Graduate Texts in Mathematics 216

For other titles in this series, go to
http://www.springer.com/series/136

Denis Serre

Matrices

Theory and Applications

Second Edition

 Springer

Denis Serre
Unité de Mathématiques Pures et Appliquées
École Normale Supérieure de Lyon
69364 Lyon Cedex 07
France
denis.serre@umpa.ens-lyon.fr

ISSN 0072-5285
ISBN 978-1-4614-2723-0   ISBN 978-1-4419-7683-3 (eBook)
DOI 10.1007/978-1-4419-7683-3
Springer New York Dordrecht Heidelberg London

Mathematics Subject Classification (2010): 15-XX, 22-XX, 47-XX, 65-XX

Printed on acid-free paper

Springer is part of Springer Science+Business Media (www.springer.com)

To Pascale, Fanny, Paul, and Joachim

Contents

Preface for the Second Edition   xi

Preface for the First Edition   xiii

1 Elementary Linear and Multilinear Algebra   1
  1.1 Vectors and Scalars   1
  1.2 Linear Maps   5
  1.3 Bilinear Maps   9

2 What Are Matrices   15
  2.1 Introduction   15
  2.2 Matrices as Linear Maps   19
  2.3 Matrices and Bilinear Forms   28

3 Square Matrices   31
  3.1   Determinant   31
  3.2   Minors   34
  3.3   Invertibility   38
  3.4   Eigenvalues and Eigenvectors   42
  3.5   The Characteristic Polynomial   43
  3.6   Diagonalization   48
  3.7   Trigonalization   49
  3.8   Rank-One Perturbations   52
  3.9   Alternate Matrices and the Pfaffian   54
  3.10 Calculating the Characteristic Polynomial   56
  3.11 Irreducible Matrices   59

4 Tensor and Exterior Products   69
  4.1 Tensor Product of Vector Spaces   69
  4.2 Exterior Calculus   72
  4.3 Tensorization of Linear Maps   77
  4.4 A Polynomial Identity in $\mathbf{M_n(K)}$ 78

5   Matrices with Real or Complex Entries 83
   5.1   Special Matrices 83
   5.2   Eigenvalues of Real- and Complex-Valued Matrices.............. 86
   5.3   Spectral Decomposition of Normal Matrices 91
   5.4   Normal and Symmetric Real-Valued Matrices 92
   5.5   Functional Calculus 94
   5.6   Numerical Range.. 98
   5.7   The Gershgorin Domain 102

6   Hermitian Matrices 109
   6.1   The Square Root over $\mathbf{HPD}_n$ 109
   6.2   Rayleigh Quotients 110
   6.3   Further Properties of the Square Root 114
   6.4   Spectrum of Restrictions 115
   6.5   Spectrum versus Diagonal 117
   6.6   The Determinant of Nonnegative Hermitian Matrices............. 119

7   Norms 127
   7.1   A Brief Review 127
   7.2   Householder's Theorem 133
   7.3   An Interpolation Inequality 135
   7.4   Von Neumann's Inequality 137

8   Nonnegative Matrices 149
   8.1   Nonnegative Vectors and Matrices 149
   8.2   The Perron–Frobenius Theorem: Weak Form 150
   8.3   The Perron–Frobenius Theorem: Strong Form 151
   8.4   Cyclic Matrices 154
   8.5   Stochastic Matrices 156

9   Matrices with Entries in a Principal Ideal Domain; Jordan Reduction 163
   9.1   Rings, Principal Ideal Domains 163
   9.2   Invariant Factors of a Matrix 167
   9.3   Similarity Invariants and Jordan Reduction 170

10   Exponential of a Matrix, Polar Decomposition, and Classical Groups 183
   10.1   The Polar Decomposition 183
   10.2   Exponential of a Matrix 184
   10.3   Structure of Classical Groups.............................. 188
   10.4   The Groups $\mathbf{U(p,q)}$.................................... 191
   10.5   The Orthogonal Groups $\mathbf{O(p,q)}$ 192
   10.6   The Symplectic Group $\mathbf{Sp_n}$ 195

11   Matrix Factorizations and Their Applications 207
   11.1 The *LU* Factorization 208
   11.2 Choleski Factorization 213
   11.3 The *QR* Factorization 214
   11.4 Singular Value Decomposition............................... 216
   11.5 The Moore–Penrose Generalized Inverse 218

12   Iterative Methods for Linear Systems 225
   12.1 A Convergence Criterion 226
   12.2 Basic Methods 227
   12.3 Two Cases of Convergence 229
   12.4 The Tridiagonal Case 231
   12.5 The Method of the Conjugate Gradient........................ 235

13   Approximation of Eigenvalues 247
   13.1 General Considerations.................................... 247
   13.2 Hessenberg Matrices 249
   13.3 The *QR* Method 253
   13.4 The Jacobi Method 259
   13.5 The Power Methods 266

References 277

Index of Notations 279

General Index 283

Cited Names 289

Preface for the Second Edition

It is only after long use that an author realizes the flaws and the gaps in his or her book. Having taught from a significant part of it, having gathered more than three hundred exercises on a public website http://www.umpa.ens-lyon.fr/serre/DPF/exobis.pdf, and having learned a lot within eight years of reading after the first edition was published, I arrived at the conclusion that a second edition of *Matrices: Theory and Applications* should be significantly different from the first one.

First of all, I felt ashamed of the very light presentation of the backgrounds in linear algebra and elementary matrix theory in Chapter 1. In French, I should say *ce n'était ni fait, ni à faire* (neither done, nor to be done). I thus began by rewriting this part completely, taking this opportunity to split pure linear algebra from the introduction to matrices. I hope that the reader is satisfied with the new Chapters 1 and 2 below.

When teaching, it was easy to recognize the lack of logical structure here and there. For the sake of a more elegant presentation, I therefore moved several statements from one chapter to another. It even happened that entire sections was displaced, such as those about singular value decomposition, the Leverrier algorithm, the Schur complement, or the square root of positive Hermitian matrices.

Next, I realized that some important material was missing in the first edition. This has led me to increase the size of this book by about forty percent. The newly added topics are

- Dunford decomposition
- Calculus with rank-one perturbations
- Improvement by Preparata and Sarwate of Leverrier's algorithm for calculating the characteristic polynomial
- Tensor calculus
- Polynomial identity of Amitsur and Levitzki
- Regularity of simple eigenvalues for complex matrices
- Functional calculus and the Dunford–Taylor formula
- Stable and unstable subspaces

- Numerical range
- Weyl inequalities
- Concavity of $(\det H)^{1/n}$ over $\mathbf{HPD}_n$
- von Neumann inequality
- Convergence of the Jacobi method with random choice (perhaps a new result).

With so many additions, the chapter on real and complex matrices extended beyond a reasonable size. This in turn led me to split it, by dedicating a specific chapter to the study of Hermitian matrices. Because tensor calculus, together with polynomial identities, also forms a new chapter, the number of chapters has increased from ten to thirteen.

The reader might wonder why I included the new fourth chapter, because it is essentially not used in the sequel. Several reasons led me to this choice. First of all, tensor and exterior calculus are fundamental in spectral analysis, in differential geometry, in theoretical physics and many other domains. Next, I think that the theorem of Amitsur and Levitzki is one of the most beautiful in matrix theory, which remained mysterious until a quite direct proof using exterior algebra was found, allowing it to be presented in a graduate textbook. The presence of this result is consistent with my philosophy that every chapter, beyond the introductory ones, should contain at least one advanced statement. My last motivation was to include more algebraic issues, because it is a permanent tendancy of matrix analysis to be too analytic.

Following this point of view, I should have loved to include a proof of A. Horn's conjecture, after W. Fulton, A. Klyachko, A. Knutson and T. Tao: the statement, which has an analytic flavor inasmuch as it consists in inequalities between eigenvalues of Hermitian matrices, is undoubtedly a great achievement of algebraists, involving, as it does, representation theory and Schubert calculus. However, the material to be included would have been far too advanced. This theory is still too fresh and will not be part of textbooks before it has been digested, and this will take perhaps a few decades. Thus I decided to be content with discussing the Weyl and Lidskiĭ inequalities, which are the lower steps in Horn's list.

Even though this new version had expanded a lot, I thought long and hard about including a fundamental and beautiful (to my opinion) issue, namely Loewner's theory of operator monotone functions. I eventually abandoned this idea because the topic would have needed too much space, thus upsetting the balance between elementary and advanced material.

Finally, I have included many new exercises, many of them offering a way to go further into the theory and practice. The website mentioned above remains available, and is expected to grow over time, but the references in it will remain unchanged, in order that it may be usable by the owners of the first edition, until I feel that it absolutely needs to be refreshed.

Lyon, France   Denis Serre
November 2009

Preface for the First Edition

The study of matrices occupies a singular place within mathematics. It is still an area of active research, and it is used by every mathematician and by many scientists working in various specialities. Several examples illustrate its versatility:

- Scientific computing libraries began growing around matrix calculus. As a matter of fact, the discretization of partial differential operators is an endless source of linear finite-dimensional problems.
- At a discrete level, the maximum principle is related to nonnegative matrices.
- Control theory and stabilization of systems with finitely many degrees of freedom involve spectral analysis of matrices.
- The discrete Fourier transform, including the fast Fourier transform, makes use of Toeplitz matrices.
- Statistics is widely based on correlation matrices.
- The generalized inverse is involved in least-squares approximation.
- Symmetric matrices are inertia, deformation, or viscous tensors in continuum mechanics.
- Markov processes involve stochastic or bistochastic matrices.
- Graphs can be described in a useful way by square matrices.
- Quantum chemistry is intimately related to matrix groups and their representations.
- The case of quantum mechanics is especially interesting. Observables are Hermitian operators; their eigenvalues are energy levels. In the early years, quantum mechanics was called "mechanics of matrices," and it has now given rise to the development of the theory of large random matrices. See [25] for a thorough account of this fashionable topic.

This text was conceived during the years 1998–2001, on the occasion of a course that I taught at the École Normale Supérieure de Lyon. As such, every result is accompanied by a detailed proof. During this course I tried to investigate all the principal mathematical aspects of matrices: algebraic, geometric, and analytic.

In some sense, this is not a specialized book. For instance, it is not as detailed as [19] concerning numerics, or as [40] on eigenvalue problems, or as [21] about

Weyl-type inequalities. But it covers, at a slightly higher than basic level, all these aspects, and is therefore well suited for a graduate program. Students attracted by more advanced material will find one or two deeper results in each chapter except the first one, given with full proofs. They will also find further information in about the half of the 170 exercises. The solutions for exercises are available on the author's site http://www.umpa.ens-lyon.fr/ ˜serre/exercises.pdf.

This book is organized into ten chapters. The first three contain the basics of matrix theory and should be known by almost every graduate student in any mathematical field. The other parts can be read more or less independently of each other. However, exercises in a given chapter sometimes refer to the material introduced in another one.

This text was first published in French by Masson (Paris) in 2000, under the title *Les Matrices: théorie et pratique*. I have taken the opportunity during the translation process to correct typos and errors, to index a list of symbols, to rewrite some unclear paragraphs, and to add a modest amount of material and exercises. In particular, I added three sections, concerning alternate matrices, singular value decomposition, and the Moore–Penrose generalized inverse. Therefore, this edition differs from the French one by about ten percent of the contents.

Acknowledgments

Many thanks to the Ecole Normale Supérieure de Lyon and to my colleagues who have had to put up with my talking to them so often about matrices. Special thanks to Sylvie Benzoni for her constant interest and useful comments.

Lyon, France   Denis Serre
December 2001

Chapter 1
Elementary Linear and Multilinear Algebra

This chapter is the only one where results are given either without proof, or with sketchy proofs. A beginner should have a close look at a textbook dedicated to linear algebra, not only reading statements and proofs, but also solving exercises in order to become familiar with all the relevant notions.

1.1 Vectors and Scalars

Scalars are elements of some field k (or K), or sometimes of a ring R. The most common fields are the field of rational numbers $\mathbb{Q}$, the field of real numbers $\mathbb{R}$, and the field of complex numbers $\mathbb{C}$. There are also finite fields, such as $\mathbb{F}_p := \mathbb{Z}/p\mathbb{Z}$ (p a prime number). Other interesting fields are $k(X)$ (rational fractions), that of formal Laurent series, or the p-adic field $\mathbb{Q}_p$. Linear algebra also makes use of the ring of integers $\mathbb{Z}$ or of those of polynomials in one or in several variables $k[X]$ and $k[X_1, \ldots, X_r]$. One encounters a lot of other rings in number theory and algebraic geometry, for instance, the Gaussian integers $\mathbb{Z}[i]$.

Inasmuch this book is about matrices, we show that square matrices form a non-commutative ring; this ring can be used as a set of scalars, when we write a large matrix blockwise. This is one of the few instances where the ring of scalars is not Abelian. Another one occurs in Section 4.4.

The digits 0 and 1 have the usual meaning in a field K, with $0 + x = 1 \cdot x = x$. The subring $\mathbb{Z} \cdot 1$, composed of all sums (possibly empty) of the form $\pm(1 + \cdots + 1)$ is isomorphic to either $\mathbb{Z}$ or a finite field $\mathbb{F}_p$. In the latter case, p is a prime number, which we call the *characteristic* of K and denote $\mathrm{charc}(K)$. In the former case, we set $\mathrm{charc}(K) = 0$.

One says that a nonzero polynomial $P \in K[X]$ *splits* over K if it can be written as a product of the form

$$a \prod_{i=1}^{r} (X - a_i)^{n_i}, \quad a, a_i \in K, \quad r \in \mathbb{N}, n_i \in \mathbb{N}^*.$$

We may assume that the a_is are pairwise distinct. Such a factorization is then unique, up to the order of the factors. A field K in which every nonconstant polynomial $P \in K[X]$ admits a root, or equivalently in which every polynomial $P \in K[X]$ splits, is *algebraically closed*. If the field K' contains the field K and if every polynomial $P \in K[X]$ splits in K', then the set $\overline{K}$ of roots in K' of polynomials in $K[X]$ is an algebraically closed field containing K, and it is the smallest such field, unique up to isomorphism. One calls $\overline{K}$ the *algebraic closure* of K. Every field K admits an algebraic closure, unique up to isomorphism. The fundamental theorem of algebra asserts that $\overline{\mathbb{R}} = \mathbb{C}$. The algebraic closure of $\mathbb{Q}$, for instance, is the set of *algebraic numbers*; it is the set of complex roots of all polynomials $P \in \mathbb{Z}[X]$.

1.1.1 Vector Spaces

Let K be a field and $(E,+)$ be a commutative group. Because E and K are distinct sets, the symbol $+$ has two meanings, depending on whether it is used for the addition in E or in K. This does not cause any confusion. Let moreover

$$(a,x) \mapsto ax,$$
$$K \times E \to E,$$

be a map such that

$$(a+b)x = ax + bx, \quad a(x+y) = ax + ay, \qquad a(bx) = (ab)x, \quad 1x = x$$

for every $x,y \in E$ and $a,b \in K$. We say that E is a *vector space* over K (or a K-vector space). The elements of E are called *vectors*. In a vector space one always has $0x = 0$ (more precisely, $0_K x = 0_E$).

When $P, Q \subset K$ and $F, G \subset E$, one denotes by PQ (respectively, $P+Q, F+G, PF$) the set of products pq as (p,q) ranges over $P \times Q$ (respectively, $p+q$, $f+g$, pf as p,q,f,g range over P,Q,F,G). A subgroup $(F,+)$ of $(E,+)$, which is stable under multiplication by scalars (i.e., such that $KF \subset F$), is again a K-vector space. One says that it is a *linear subspace* of E, or just a subspace. Observe that F, as a subgroup, is nonempty, because it contains 0_E. The intersection of any family of linear subspaces is a linear subspace. The sum $F + G$ of two linear subspaces is again a linear subspace. The trivial formula $(F+G)+H = F+(G+H)$ allows us to define unambiguously $F+G+H$ and, by induction, the sum of any finite family of subsets of E. When these subsets are linear subspaces, their sum is also a linear subspace.

Let I be a set. One denotes by K^I the set of maps $a = (a_i)_{i \in I} : I \to K$ where only finitely many of the a_is are nonzero. This set is naturally endowed with a K-vector space structure, with the addition and product laws

$$(a+b)_i := a_i + b_i, \quad (\lambda a)_i := \lambda a_i.$$

Let E be a vector space and let $i \mapsto f_i$ be a map from I to E. A *linear combination* of $(f_i)_{i \in I}$ is a sum

$$\sum_{i \in I} a_i f_i,$$

where the a_is are scalars, only finitely many of them being nonzero (in other words, $(a_i)_{i \in I} \in K^I$). This sum involves only finitely many nonzero terms, thus makes sense. It is a vector of E. The family $(f_i)_{i \in I}$ is *free*, or *linearly independent*, if every linear combination but the trivial one (when all coefficients are zero) is nonzero. It is a *generating* family if every vector of E is a linear combination of its elements. In other words, $(f_i)_{i \in I}$ is free (respectively, generating) if the map

$$K^I \to E,$$
$$(a_i)_{i \in I} \mapsto \sum_{i \in I} a_i f_i,$$

is injective (respectively, onto). Finally, one says that $(f_i)_{i \in I}$ is a *basis* of E if it is both free and generating. In that case, the above map is bijective, and it is actually an isomorphism between vector spaces.

If $\mathscr{G} \subset E$, one often identifies $\mathscr{G}$ and the associated family $(g)_{g \in \mathscr{G}}$. The set G of linear combinations of elements of $\mathscr{G}$ is a linear subspace E, called the linear subspace *spanned* by $\mathscr{G}$. It is the smallest linear subspace E containing $\mathscr{G}$, equal to the intersection of all linear subspaces containing $\mathscr{G}$. The subset $\mathscr{G}$ is generating when $G = E$.

1.1.1.1 Dimension of a Vector Space

Let us mention an abstract result.

Theorem 1.1 *Every K-vector space admits at least one basis. Every free family is contained in a basis. Every generating family contains a basis. All the bases of E have the same cardinality (which is called the* dimension *of E).*

For general vector spaces, this statement is a consequence of the axiom of choice. As such, it is overwhelmingly (but not universally) accepted by mathematicians. Because we are interested throughout this book in finite-dimensional spaces, for which the existence of bases follows from elementary considerations, we prefer to start with the following.

Definition 1.1 *The dimension of a vector space E, denoted by $\dim E$, is the upper bound of the cardinality of free families in E. It may be infinite. If $E = \{0\}$, the dimension is zero.*

If $\dim E < +\infty$, we say that E is finite-dimensional.

When E is finite-dimensional, every free family of cardinal $\dim E$ is contained in a free family that is maximal for the inclusion (obvious); the maximality implies that the latter family is generating, hence is a basis. Next, given a generating family β in E, one may consider free families contained in β. Again, a maximal one is a basis.

Thus β contains a basis. The fact that two bases have the same cardinality is less easy, but still elementary.

The dimension is monotone with respect to inclusion: if $F \subset E$, then $\dim F \leq \dim E$. The equality case is useful.

Proposition 1.1 *Let E be a finite-dimensional vector space, and F be a linear subspace of E.*

If $\dim F = \dim E$, then $F = E$.

Proposition 1.2 *If F, G are two linear subspaces of E, the following formula holds,*

$$\dim F + \dim G = \dim F \cap G + \dim(F + G).$$

Corollary 1.1 *In particular,*

$$\dim F \cap G \geq \dim F + \dim G - \dim E.$$

If $F \cap G = \{0\}$, one writes $F \oplus G$ instead of $F + G$, and one says that the sum of F and G is *direct*. Proposition 1.2 gives

$$\dim F \oplus G = \dim F + \dim G.$$

Let E and F be vector spaces over K. One builds the abstract direct sum of E and F as follows, and one denotes it again $E \oplus F$. Its vectors are those of the Cartesian product $E \times F$, whereas the sum and the multiplication by a scalar are defined by

$$(e, f) + (e', f') = (e + e', f + f'), \qquad \lambda(e, f) = (\lambda e, \lambda f).$$

The spaces E and F can be identified with the subspaces $E \times \{0_F\}$ and $\{0_E\} \times F$ of $E \oplus F$, respectively.

Given a set I, the family $(\mathbf{e}^i)_{i \in I}$, defined by

$$(\mathbf{e}^i)_j = \begin{cases} 0, \ j \neq i, \\ 1, \ j = i, \end{cases}$$

is a basis of K^I, called the *canonical basis*. The dimension of K^I is therefore equal to the cardinality of I.

1.1.1.2 Extension of the Scalars

Let L be a field and K a subfield of L. If F is an L-vector space, then F is also a K-vector space. As a matter of fact, L is itself a K-vector space, and one has

$$\dim_K F = \dim_L F \cdot \dim_K L.$$

The most common example (the only one that we consider) is $K = \mathbb{R}$, $L = \mathbb{C}$, for which we have

$$\dim_{\mathbb{R}} F = 2 \dim_{\mathbb{C}} F.$$

Conversely, if G is an $\mathbb{R}$-vector space, one builds its *complexification* $G \otimes_{\mathbb{R}} \mathbb{C}$ (read *G tensor* $\mathbb{C}$) as follows:

$$G \otimes_{\mathbb{R}} \mathbb{C} = G \times G,$$

with the induced structure of the additive group. An element (x,y) of $G \otimes_{\mathbb{R}} \mathbb{C}$ is also denoted $x + iy$. One defines multiplication by a complex number by

$$(\lambda = a + ib, z = x + iy) \mapsto \lambda z := (ax - by, ay + bx).$$

One verifies easily that $G \otimes_{\mathbb{R}} \mathbb{C}$ is a $\mathbb{C}$-vector space, with

$$\dim_{\mathbb{C}} G \otimes_{\mathbb{R}} \mathbb{C} = \dim_{\mathbb{R}} G.$$

Furthermore, G may be identified with an $\mathbb{R}$-linear subspace of $G \otimes_{\mathbb{R}} \mathbb{C}$ by

$$x \mapsto (x, 0).$$

Under this identification, one has $G \otimes_{\mathbb{R}} \mathbb{C} = G \oplus iG$. In a more general setting, one may consider two fields K and L with $K \subset L$, instead of $\mathbb{R}$ and $\mathbb{C}$. The extension of scalars from K to L yields the space $G \otimes_K L$, a *tensor product*. We construct the tensor product of arbitrary vector spaces in Section 4.1.

1.2 Linear Maps

Let E, F be two finite-dimensional K-vector spaces. A map $u : E \rightarrow F$ is *linear* (one also speaks of a *homomorphism*) if $u(x + y) = u(x) + u(y)$ and $u(ax) = au(x)$ for every $x, y \in E$ and $a \in K$. One then has $u(0_E) = 0_F$. The preimage $u^{-1}(0_F)$, denoted by $\ker u$, is the *kernel* of u. It is a linear subspace of E. The *range* $u(E)$ is also a linear subspace of F, whose dimension is called the *rank* of u, and denoted by $\mathrm{rk}\, u$. It is often denoted $R(u)$. Taking a basis $(u(x_i))_{i \in I}$ of $u(E)$, together with a basis $(y_j)_{j \in J}$ of $\ker u$, the xs and the ys form a basis of E, hence comes the following.

Theorem 1.2 *If $u : E \rightarrow F$ is linear, then*

$$\dim E = \dim \ker u + \mathrm{rk}\, u.$$

The set of homomorphisms from E to F is naturally a K-vector space, with operations

$$(u + v)(x) = u(x) + v(x), \qquad (\lambda u)(x) = \lambda u(x).$$

It is denoted $\mathscr{L}(E; F)$. Its dimension equals the product of $\dim E$ and $\dim F$.

If $u \in \mathscr{L}(E; F)$ and $v \in \mathscr{L}(F; G)$ are given, the composition $v \circ u$ is well defined and is linear: $v \circ u \in \mathscr{L}(E; G)$.

1.2.1 Eigenvalues, Eigenvectors

If $F = E$, one denotes $\mathrm{End}(E) := \mathscr{L}(E;E)$; its elements are the *endomorphisms* of E. Therefore $\mathrm{End}(E)$ is an algebra, that is a ring under the laws $(+, \circ)$, a K-vector space, with the additional property that $\lambda(u \circ v) = (\lambda u) \circ v$.

For an endomorphism, Theorem 1.2 reads $n = \dim \ker u + \mathrm{rk}\, u$. A subspace of E of dimension n equals E itself, therefore we infer the equivalence

$$u \text{ is bijective} \Longleftrightarrow u \text{ is injective} \Longleftrightarrow u \text{ is surjective.}$$

Replacing u by $\lambda \mathrm{id}_E - u$, we thus have

$$\lambda \mathrm{id}_E - u \text{ is bijective} \Longleftrightarrow \lambda \mathrm{id}_E - u \text{ is injective} \Longleftrightarrow \lambda \mathrm{id}_E - u \text{ is surjective.}$$

In other words, we face the alternative

- Either there exists $x \in E \setminus \{0\}$ such that $u(x) = \lambda x$. Such a vector is called an *eigenvector* and λ is called an *eigenvalue*,
- Or, for every $b \in E$, the following equation admits a unique solution $y \in E$,

$$u(y) - \lambda y = b.$$

The set of eigenvalues of u is denoted $\mathrm{Sp}(u)$. We show later on that it is a finite set. Notice that an eigenvector is always a nonzero vector.

1.2.2 Linear Forms and Duality

When the target space is the field of scalars, a linear map (i.e., $u : E \to K$) is called a *linear form*. The set of linear forms is the *dual space* of E, denoted by E':

$$E' = \mathscr{L}(E;K).$$

The dimension of E equals that of E'. If $\mathscr{B} = \{v^1, \ldots, v^n\}$ is a basis of E, then the *dual basis* of E' is $\{\ell_1, \ldots, \ell_n\}$ defined by

$$\ell_i(v^j) := \begin{cases} 1, & \text{if } j = i, \\ 0, & \text{if } j \neq i. \end{cases}$$

The ℓ_is are the coordinate maps over in E in the basis $\mathscr{B}$, inasmuch as we have

$$x = \sum_{i=1}^{n} \ell_i(x) v^i, \qquad \forall x \in E.$$

In other words, the identity map id_E decomposes as

$$\mathrm{id}_E = \sum_{i=1}^{n} v^i \ell_i.$$

In the above equality, the ℓ_is play the role of functions, and the v^is can be viewed as coefficients. The image of x under id_E is a vector (here, itself), therefore these coefficients are vectors. The same comment applies below.

Every linear map $u : E \to F$ decomposes as a finite sum $w^1 m_1 + \cdots + w^r m_r$, where the w^js are vectors of F and the m_is are linear forms on E. In other words,

$$u(x) = \sum_{i=1}^{r} m_i(x) w^i, \qquad \forall x \in E, \qquad \left(\text{equivalently } u = \sum_{i=1}^{r} w^i m_i \right).$$

This decomposition is highly non unique. The minimal number r in such a sum equals the rank of u. In terms of the tensor product introduced in Section 4.1, we identify $\mathcal{L}(E;F)$ with $F \otimes E'$ and this decomposition reads

$$u = \sum_{i=1}^{r} w^i \otimes m_i.$$

1.2.2.1 Bidual

The *bidual* of E is the dual space of E'. A vector space E can be identified canonically with a subspace of its bi-dual: given $x \in E$, one defines a linear form over E' by

$$\ell \overset{\delta_x}{\mapsto} \ell(x).$$

The map $\delta : x \mapsto \delta_x$ is linear and one-to-one from E to $(E')'$. These spaces have the same dimension and thus δ is an isomorphism. Because it is canonically defined, we identify E with its bi-dual.

1.2.2.2 Polarity

Given a subset S of E, the set of linear forms vanishing identically over S is the *polar set* of S, denoted by S^0. The following properties are obvious:

- A polar set is a linear subspace of E'.
- If $S \subset T$, then $T^0 \subset S^0$.
- $S^0 = (\mathrm{Span}(S))^0$.

Proposition 1.3 *If F is a subspace of E, then*

$$\dim F + \dim F^0 = \dim E.$$

Proof. Let $\{v^1, \ldots, v^r\}$ be a basis of F, which we extend as a basis $\mathscr{B}$ of E. Let $\{\ell^1, \ldots, \ell^n\}$ be the dual basis of E'. Then F^0 equals $\mathrm{Span}(\ell^{r+1}, \ldots, \ell^n)$ has dimension $n - r$. $\square$

Corollary 1.2 *A subspace F of E equals its bipolar $(F^0)^0$. Rigorously speaking,*
$(F^0)^0 = \delta(F)$.

The bipolar of a subset $S \subset E$ equals Span(S).

Proof. Obviously, $(F^0)^0 \subset F$. In addition, their dimensions are equal to $n - \dim F^0$ because of Proposition 1.3. Therefore they coincide.

If $S \subset E$, we know that $S^0 = (\mathrm{Span}(S))^0$. Therefore

$$(S^0)^0 = \left((\mathrm{Span}(S))^0\right)^0 = \mathrm{Span}(S).$$

$\square$

1.2.2.3 Adjoint Linear Map

Let $u \in \mathcal{L}(E;F)$ be given. If ℓ is a linear form over F, then $\ell \circ u$ is a linear form over E, thus an element of E'. The map $\ell \mapsto \ell \circ u$ is linear and is denoted by u^*. It is an element of $\mathcal{L}(F';E')$, called the *adjoint* of u. One has

$$u^*(\ell) = \ell \circ u.$$

Because E and F have finite dimensions, then $(u^*)^*$, an element of $\mathcal{L}(E'';F'')$, is an element of $\mathcal{L}(E;F)$, after the identification of E and F with their bi-duals $(E')' = E$. We prove below that it coincides with u, or more accurately, that the following diagram is commutative.

$$
\begin{array}{ccc}
E & \xrightarrow{\;u\;} & F \\
{\scriptstyle \delta_E}\downarrow & & \downarrow{\scriptstyle \delta_F} \\
E'' & \xrightarrow{\;u^{**}\;} & F''
\end{array}
$$

We list below the main facts about adjunction.

Proposition 1.4 *We recall that E is a finite-dimensional vector space. Then*

- $(u^*)^* = u$.
- $(\ker u)^0 = R(u^*)$ *and* $R(u)^0 = \ker u^*$.
- u *is injective if and only if* u^* *is surjective.*
- u *is surjective if and only if* u^* *is injective.*
- *For every* $u : E \to F$ *and* $v : F \to G$, *one has*

$$(v \circ u)^* = u^* \circ v^*.$$

Proof. Let $x \in E$ and $L \in F'$ be given. Then

$$
\begin{aligned}
(u^{**} \circ \delta_E(x))(L) = (\delta_x \circ u^*)(L) &= \delta_x(u^*(L)) = \delta_x(L \circ u) \\
&= (L \circ u)(x) = L(u(x)) = \delta_{u(x)}(L).
\end{aligned}
$$

We therefore have

$$(u^{**} \circ \delta_E)(x) = \delta_{u(x)} = \delta_F(u(x)) = \delta_F \circ u(x),$$

whence $u^{**} \circ \delta_E = \delta_F \circ u$.

If $x \in \ker u$ and $\ell \in F'$, then $u^*(\ell)(x) = \ell \circ u(x) = \ell(0) = 0$, whence $R(u^*) \subset (\ker u)^0$. Conversely, let $m \in (\ker u)^0$ be given, that is, a linear form on E, vanishing over $\ker u$. When y is in $R(u)$, m is constant over $u^{-1}(y)$. We thus define a map

$$R(u) \to K$$
$$y \mapsto m(x),$$

where x is any element in $u^{-1}(y)$. We extend this map as a linear form ℓ over F. It satisfies $m = \ell \circ u = u^*(\ell)$, hence the converse inclusion $(\ker u)^0 \subset R(u^*)$. We deduce immediately that u is injective if and only if u^* is surjective.

Again, if $x \in u$ and $\ell \in \ker u^*$, then $\ell(u(x)) = u^*(\ell)(x) = 0(x) = 0$ shows that $\ker u^* \subset R(u)^0$. Conversely, let $\ell \in F'$ vanish over $R(u)$. Then $u^*(\ell) = u \circ \ell \equiv 0$, hence the equality $R(u)^0 = \ker u^*$. It follows immediately that u is surjective if and only if u^* is injective.

Finally, if $\ell \in G'$, then

$$(v \circ u)^*(\ell) = \ell \circ (v \circ u) = (\ell \circ v) \circ u = v^*(\ell) \circ u = u^*(v^*(\ell)) = (u^* \circ v^*)(\ell).$$

$\square$

1.3 Bilinear Maps

Let E, F, and G be three K-vector spaces. A map $b : E \times F \to G$ is *bilinear* if the partial maps $x \mapsto b(x,y)$ and $y \mapsto b(x,y)$ are linear from E (respectively, from F) into G. The set of bilinear maps from $E \times F$ into G is a vector space, denoted by **Bil**$(E \times F; G)$.

If the target space is K itself, then one speaks of a bilinear form. The set of bilinear forms over $E \times F$ is a vector space, denoted by **Bil**$(E \times F)$. Its dimension equals the product of $\dim E$ and $\dim F$. If $F = E$, we simply write **Bil**(E).

1.3.1 Bilinear Forms When F = E

Let $b \in$ **Bil**(E) be given. We say that b is *symmetric* if

$$b(x,y) = b(y,x), \qquad \forall x,y \in E.$$

Likewise, we say that b is *skew-symmetric* if

$$b(x,y) = -b(y,x), \qquad \forall x,y \in E.$$

Finally, we say that b is *alternating* if it satisfies

$$b(x,x) = 0, \qquad \forall x \in E.$$

An alternating form is skew-symmetric, because

$$b(x,y) + b(y,x) = b(x+y,x+y) - b(x,x) - b(y,y).$$

If the characteristic of K is different from 2, the converse is true, because the definition contains the identity $2b(x,x) = 0$. Notice that in characteristic 2, skew-symmetry is equivalent to symmetry. To summarize, there are basically two distinct classes, those of symmetric forms and of alternating forms. The third class of skew-symmetric forms equals either the latter if $\mathrm{charc}(K) \neq 2$ or the former if $\mathrm{charc}(K) = 2$.

1.3.2 Degeneracy versus Nondegeneracy

We assume that E has finite dimension. Let $b \in \mathbf{Bil}(E)$ be given. We may define two linear maps $b_0, b_1 : E \to E'$ by the formulæ

$$b_0(x)(y) = b_1(y)(x) := b(x,y), \qquad \forall x,y \in E.$$

Recalling that $(E')'$ is identified with E through $y(\ell) := \ell(y)$, the following calculation shows that b_1 is the adjoint of b_0, and conversely:

$$b_0^*(y)(x) = (y \circ b_0)(x) = y(b_0(x)) = b_0(x)(y) = b(x,y) = b_1(y)(x),$$

whence $b_0^*(y) = b_1(y)$ for all $y \in E$.

Because $\dim E' = \dim E$, and thanks to Proposition 1.4, we thus have an equivalence among the injectivity, surjectivity, and bijectivity of b_0 and b_1. We say that b is *nondegenerate* if b_0, or equivalently b_1, is one-to-one. It is degenerate otherwise. Degeneracy means that there exists a nonzero vector $\bar{x} \in E$ such that $b(\bar{x}, \cdot) \equiv 0$.

When b is symmetric (respectively, skew-symmetric), the maps b_0 and b_1 are identical (respectively, opposite), thus their kernels coincide. It is then called the *kernel* of b. A (skew-)symmetric bilinear form is nondegenerate if and only if $\ker b = \{0\}$.

1.3.3 Bilinear Spaces

Let a nondegenerate bilinear form b be given on a space E, either symmetric or skew-symmetric. We say that (E, b) is a *bilinear space*. If $u \in \mathbf{End}(E)$ is given, we may define an *adjoint* u^*, still an element of $\mathbf{End}(E)$, by the formula

$$b(u(x), y) = b(x, u^*(y)), \qquad \forall x, y \in E.$$

An accurate expression of $u^*(y)$ is $(b_1)^{-1}(b_1(y) \circ u)$. The following properties are obvious

$$(\lambda u + v)^* = \lambda u^* + v^*, \qquad (v \circ u)^* = u^* \circ v^*, \qquad (u^*)^* = u.$$

Mind that this adjoint depends on the bilinear structure; another bilinear form yields another adjoint.

We also say that $u \in \mathbf{End}(E)$ is an *isometry* if

$$b(u(x), u(y)) = b(x, y), \qquad \forall x, y \in E.$$

This is equivalent to saying that $u^* \circ u = \mathrm{id}_E$. In particular, an isometry is one-to-one. One easily checks that the set of isometries is a group for the composition of linear maps.

1.3.4 Quadratic Forms

A *quadratic form* over E is a function $q : E \to K$ given by a formula

$$q(x) = b(x, x), \qquad \forall x \in E,$$

where b is a symmetric bilinear form over E.

When the characteristic of K is different from 2, there is a one-to-one correspondence between quadratic forms and symmetric bilinear forms, because of the reciprocal formula

$$b(x, y) = \frac{1}{2}(q(x + y) - q(x) - q(y)),$$

or as well

$$b(x, y) = \frac{1}{4}(q(x + y) - q(x - y)).$$

We then say that b is the *polar form* of q.

The *kernel* of q is by definition that of b, and q is nondegenerate when b is so. We warn the beginner that the kernel of q is usually different from the set

$$\Gamma(q) := \{x \in E \mid q(x) = 0\}.$$

The latter is a cone, that is a set invariant under the multiplication by scalars, whereas the former is a vector space. For this reason, $\Gamma(q)$ is called the *isotropic cone* of q. The kernel is obviously contained in $\Gamma(q)$ and we have the stronger property that

$$\Gamma(q) + \ker q = \Gamma(q).$$

When q is nondegenerate and $\operatorname{charc}(K) \neq 2$, $u \in \mathbf{End}(E)$ is a b-isometry if and only if $q \circ u = q$ (use the correspondence $q \leftrightarrow b$ above). We also say that u is a *q-isometry*.

1.3.5 Euclidean Spaces

When $K = \mathbb{R}$ is the field of real numbers, the range of a quadratic form may be either $\mathbb{R}$ or $\mathbb{R}^+$. The latter situation is especially interesting. We say that q (or b as well) is *positive semidefinite* if $q \geq 0$ over E. We say that q is *positive definite* if moreover $q(x) = 0$ implies $x = 0$; that is, $q(y) > 0$ for every nonzero vector. Then the polar form b is called a *scalar product*. A positive-definite form is always nondegenerate, but the converse statement is false.

Definition 1.2 *A pair* (E, q) *where* E *is a real vector space and* q *is a positive definite quadratic form on* E *is called a* Euclidean space.

Proposition 1.5 *A scalar product satisfies the* Cauchy–Schwarz[1] *inequality*

$$b(x,y)^2 \leq q(x)q(y), \qquad \forall x, y \in E.$$

The equality holds true if and only if x *and* y *are colinear.*

Proof. The polynomial

$$t \mapsto q(tx+y) = q(x)t^2 + 2b(x,y)t + q(y)$$

takes nonnegative values for $t \in \mathbb{R}$. Hence its discriminant $4(b(x,y)^2 - q(x)q(y))$ is nonpositive. When the latter vanishes, the polynomial has a real root t_0, which implies that $t_0 x + y = 0$. $\square$

The Cauchy–Schwarz inequality implies immediately

$$q(x+y) \leq \left(\sqrt{q(x)} + \sqrt{q(y)}\right)^2,$$

which means that the square root $\|\cdot\| := q^{1/2}$ satisfies the triangle inequality

$$\|x+y\| \leq \|x\| + \|y\|.$$

---

[1] In *Cauchy–Schwarz*, the name Schwarz (1843–1921) is spelled without a *t*.

Because $\| \cdot \|$ is positively homogeneous, it is thus a *norm* over E: every Euclidean space is a normed space. The converse is obviously false.

The space $\mathbb{R}^n$ is endowed with a canonical scalar product

$$\langle x, y \rangle := x_1 y_1 + \cdots + x_n y_n.$$

The corresponding norm is

$$\|x\| = \left(x_1^2 + \cdots + x_n^2 \right)^{1/2}.$$

It is denoted $\| \cdot \|_2$ in Chapter 7.

1.3.6 Hermitian Spaces

When the scalar field is that of complex numbers $\mathbb{C}$, the complex conjugation yields an additional structure.

Definition 1.3 *Let E be a complex space, and $\phi : E \times E \to \mathbb{C}$ be a scalar-valued map. We say that ϕ is a* sesquilinear *form if it satisfies the following*

Linearity: *For every $x \in E$, $y \mapsto \phi(x,y)$ is linear,*
Anti-linearity: *For every $y \in E$, $x \to \phi(x,y)$ is antilinear, meaning*

$$\phi(\lambda x + x', y) = \bar{\lambda} \phi(x,y) + \phi(x',y).$$

Given a sesquilinear form ϕ, the formula $\psi(x,y) := \overline{\phi(y,x)}$ defines another sesquilinear form, in general different from ϕ. The equality case is especially interesting:

Definition 1.4 *An Hermitian form is a sesquilinear form satisfying in addition*

$$\phi(y,x) = \overline{\phi(x,y)}, \qquad \forall x,y \in E.$$

For an Hermitian form, the function $q(x) := \phi(x,x)$ is real-valued and satisfies

$$q(\lambda x) = |\lambda|^2 q(x).$$

The form ϕ can be retrieved from q via the formula

$$\phi(x,y) = \frac{1}{4} (q(x+y) - q(x-y) - iq(x+iy) + iq(x-iy)). \qquad (1.1)$$

Definition 1.5 *An Hermitian form is said to be* positive definite *if $q(x) > 0$ for every $x \neq 0$.*

There are also semipositive-definite Hermitian forms, satisfying $q(x) \geq 0$ for every $x \in E$. A semipositive-definite form satisfies the Cauchy–Schwarz inequality

$$|\phi(x,y)|^2 \le q(x)q(y), \qquad \forall x,y \in E.$$

In the positive-definite case, the equality holds if and only if x and y are colinear.

Definition 1.6 *An Hermitian space is a pair (E,ϕ) where E is a complex space and ϕ is a positive-definite Hermitian form.*

As in the Euclidean case, an Hermitian form is called a *scalar product*. An Hermitian space is a normed space, where the norm is given by

$$\|x\| := \sqrt{q(x)}.$$

The space $\mathbb{C}^n$ is endowed with a canonical scalar product

$$\langle x,y \rangle := \bar{x}_1 y_1 + \cdots + \bar{x}_n y_n.$$

The corresponding norm is

$$\|x\| = \left(|x_1|^2 + \cdots + |x_n|^2\right)^{1/2}.$$

Chapter 2
What Are Matrices

2.1 Introduction

In real life, a *matrix* is a rectangular array with prescribed numbers n of rows and m of columns ($n \times m$ matrix). To make this array as clear as possible, one encloses it between delimiters; we choose parentheses in this book. The position at the intersection of the ith row and jth column is labeled by the pair (i, j). If the name of the matrix is M (respectively, A, X, etc.), the entry at the (i, j)th position is usually denoted m_{ij} (respectively, a_{ij}, x_{ij}). An entry can be anything provided it gives the reader information. Here is a the real-life example.

$$M = \begin{pmatrix} 11 & 27 & 83 \\ \text{blue} & \text{green} & \text{yellow} \\ \text{undefined} & \text{Republican} & \text{Democrat} \end{pmatrix}.$$

Perhaps this matrix gives the age, the preferred color, and the political tendency of three people. In the present book, however, we restrict to matrices whose entries are mathematical objects. In practice, they are elements of a ring A. In most cases, this ring is Abelian; if it is a field, then it is denoted k or K, unless it is one of the classical number fields $\mathbb{Q}, \mathbb{R}, \mathbb{C}, \mathbb{F}_p$. When writing a matrix blockwise, it becomes a smaller matrix whose elements are themselves matrices, and thus belong to some spaces that are not even rings; having possibly different sizes, these submatrices may even belong to distinct sets.

In some circumstances (extraction of matrices or minors, e.g.) the rows and the columns can be numbered in a different way, using nonconsecutive numbers i and j. In general one needs only two finite sets I and J, one for indexing the rows and the other for indexing the columns. For instance, the following extraction from a 4×5 matrix M corresponds to the choice $I = (1, 3)$, $J = (2, 5, 3)$.

$$M_I^J = \begin{pmatrix} m_{12} & m_{15} & m_{13} \\ m_{32} & m_{35} & m_{33} \end{pmatrix}.$$

Notice that the indices need not be taken in increasing order.

2.1.1 Addition of Matrices

The set of matrices of size $n \times m$ with entries in A is denoted by $\mathbf{M}_{n \times m}(A)$. It is an additive group, where $M + M'$ denotes the matrix M'' whose entries are given by $m''_{ij} = m_{ij} + m'_{ij}$.

2.1.2 Multiplication by a Scalar

One defines the multiplication by a scalar $a \in A$: $M' := aM$ by $m'_{ij} = am_{ij}$. One has the formulæ $a(bM) = (ab)M$, $a(M + M') = (aM) + (aM')$, and $(a + b)M = (aM) + (bM)$. Likewise we define $M'' = Ma$ by $m''_{ij} := m_{ij}a$ and we have similar properties, together with $(aM)b = a(Mb)$.

With these operations, $\mathbf{M}_{n \times m}(A)$ is a left and right A-module. If A is Abelian, then $aM = Ma$. When the set of scalars is a field K, $\mathbf{M}_{n \times m}(K)$ is a K-vector space. The zero matrix is denoted by 0, or 0_{nm} when one needs to avoid ambiguity:

$$0_{n \times m} = \begin{pmatrix} 0 & \cdots & 0 \\ \vdots & & \vdots \\ 0 & \cdots & 0 \end{pmatrix}.$$

When $m = n$, one writes simply $\mathbf{M}_n(K)$ instead of $\mathbf{M}_{n \times n}(K)$, and 0_n instead of 0_{nn}. The matrices of sizes $n \times n$ are called *square* matrices of size n. When A has a unit 1, one writes I_n for the *identity* matrix, a square matrix of order n defined by

$$m_{ij} = \delta_i^j = \begin{cases} 0, & \text{if } i \neq j, \\ 1, & \text{if } i = j. \end{cases}$$

In other words,

$$I_n = \begin{pmatrix} 1 & 0 & \cdots & 0 \\ 0 & \ddots & \ddots & \vdots \\ \vdots & \ddots & \ddots & 0 \\ 0 & \cdots & 0 & 1 \end{pmatrix}.$$

2.1.3 Special Matrices

The identity matrix is a special case of a *permutation matrix*, which is a square matrix having exactly one nonzero entry in each row and each column, that entry being a 1. In other words, a permutation matrix M reads

$$m_{ij} = \delta_i^{\sigma(j)}$$

for some permutation $\sigma \in S_n$.

A square matrix for which $i < j$ implies $m_{ij} = 0$ is called a *lower-triangular* matrix. It is *upper*-triangular if $i > j$ implies $m_{ij} = 0$. It is *strictly* upper- (respectively, *lower*)-triangular if $i \geq j$ (respectively, $i \leq j$) implies $m_{ij} = 0$. It is *diagonal* if m_{ij} vanishes for every pair (i, j) such that $i \neq j$. When $d_1, \ldots, d_n \in A$ are given, one denotes by $\mathrm{diag}(d_1, \ldots, d_n)$ the diagonal matrix M whose diagonal term m_{ii} equals d_i for every index i. See below typical triangular and diagonal matrices.

$$L = \begin{pmatrix} * & 0 & \cdots & 0 \\ * & \ddots & \ddots & \vdots \\ \vdots & \ddots & \ddots & 0 \\ * & \cdots & * & * \end{pmatrix}, \quad U = \begin{pmatrix} * & * & \cdots & * \\ 0 & \ddots & \ddots & \vdots \\ \vdots & \ddots & \ddots & * \\ 0 & \cdots & 0 & * \end{pmatrix}, \quad D = \begin{pmatrix} * & 0 & \cdots & 0 \\ 0 & \ddots & \ddots & \vdots \\ \vdots & \ddots & \ddots & 0 \\ 0 & \cdots & 0 & * \end{pmatrix}.$$

When $m = 1$, a matrix M of size $n \times 1$ is called a *column vector*. One identifies it with the vector of A^n whose ith coordinate in the canonical basis is m_{i1}. This identification is an isomorphism between $\mathbf{M}_{n \times 1}(A)$ and A^n. Likewise, the matrices of size $1 \times m$ are called *row vectors*.

A matrix $M \in \mathbf{M}_{n \times m}(A)$ may be viewed as the ordered list of its columns $M^{(j)}$ ($1 \leq j \leq m$). When the set of scalars is a field, the dimension of the linear subspace spanned by the $M^{(j)}$s in K^n is called the *rank* of M and denoted by $\mathrm{rk}\, M$.

Here are examples of row and column matrices, and of an $n \times m$ matrix written rowwise and columnwise:

$$R = \begin{pmatrix} * & \cdots & * \end{pmatrix}, \quad C = \begin{pmatrix} * \\ \vdots \\ * \end{pmatrix}, \quad M = \begin{pmatrix} R_1 \\ \hline \vdots \\ \hline R_m \end{pmatrix} = \begin{pmatrix} C_1 \mid \cdots \mid C_m \end{pmatrix}.$$

2.1.4 Transposition

Definition 2.1 *The transpose matrix of* $M \in \mathbf{M}_{n \times m}(A)$ *is the matrix* $M^T \in \mathbf{M}_{m \times n}(A)$ *defined as*

$$m_{ij}^T := m_{ji}, \qquad \forall 1 \leq i \leq m, 1 \leq j \leq n.$$

Mind that the numbers of rows and columns are exchanged.

For instance, the transpose of a column is a row, and conversely.
  The following formulæ are obvious.

$$(aM+N)^T = aM^T + N^T, \qquad \left(M^T\right)^T = M.$$

When M is a square matrix, M and M^T have the same size and we can compare them. We thus say that $M \in \mathbf{M}_n(A)$ is *symmetric* if $M^T = M$, and *skew-symmetric* if $M^T = -M$ (notice that these two notions coincide when K has characteristic 2). We denote by $\mathbf{Sym}_n(K)$ the subset of symmetric matrices in $\mathbf{M}_n(K)$. It is a linear subspace of $\mathbf{M}_n(K)$.

2.1.5 Writing a Matrix Blockwise

The size $n \times m$ of a matrix can be quite large, and its entries may have nice patterns such as repetitions or lots of zeroes. It often helps to partition a matrix into blocks, in order to better understand its overall structure. For this purpose, we may write a matrix blockwise. The standard way to do so is to choose partitions of n and m:

$$n = n_1 + \cdots + n_r, \qquad m = m_1 + \cdots + m_s,$$

with $0 \leq n_p, m_q$. For each $1 \leq p \leq r$ and $1 \leq q \leq s$, let us form the submatrix $M_{pq} \in \mathbf{M}_{n_p \times m_q}(A)$ whose entries are

$$m_{pq,ij} := m_{\nu_{p-1}+i,\mu_{q-1}+j}, \qquad \nu_{p-1} := n_1 + \cdots + n_{p-1}, \ \mu_{q-1} := m_1 + \cdots + m_{q-1}.$$

As usual, $\nu_0 = \mu_0 = 0$. Then M is nothing but an $r \times s$ matrix whose (p,q)-entry is M_{pq}. Mind that these entries belong to distinct rings, inasmuch as the numbers n_p (respectively, m_q) need not be equal. Here is an example with $r = s = 2$, where we have indicated the partitions

$$M = \begin{pmatrix} 0 & 1 & 2 & 3 & | & 4 \\ \hline 5 & 6 & 7 & 8 & | & 9 \\ 10 & 11 & 12 & 13 & | & 14 \\ 15 & 16 & 17 & 18 & | & 19 \end{pmatrix}.$$

In this example, M_{11} is a row, M_{22} a column, and M_{12} is just a 1×1 matrix, that is, a scalar!

Multiplication of a matrix by a scalar can be done blockwise: we have $(aM)_{pq} = aM_{pq}$. The same remark holds true for the addition of two $n \times m$ matrices M and N, provided we choose the same partitions for each: $(M+N)_{pq} = M_{pq} + N_{pq}$.

2.1.6 Writing Blockwise Square Matrices

When $n = m$, it is often useful to choose the same partition for columns and rows: $r = s$ and $m_p = n_p$ for every $1 \leq p \leq r$. We say that M is *blockwise upper-* (respectively, lower)-*triangular* if $p > q$ (respectively, $p < q$) implies $M_{pq} = 0_{n_p \times n_q}$. We also speak of *block-triangular* matrices. A block-triangular matrix need not be triangular; after all, it is not necessarily a square matrix. Likewise, M is *blockwise diagonal* if $p \neq q$ implies $M_{pq} = 0_{n_p \times n_q}$. Again, a blockwise diagonal (or *block-diagonal*) matrix need not be diagonal. If $n_p \times n_p$ matrices M_{pp} are given, we form the block-diagonal matrix $\mathrm{diag}(M_{11}, \ldots, M_{rr})$.

2.2 Matrices as Linear Maps

2.2.1 Matrix of a Linear Map

Let K be a field and E, F be finite-dimensional vector spaces over K. Let us choose a basis $\mathscr{B}_E = \{\mathbf{e}^1, \ldots, \mathbf{e}^m\}$ of E and a basis $\mathscr{B}_F = \{\mathbf{f}^1, \ldots, \mathbf{f}^n\}$ of F. Thus $\dim E = m$ and $\dim F = n$.

A linear map $u \in \mathscr{L}(E;F)$ can be described by its action over $\mathscr{B}_E$: let m_{ij} be the coordinate of $u(\mathbf{e}^j)$ in the basis $\mathscr{B}_F$; that is,

$$u(\mathbf{e}^j) = \sum_{i=1}^{n} m_{ij}\mathbf{f}^i.$$

The numbers m_{ij} are the entries of an $n \times m$ matrix which we call M. By linearity, one finds the image of a general vector $x \in E$:

$$u\left(\sum_{j=1}^{m} x_j \mathbf{e}^j\right) = \sum_{i=1}^{n}\left(\sum_{j=1}^{m} m_{ij}x_j\right)\mathbf{f}^i. \tag{2.1}$$

Conversely, given a matrix $M \in \mathbf{M}_{n \times m}(K)$, the formula (2.1) defines a linear map u. We therefore have a one-to-one correspondance $u \leftrightarrow M$ between $\mathscr{L}(E;F)$ and $\mathbf{M}_{n \times m}(K)$. We say that M is the *matrix of u in the bases* $\mathscr{B}_E$ *and* $\mathscr{B}_F$. We warn the reader that this bijection is by no means canonical, because it depends upon the choice of the bases. We sometimes employ the notation M_u for the matrix associated with u, and u_M for the linear map associated with M, but this is dangerous because the bases are not indicated explicitly; this notation is recommended only when it is clear for both the writer and the reader what the bases of the underlying spaces are.

The addition of matrices is nothing but the addition of the linear maps, and the same can be said for multiplication by a scalar:

$$M_u + M_v = M_{u+v}, \qquad u_M + u_{M'} = u_{M+M'}, \qquad M_{\lambda u} = \lambda M_u, \qquad u_{\lambda M} = \lambda u_M.$$

The bijection above is thus an isomorphism between the vector spaces $\mathscr{L}(E;F)$ and $\mathbf{M}_{n \times m}(K)$.

The jth column of M is the representation of $u_M(\mathbf{e}^j)$ in the basis $\mathscr{B}_F$. The space spanned by the $M^{(j)}$s is thus in one-to-one correspondence with the space spanned by the $u_M(\mathbf{e}^j)$s, which is nothing but the range of u_M. Thus the rank of M equals that of u_M.

2.2.1.1 Transposition versus Duality

Let $u \in \mathscr{L}(E;F)$ be given and let us choose bases $\mathscr{B}_E$ and $\mathscr{B}_F$. We recall that the dual basis of $\mathscr{B}_E$ is a basis of the dual space E'. Likewise, the dual basis of $\mathscr{B}_F$ is a basis of the dual space F'.

Proposition 2.1 *Let M be the matrix associated with u in the bases $\mathscr{B}_E$ and $\mathscr{B}_F$. Then the matrix of the adjoint u^* in the dual bases is M^T.*

Proof. Let v^j be the elements of $\mathscr{B}_E$, w^k those of $\mathscr{B}_F$, and α^j, β^k those of the dual bases. We have $\alpha^j(v^i) = \delta_i^j$ and $\beta^\ell(w^k) = \delta_\ell^k$.

Let M' be the matrix of u^*. We have

$$u^*(\beta^\ell)(v^j) = \beta^\ell(u(v^j)) = \beta^\ell \left(\sum_i m_{ij} w^i \right) = m_{\ell j}.$$

Therefore

$$u^*(\beta^\ell) = \sum_j m_{\ell j} \alpha^j,$$

showing that $m'_{j\ell} = m_{\ell j}$. $\square$

2.2.2 Multiplication of Matrices

Let E, F, and G be three vector spaces over K, of respective dimensions p, m, n. Let $\mathscr{B}_E$, $\mathscr{B}_F$, and $\mathscr{B}_G$ be respective bases. Using the isomorphism above, we can define a product of matrices by using the composition of maps. If $M \in \mathbf{M}_{n \times m}(K)$ and $M' \in \mathbf{M}_{m \times p}(K)$, then we have two linear maps

$$u_M \in \mathscr{L}(F;G), \qquad u_{M'} \in \mathscr{L}(E;F).$$

We define MM' as the matrix of $u_M \circ u_{M'}$.

At first glance, this definition depends heavily on the choice of three bases. But the following calculation shows that it does not at all. Denote M'' the product MM'. Then

$$\sum_{i=1}^{n} m''_{ik}\mathbf{g}^i = u_M \circ u_{M'}(\mathbf{e}^k) = u_M \left(\sum_{j=1}^{n} m'_{jk}\mathbf{f}^j \right)$$

$$= \sum_{j=1}^{n} m'_{jk} u_M(\mathbf{f}^j) = \sum_{j=1}^{n} m'_{jk} \left(\sum_{i=1}^{n} m_{ij}\mathbf{g}^i \right)$$

$$= \sum_{i=1}^{n} \left(\sum_{j=1}^{n} m_{ij}m'_{jk} \right) \mathbf{g}^i.$$

Finally, the matrix $M'' = MM'$ is given by the formula

$$m''_{ij} = \sum_{k=1}^{m} m_{ik}m'_{kj}, \quad 1 \le i \le n, 1 \le j \le p, \tag{2.2}$$

which is clearly independent of the chosen bases.

We point out that a product of matrices MM' makes sense as long as the number of columns of M equals the number of rows of M', and only in this situation. If MN makes sense, then $N^T M^T$ does too, and we have the obvious formula

$$(MN)^T = N^T M^T,$$

where we warn the reader that the positions of M and N are flipped under transposition.

Thanks to the associativity of the composition, the product is associative:

$$(MP)Q = M(PQ),$$

whenever the sizes agree. Likewise, the product is distributive with respect to the addition, and associates with the scalar multiplication:

$$M(P+Q) = MP + MQ, \qquad (P+Q)M = PM + QM, \qquad (aM)M' = a(MM').$$

The following formula extends that for linear maps (see Exercise 2)

$$\mathrm{rk}(MM') \le \min\{\mathrm{rk}\, M, \mathrm{rk}\, M'\}.$$

In particular, we have the following.

Proposition 2.2 *The rank of a submatrix of M is not larger than that of M.*

Proof. Just remark that the submatrix M' formed by retaining only the rows of indices $i_1 < \cdots < i_r$ and the columns of indices $j_1 < \cdots < j_r$ is given by a formula $M' = PMQ$ where P is the matrix of projection from K^n over the space spanned by $\mathbf{f}^{i_1}, \ldots, \mathbf{f}^{i_r}$, and Q is the embedding matrix from the space spanned by $\mathbf{e}^{j_1}, \ldots, \mathbf{e}^{j_r}$ over K^m. Then

$$\mathrm{rk}(M') = \mathrm{rk}(PMQ) \le \mathrm{rk}(MQ) \le \mathrm{rk}(M).$$

□

2.2.2.1 Matrices with Entries in a Ring

When the scalar set is a ring A, the formula (2.2) still makes sense and lets us define a product MN when $M \in \mathbf{M}_{n \times m}(A)$ and $N \in \mathbf{M}_{m \times p}(A)$. Associativity is straightforward. In particular $\mathbf{M}_n(A)$ is itself a ring, although a noncommutative one, even if A is Abelian.

2.2.2.2 The Case of Square Matrices

Square matrices of a given size can be multiplied together, which makes $\mathbf{M}_n(K)$ an algebra. We cannot emphasize enough that *the multiplication of matrices is not commutative*: in general, MM' differs from $M'M$. This is reminiscent of the lack of commutativity of the composition of endomorphisms. It is an endless source of interesting questions regarding matrices. For instance,

$$\begin{pmatrix} 0 & 1 \\ 0 & 0 \end{pmatrix} \begin{pmatrix} 0 & 0 \\ 1 & 0 \end{pmatrix} = \begin{pmatrix} 1 & 0 \\ 0 & 0 \end{pmatrix} \neq \begin{pmatrix} 0 & 0 \\ 0 & 1 \end{pmatrix} = \begin{pmatrix} 0 & 0 \\ 1 & 0 \end{pmatrix} \begin{pmatrix} 0 & 1 \\ 0 & 0 \end{pmatrix}.$$

We say that two matrices $M, N \in \mathbf{M}_n(K)$ *commute* to each other if $MN = NM$. To quantify the lack of commutativity, we define the *commutator* of square matrices M, N by

$$[M, N] := MN - NM.$$

Section 4.4 discusses the amount of noncommutativity in $\mathbf{M}_n(K)$.

In $\mathbf{M}_n(K)$, the unit matrix I_n is a neutral element for the multiplication:

$$I_n M = M I_n = M.$$

We thus have the standard notion of *inverse*: a matrix $M \in \mathbf{M}_n(K)$ is invertible if there exists $N \in \mathbf{M}_n(K)$ such that $NM = MN = I_n$. We say that N is the *inverse* of M and we denote it M^{-1}. We could as well define right-inverse and left-inverse, but we show (Proposition 3.5) that the three notions coincide. We say that M is *invertible* or *nonsingular*. A characterization of invertible matrices is given in Chapter 3. As in every algebra, the product of nonsingular matrices is nonsingular, and we have

$$(MN)^{-1} = N^{-1} M^{-1}, \qquad \left(M^{-1}\right)^{-1} = M.$$

The subset of nonsingular matrices of size n is a multiplicative group, which is denoted by $\mathbf{GL}_n(K)$.

Powers of a square matrix M are defined inductively by $M^2 = MM$, $M^3 = MM^2 = M^2 M$ (from associativity), ..., $M^{k+1} = M^k M$. We complete this notation by $M^1 = M$ and $M^0 = I_n$, so that $M^j M^k = M^{j+k}$ for all $j, k \in \mathbb{N}$. The powers of a square matrix M commute pairwise. In particular, the set $K[M]$ formed by polynomials in M, which consists of matrices of the form

$$a_0 I_n + a_1 M + \cdots + a_r M^r, \quad a_0, \ldots, a_r \in K, \quad r \in \mathbb{N},$$

is a commutative algebra.

If M is nonsingular, we define $M^{-k} := (M^k)^{-1} = (M^{-1})^k$, which yields $M^j M^k = M^{j+k}$ for all $j, k \in \mathbb{Z}$.

If $M^k = 0_n$ for some integer $k \in \mathbb{N}$, we say that M is *nilpotent*. We say that M is *idempotent* if $I_n - M$ is nilpotent.

A matrix $M \in \mathbf{M}_n(K)$ is *orthogonal* if $M^T M = M M^T = I_n$. It is equivalent to saying that M is nonsingular and M^{-1} is the transpose of M. The set of orthogonal matrices is a multiplicative group in $\mathbf{M}_n(K)$, called the *orthogonal* and denoted $\mathbf{O}_n(K)$.

2.2.2.3 Multiplication of a Vector and a Matrix

Another interesting case is that of multiplication with a column vector. If $M \in \mathbf{M}_{n \times m}(K)$ and $X \in K^m$, the product MX makes sense because X can be viewed as an $m \times 1$ matrix. The result is an $n \times 1$ matrix, that is, a vector Y in K^n, given by

$$y_i = \sum_{j=1}^m m_{ij} x_j.$$

In the terminology of Section 2.2.1, M induces a linear map $u_M \in \mathscr{L}(K^m; K^n)$, which refers to the choice of the canonical bases; this correspondence is thus canonical somehow. When $n = m$, $\mathbf{M}_n(K)$ operates over K^n and is canonically isomorphic to $\mathbf{End}(K^n)$.

The above action of a given matrix is the straightforward translation of that of its associated linear map: if x and y are the vectors associated with the columns X and Y, then $y = u_M(x)$. This leads us to extend several notions already encountered for linear maps, such as the kernel and the range:

$$\ker M = \{X \in K^m \,|\, MX = 0\}, \qquad R(M) = \{MX \,|\, X \in K^m\}.$$

The *rank* is the dimension of $R(M)$ and is denoted by $\mathrm{rk}\, M$. Theorem 1.2 becomes the following.

Proposition 2.3 *Let K be a field. If $M \in \mathbf{M}_{n \times m}(K)$, then*

$$m = \dim \ker M + \mathrm{rk}\, M.$$

2.2.2.4 Scalar Product as a Matrix Product

When ℓ is a row vector and y a column vector with the same number of entries, then ℓy is a 1×1 matrix, that is, a scalar. This can be interpreted simply in terms of linear algebra: y is the matrix of an element of K^n (which we still denote y) in the canonical basis, and ℓ is the matrix of a linear form f over K^n, in the dual basis. Then ℓy is nothing but $f(y)$. We notice that $\ell = 0$ if and only if $\ell y = 0$ for every y. Likewise, $y = 0$ if and only if $\ell y = 0$ for every ℓ.

When x and y are both in K^n, then x^T is a row vector. We can form their product $x^T y$, which is the *canonical scalar product* over K^n:

$$x^T y = x_1 y_1 + \cdots + x_n y_n.$$

We notice that $x = 0$ if and only if $x^T y = 0$ for every y. Thus the scalar product is a nondegenerate bilinear form over K^n. When $x^T y = 0$, we say that x and y are *orthogonal* (to each other) and we denote $x \perp y$. If $x \perp x$, we say that x is *isotropic*. We warn the reader that for many fields K, there are nonzero isotropic vectors, even though there is not if $K = \mathbb{R}$. For instance, if $K = \mathbb{C}$ the vector

$$x = \begin{pmatrix} 1 \\ i \end{pmatrix}$$

is isotropic.

If $x \in K^n$, the map $y \mapsto \ell_x(y) := x^T y$ is a linear form over K^n. In addition, the map $x \mapsto \ell_x$ is one-to-one. Because $(K^n)'$ has the same dimension n, this map is an isomorphism. We thus identify K^n with its dual space in a natural way.

Two subsets A and B of K^n are *orthogonal* if every vector of A is orthogonal to every vector of B. The *orthogonal* of a subset A is the set of all vectors in K^n that are orthogonal to A; it is denoted $A^\perp$. Because of the linearity of the scalar product with respect to each argument, the orthogonal $A^\perp$ is a subspace of K^n. For the same reason, we have

$$(\mathrm{Span}(A))^\perp = A^\perp. \tag{2.3}$$

Obviously, $A \subset B$ implies $B^\perp \subset A^\perp$.

When identifiying K^n with its dual, the orthogonal of S identifies to the polar set S^0. We therefore rephrase the results obtained in Paragraph 1.2.2:

Proposition 2.4 *If E is a subspace of K^n, then*

$$\dim E^\perp + \dim E = n.$$

Proposition 2.5 *Given a subset A of K^n, its biorthogonal is the subspace spanned by A:*

$$\left(A^\perp \right)^\perp = \mathrm{Span}(A).$$

2.2.2.5 Range, Kernel, and Duality

Let $M \in \mathbf{M}_{n \times m}(K)$ and $x \in \ker M^T$. Then $x^T M = (M^T x)^T = 0$. If $y \in K^m$, there follows $x^T M y = 0$. In other words, x is orthogonal to the range of M.

Conversely, let x be orthogonal to $R(M)$. Then $(M^T x)^T y = x^T (My) = 0$ for every $y \in K^m$. This tells us that $M^T x = 0$, and proves that the orthogonal of $R(M)$

is $\ker M^T$. Applying Proposition 2.5, we find also that the orthogonal of $\ker M^T$ is $R(M)$. Exchanging the roles of M and M^T leads to the following.

Proposition 2.6 *The orthogonal of $R(M)$ is $\ker M^T$ and that of $\ker M$ is $R(M^T)$.*

The following consequence is sometimes called the *Fredholm principle*.

Corollary 2.1 *Let $M \in \mathbf{M}_{n\times m}(K)$ and $b \in K^n$. In order that the linear equation $Mx = b$ be solvable, it is necessary and sufficient that $z^T b = 0$ for every $z \in \ker(M^T)$.*

Assembling Propositions 2.3, 2.4, and 2.6, we obtain the following identities for a matrix $M \in \mathbf{M}_{n\times m}(K)$:

$$m = \dim \ker M + \mathrm{rk}\,M, \qquad n = \dim \ker M^T + \mathrm{rk}\,M^T,$$
$$n = \dim \ker M^T + \mathrm{rk}\,M, \qquad m = \dim \ker M + \mathrm{rk}\,M^T.$$

Besides some redundancy, this list has an interesting consequence:

Proposition 2.7 *For every $M \in \mathbf{M}_{n\times m}(K)$, there holds*

$$\mathrm{rk}\,M^T = \mathrm{rk}\,M.$$

The kernels, however, do not have the same dimension if $m \neq n$. Only for square matrices, we deduce the following.

Proposition 2.8 *If $M \in \mathbf{M}_n(K)$ is a square matrix, then*

$$\dim \ker M^T = \dim \ker M.$$

Corollary 2.2 *If $M \in \mathbf{M}_n(K)$, then*

$$(M : K^n \to K^n \text{ is bijective}) \Longleftrightarrow \ker M = \{0\} \Longleftrightarrow \mathrm{rk}\,M = n.$$

In particular, there is a well-defined notion of inverse in $\mathbf{M}_n(K)$: a left-inverse exists if and only if a right-inverse exists, and then they are equal to each other. In particular, this inverse is unique.

Going back to $n \times m$ matrices, we say that M is a *rank-one* matrix if $\mathrm{rk}\,M = 1$. A rank-one matrix decomposes as xy^T where $x \in K^n$ spans $R(M)$ and $y \in K^m$ spans $R(M^T)$ (remark that M^T is rank-one too, because of Proposition 2.7).

2.2.3 Change of Basis

Let E be a K-vector space, in which one chooses a basis $\beta = \{e_1, \ldots, e_n\}$. Choosing another n-tuple $\beta' = \{e'_1, \ldots, e'_n\}$ in E amounts to prescribing the coordinates of each e'_i in the basis β:

$$e'_i = \sum_{j=1}^n p_{ji} e_j.$$

It is thus equivalent to selecting a matrix $P \in \mathbf{M}_n(K)$. Whether β' is a basis of E depends on whether P is nonsingular: If P is nonsingular and $Q := P^{-1}$, then a straightforward calculation yields

$$e_j = \sum_{i=1}^{n} q_{ji} e_i',$$

which shows that β' is generating, thus a basis, because of cardinality. Conversely, if β' is a basis, then the coordinates of each e_i in β' provide a matrix Q which is nothing but the inverse of P.

Proposition 2.9 *The matrix P above is nonsingular if and only if β' is another basis of E.*

Definition 2.2 *The matrix P above is the matrix of the change of basis from β to β'.*

The matrix M_u of a linear map $u \in \mathscr{L}(E; F)$ depends upon the choice of the bases of E and F. Therefore it must be modified when they are changed. The following formula describes this modification. Let β, β' be two bases of E, and γ, γ' two bases of F. Let M be the matrix of u associated with the bases (β, γ), and M' be that associated with (β', γ'). Finally, let P be the matrix of the change of basis $\beta \mapsto \beta'$ and Q that of $\gamma \mapsto \gamma'$. We have $P \in \mathbf{GL}_m(K)$ and $Q \in \mathbf{GL}_n(K)$.

With obvious notations, we have

$$f_k' = \sum_{i=1}^{n} q_{ik} f_i, \qquad e_j' = \sum_{\ell=1}^{m} p_{\ell j} e_\ell.$$

We have

$$u(e_j') = \sum_{k=1}^{n} m_{kj}' f_k' = \sum_{i,k=1}^{n} m_{kj}' q_{ik} f_i.$$

On the other hand, we have

$$u(e_j') = u\left(\sum_{\ell=1}^{m} p_{\ell j} e_\ell \right) = \sum_{\ell=1}^{m} p_{\ell j} \sum_{i=1}^{m} m_{i\ell} f_i.$$

Comparing the two formulæ, we obtain

$$\sum_{\ell=1}^{m} m_{i\ell} p_{\ell j} = \sum_{k=1}^{n} q_{ik} m_{kj}', \qquad \forall 1 \leq i \leq n, \, 1 \leq j \leq m.$$

This exactly means the formula

$$MP = QM'. \tag{2.4}$$

Definition 2.3 *Two matrices $M, M' \in \mathbf{M}_{n \times m}(K)$ are equivalent if there exist two matrices $P \in \mathbf{GL}_m(K)$ and $Q \in \mathbf{GL}_n(K)$ such that equality (2.4) holds true.*

Thus equivalent matrices represent the same linear map in different bases.

2.2.3.1 The Situation for Square Matrices

When $F = E$ and thus $m = n$, it is natural to represent $u \in \mathbf{End}(E)$ by using only one basis, that is, choosing $\beta' = \beta$ with the notations above. In a change of basis, we have likewise $\gamma' = \gamma$, which means that $Q = P$. We now have

$$MP = PM',$$

or equivalently

$$M' = P^{-1}MP. \tag{2.5}$$

Definition 2.4 *Two matrices* $M, M' \in \mathbf{M}_n(K)$ *are* similar *if there exists a matrix* $P \in \mathbf{GL}_n(K)$ *such that equality* (2.5) *holds true.*

Thus similar matrices represent the same endomorphism in different bases.

The equivalence and the similarity of matrices both are equivalence relations. They are studied in detail in Chapter 9.

2.2.4 Multiplying Blockwise

Let $M \in \mathbf{M}_{n \times m}(K)$ and $M' \in \mathbf{M}_{m' \times p}(K)$ be given. We assume that partitions

$$n = n_1 + \cdots + n_r, \qquad m = m_1 + \cdots + m_s,$$
$$m' = m_1' + \cdots + m_s', \qquad p = p_1 + \cdots + p_t$$

have been chosen, so that M and M' can be written blockwise with blocks $M_{\alpha\beta}$ and $M'_{\beta\gamma}$ with $\alpha = 1, \ldots, r$, $\beta = 1, \ldots, s$, $\gamma = 1, \ldots, t$. We can make the product MM', which is an $n \times p$ matrix, provided that $m' = m$. On the other hand, we wish to use the block form to calculate this product more concisely. Let us write blockwise MM' by using the partitions

$$n = n_1 + \cdots + n_r, \qquad p = p_1 + \cdots + p_t.$$

We expect that the blocks $(MM')_{\alpha\gamma}$ obey a simple formula, say

$$(MM')_{\alpha\beta} = \sum_{\beta=1}^{s} M_{\alpha\beta}M'_{\beta\gamma}. \tag{2.6}$$

The block products $M_{\alpha\beta}M'_{\beta\gamma}$ make sense provided $m'_\beta = m_\beta$ for every $\beta = 1, \ldots, s$ (which in turn necessitates $m' = m$). Once this requirement is fulfilled, it is easy to see that formula (2.6) is correct. We leave its verification to the reader as an exercise.

In conclusion, multiplication of matrices written blockwise follows the same rule as when the matrices are given entrywise. The multiplication is done in two stages: one level using block multiplication, the other one using multiplication in K. Actually, we may have as many levels as wished, by writing blocks blockwise (using subblocks), and so on. This recursive strategy is employed in Section 11.1.

2.3 Matrices and Bilinear Forms

Let E, F be two K-vector spaces. One chooses two respective bases $\beta = \{e_1, \ldots, e_n\}$ and $\gamma = \{f_1, \ldots, f_m\}$. If $B : E \times F \to K$ is a bilinear form, then

$$B(x,y) = \sum_{i,j} B(e_i, f_j) x_i y_j,$$

where the x_is are the coordinates of x in β and the y_js are those of y in γ. Let us define a matrix $M \in \mathbf{M}_{n \times m}(K)$ by $m_{ij} = B(e_i, f_j)$. Then B can be recovered from the formula

$$B(x,y) := x^T M y = \sum_{i,j} m_{ij} x_i y_j. \tag{2.7}$$

Conversely, if $M \in \mathbf{M}_{n \times m}(K)$ is given, one can construct a bilinear form on $E \times F$ by applying (2.7). We say that M is the *matrix of the bilinear form B*, or that B is the bilinear form associated with M. We warn the reader that once again, the correspondence $B \leftrightarrow M$ depends upon the choice of the bases.

This correspondence is a (noncanonical) isomorphism between $\mathbf{Bil}(E, F)$ and $\mathbf{M}_{n \times m}(K)$. We point out that, opposite to the isomorphism with $\mathscr{L}(E; F)$, n is now the dimension of E and m that of F.

If M is associated with B, its transpose M^T is associated with the bilinear form B_T defined on $F \times E$ by

$$B_T(y,x) := B(x,y).$$

When $F = E$, it makes sense to assume that $\gamma = \beta$. Then M is symmetric if and only if B is symmetric: $B(x,y) = B(y,x)$. Likewise, one says that M is *alternate* if B itself is an alternate form. This is equivalent to saying that

$$m_{ij} + m_{ji} = 0, \qquad m_{ii} = 0, \qquad \forall 1 \le i, j \le n.$$

An alternate matrix is skew-symmetric, and the converse is true if $\mathrm{charc}(K) \ne 2$. If $\mathrm{charc}(K) = 2$, an alternate matrix is a skew-symmetric matrix whose diagonal vanishes.

2.3.1 Change of Bases

As for matrices associated with linear maps, we need a description of the effect of a change of bases for the matrix associated with a bilinear form.

Denoting again by P, Q the matrices of the changes of basis $\beta \mapsto \beta'$ and $\gamma \mapsto \gamma'$, and by M, M' the matrices of B in the bases (β, γ) or (β', γ'), respectively, one has

$$m'_{ij} = B(e'_i, f'_j) = \sum_{k,l} p_{ki} q_{lj} B(e_k, f_\ell) = \sum_{k,l} p_{ki} q_{lj} m_{kl}.$$

Therefore,

$$M' = P^T M Q.$$

The case $F = E$

When $F = E$ and $\gamma = \beta$, $\gamma' = \beta'$, the change of basis has the effect of replacing M by $M' = P^T M P$. We say that M and M' are *congruent*. If M is symmetric, then M' is too. This was expected, inasmuch as one expresses the symmetry of the underlying bilinear form B.

If the characteristic of K is distinct from 2, there is an isomorphism between $\mathbf{Sym}_n(K)$ and the set of quadratic forms on K^n. This isomorphism is given by the formula

$$Q(e_i + e_j) - Q(e_i) - Q(e_j) = 2m_{ij}.$$

In particular, $Q(e_i) = m_{ii}$.

Exercises

1. Let G be an $\mathbb{R}$-vector space. Verify that its complexification $G \otimes_{\mathbb{R}} \mathbb{C}$ is a $\mathbb{C}$-vector space and that $\dim_{\mathbb{C}} G \otimes_{\mathbb{R}} \mathbb{C} = \dim_{\mathbb{R}} G$.
2. Let $M \in \mathbf{M}_{n \times m}(K)$ and $M' \in \mathbf{M}_{m \times p}(K)$ be given. Show that

$$\mathrm{rk}(MM') \leq \min\{\mathrm{rk}\, M,\ \mathrm{rk}\, M'\}.$$

   First show that $\mathrm{rk}(MM') \leq \mathrm{rk}\, M$, and then apply this result to the transpose matrix.
3. Let K be a field and let A, B, C be matrices with entries in K, of respective sizes $n \times m$, $m \times p$, and $p \times q$.

   a. Show that $\mathrm{rk}\, A + \mathrm{rk}\, B \leq m + \mathrm{rk}\, AB$. It is sufficient to consider the case where B is onto, by considering the restriction of A to the range of B.

   b. Show that $\mathrm{rk}\, AB + \mathrm{rk}\, BC \leq \mathrm{rk}\, B + \mathrm{rk}\, ABC$. One may use the vector spaces $K^p/\ker B$ and $R(B)$, and construct three homomorphisms u, v, w, with v being onto.
4. a. Let $n, n', m, m' \in \mathbb{N}^*$ and let K be a field. If $B \in \mathbf{M}_{n \times m}(K)$ and $C \in \mathbf{M}_{n' \times m'}(K)$, one defines a matrix $B \otimes C \in \mathbf{M}_{nn' \times mm'}(K)$, the tensor product, whose block form is

$$B \otimes C = \begin{pmatrix} b_{11}C & \cdots & b_{1m}C \\ \vdots & & \vdots \\ b_{n1}C & \cdots & b_{nm}C \end{pmatrix}.$$

Show that $(B,C) \mapsto B \otimes C$ is a bilinear map and prove that its range spans $\mathbf{M}_{nn' \times mm'}(K)$. Is this map onto?

b. If $p, p' \in \mathbb{N}^*$ and $D \in \mathbf{M}_{m \times p}(K)$, $E \in \mathbf{M}_{m' \times p'}(K)$, then compute $(B \otimes C)(D \otimes E)$.

c. Show that for every bilinear form $\phi : \mathbf{M}_{n \times m}(K) \times \mathbf{M}_{n' \times m'}(K) \to K$, there exists one and only one linear form $L : \mathbf{M}_{nn' \times mm'}(K) \to K$ such that $L(B \otimes C) = \phi(B,C)$.

Chapter 3
Square Matrices

The essential ingredient for the study of square matrices is the determinant. For reasons given in Section 3.5, as well as in Chapter 9, it is useful to consider matrices with entries in a ring. This allows us to consider matrices with entries in $\mathbb{Z}$ (rational integers) as well as in $K[X]$ (polynomials with coefficients in K). We assume that the ring of scalars A is a commutative (meaning that the multiplication is commutative) integral domain (meaning that it does not have divisors of zero: $ab = 0$ implies either $a = 0$ or $b = 0$), with a unit denoted by 1, that is, an element satisfying $1x = x1 = x$ for every $x \in A$.

An element a of A is *invertible* if there exists $b \in A$ such that $ab = 1$. The element b is unique (because A is an integral domain), and is called the *inverse* of a, with the notation $b = a^{-1}$. The set of invertible elements of A is a multiplicative group, denoted by A^*. One has

$$(ab)^{-1} = b^{-1}a^{-1} = a^{-1}b^{-1}.$$

3.1 Determinant

We recall that S_n, the *symmetric group*, denotes the group of permutations over the set $\{1, \ldots, n\}$. We denote by $\varepsilon(\sigma) = \pm 1$ the signature of a permutation σ, equal to $+1$ if σ is the product of an even number of transpositions, and -1 otherwise. Two explicit formulæ are

$$\varepsilon(\sigma) = \prod_{i<j} \frac{\sigma(j) - \sigma(i)}{j - i},$$

or

$$\varepsilon(\sigma) = (-1)^{\#\{(i,j) \mid 1 \leq i < j \leq n \text{ and } \sigma(j) < \sigma(i)\}}. \tag{3.1}$$

Recall that $\varepsilon(\sigma\sigma') = \varepsilon(\sigma)\varepsilon(\sigma')$.

Definition 3.1 *Let $M \in \mathbf{M}_n(A)$ be a square matrix. Its determinant is defined as*

$$\det M := \sum_{\sigma \in S_n} \varepsilon(\sigma) m_{1\sigma(1)} \cdots m_{n\sigma(n)},$$

where the sum ranges over all the permutations of the integers $1, \ldots, n$.

In the sequel, we use the notation

$$\pi_\sigma(M) := m_{1\sigma(1)} \cdots m_{n\sigma(n)},$$

so that we may write

$$\det M = \sum_{\sigma \in S_n} \varepsilon(\sigma) \pi_\sigma(M).$$

When M is given entrywise, we sometimes write its determinant by replacing the parentheses by vertical bars:

$$\det M =: \begin{vmatrix} m_{11} & \ldots & m_{1n} \\ \vdots & \ddots & \vdots \\ m_{n1} & \ldots & m_{nn} \end{vmatrix}.$$

Here are the formulæ for the determinant of 2×2 and 3×3 matrices:

$$\begin{vmatrix} m_{11} & m_{12} \\ m_{21} & m_{22} \end{vmatrix} = m_{11}m_{22} - m_{12}m_{21},$$

$$\begin{vmatrix} m_{11} & m_{12} & m_{13} \\ m_{21} & m_{22} & m_{23} \\ m_{31} & m_{32} & m_{33} \end{vmatrix} = m_{11}m_{22}m_{33} - m_{11}m_{23}m_{32} - m_{13}m_{22}m_{31}$$

$$- m_{12}m_{21}m_{33} + m_{12}m_{23}m_{31} + m_{21}m_{13}m_{32}.$$

3.1.1 Elementary Properties

3.1.1.1 Determinant of a Triangular Matrix

If M is upper-triangular, then a nonvanishing product $\pi_\sigma(M)$ must satisfy $i \le \sigma(i)$ for every $i = 1, \ldots, n$. This implies

$$1 + \cdots + n \le \sigma(1) + \cdots + \sigma(n) = 1 + \cdots + n,$$

where the latter equality comes from the fact that σ is a bijection. Therefore each inequality $i \le \sigma(i)$ turns out to be an equality, which means that σ is the identity. Finally, $\det M$ reduces to $\pi_{\text{id}}(M)$. An analogous argument works for a lower-triangular matrix.

Proposition 3.1 *If* $M \in \mathbf{M}_n(A)$ *is triangular, then*

$$\det M = m_{11} \cdots m_{nn}. \tag{3.2}$$

In particular, $\det I_n = 1$ and $\det 0_n = 0$. We leave to the reader the slightly more involved, but still easy, proof that the determinant of a block-triangular matrix is equal to the product of the determinants of the diagonal blocks M_{jj}. An alternate proof of this fact follows from the Schur complement formula (3.6).

3.1.1.2 The Determinant of the Transpose Matrix

The map $\sigma \mapsto \rho := \sigma^{-1}$ is an involution of S_n. Rearranging the monomials, we have

$$\pi_\sigma(M^T) = \pi_{\sigma^{-1}}(M).$$

Because $\varepsilon(\sigma^{-1}) = \varepsilon(\sigma)$, we infer

$$\det M^T = \sum_{\sigma \in S_n} \varepsilon(\sigma) \pi_\sigma(M^T) = \sum_{\sigma \in S_n} \varepsilon(\sigma^{-1}) \pi_{\sigma^{-1}}(M) = \sum_{\rho \in S_n} \varepsilon(\rho) \pi_\rho(M),$$

hence the identity

$$\det M^T = \det M. \tag{3.3}$$

3.1.1.3 Multilinearity

Viewing $M = \left(M^{(1)}, \ldots, M^{(n)} \right)$ as a row matrix with entries in A^n, one may view the determinant as a function of the n columns of M:

$$\det M = \Delta \left(M^{(1)}, \ldots, M^{(n)} \right).$$

This expression can be used to define the determinant of n vectors $M^{(1)}, \ldots, M^{(n)}$ taken in A^n: just form the matrix M from these vectors, and then take its determinant.

The function Δ is a multilinear form: each partial map $M^{(j)} \mapsto \Delta \left(M^{(1)}, \ldots, M^{(n)} \right)$ is a linear form, because this is true for each monomial $\pi_\sigma(M)$. In addition, Δ is an *alternate* form:

Proposition 3.2 *If two columns of M are equal, then $\det M = 0$.*

Proof. Let us assume that the kth and the ℓth columns are equal, with $k < \ell$. The symmetric group S_n is the disjoint union of the alternate group A_n, made of even permutations (those with $\varepsilon(\sigma) = +1$) and of τA_n, where τ is the transposition (k, ℓ). We thus have

$$\det M = \sum_{\sigma \in A_n} \pi_\sigma(M) - \sum_{\sigma \in A_n} \pi_{\tau\sigma}(M).$$

There remains to observe that

$$\pi_\sigma(M) = \pi_{\tau\sigma}(M),$$

this because of $m_{i\sigma(i)} = m_{i\tau(\sigma(i))}$ for every $i = 1,\ldots,n$. □

More generally, let us assume that the columns of M are linearly dependent; that is,

$$a_1 M^{(1)} + \cdots + a_n M^{(n)} = 0,$$

where at least one of $a_1,\ldots,a_n$ does not vanish. Then using linearity, plus Proposition 3.2, we have

$$
\begin{aligned}
0 &= \Delta(\ldots,M^{(j-1)},0,M^{(j+1)},\ldots) \\
  &= \Delta(\ldots,M^{(j-1)},a_1 M^{(1)} + \cdots + a_n M^{(n)},M^{(j+1)},\ldots) \\
  &= \sum_{k=1}^{n} a_k \Delta(\ldots,M^{(j-1)},M^{(k)},M^{(j+1)},\ldots) \\
  &= a_j \Delta(\ldots,M^{(j-1)},M^{(j)},M^{(j+1)},\ldots) = a_j \det M.
\end{aligned}
$$

Taking an index j such that $a_j \neq 0$, and using that A is an integral domain, we obtain the following.

Corollary 3.1 *If the columns of M are linearly dependent, then* $\det M = 0$.

3.1.1.4 The Determinant as a Polynomial

Let $\{x_{ij} \mid 1 \leq i,j \leq n\}$ be indeterminates. The polynomial ring $A[x_{11},\ldots,x_{nn}]$ is an integral domain with a unit, in which we may apply the previous results. Let us define the matrix $X = (x_{ij})_{1 \leq i,j \leq n}$, which belongs to $\mathbf{M}_n(A[x_{11},\ldots,x_{nn}])$. Its determinant is a polynomial, which we denote Det_A. We observe that Det_A does not really depend on the ring A, in the sense that it is the image of $\mathrm{Det}_\mathbb{Z}$ through the canonical ring homomorphism $\mathbb{Z} \to A$. For this reason, we simply write Det.

Given $M \in \mathbf{M}_n(A)$, the determinant of M is obtained by substituting the entries m_{ij} to the indeterminates x_{ij} in Det. The determinant is thus a polynomial function.

3.2 Minors

For a matrix $M \in \mathbf{M}_{n \times m}(A)$, not necessarily a square one, and for $p \geq 1$ an integer with $p \leq m,n$, one may extract a $p \times p$ matrix $M' \in \mathbf{M}_p(A)$ by retaining only p rows and p columns of M. The determinant of such a matrix M' is called a *minor of order* p. Once the choice of the row indices $i_1 < \cdots < i_p$ and column indices $j_1 < \cdots < j_p$ has been made, one denotes by

$$M \begin{pmatrix} i_1 & i_2 & \cdots & i_p \\ j_1 & j_2 & \cdots & j_p \end{pmatrix}$$

the corresponding minor. A *principal minor* is a minor with equal row and column indices, that is, of the form

$$M \begin{pmatrix} i_1 \; i_2 \; \cdots \; i_p \\ i_1 \; i_2 \; \cdots \; i_p \end{pmatrix}.$$

In particular, the *leading* principal minor of order p is

$$M \begin{pmatrix} 1 \; 2 \; \cdots \; p \\ 1 \; 2 \; \cdots \; p \end{pmatrix}.$$

3.2.1 Cofactors

Given $i \in \{1, \ldots, n\}$, we adopt the notation $\hat{i} := (\ldots, i-1, i+1, \ldots)$, which is the $(n-1)$-uplet obtained from $(1, \ldots, n)$ by removing i. If $M \in \mathbf{M}_n(A)$ and i, j are two indices, we define the *cofactor*

$$\hat{m}_{ij} := (-1)^{i+j} M \begin{pmatrix} \hat{i} \\ \hat{j} \end{pmatrix}.$$

The matrix $\hat{M}$ of entries $\hat{m}_{ij}$ is the matrix of cofactors. Especially important is its transpose

$$\operatorname{adj} M := \hat{M}^T,$$

called the *adjugate* matrix of M.

Viewing the determinant as a polynomial in the entries of M, we can differentiate it with respect to any one. We have

Lemma 1. *For every* $1 \leq i, j \leq n$,

$$\frac{\partial \operatorname{Det}}{\partial m_{ij}}(M) = \hat{m}_{ij}.$$

Proof. Let N be the matrix obtained from M by removing the ith row and the jth column. The partial derivative of π_σ is zero, unless $\sigma(i) = j$. In the latter case, we may write

$$\pi_\sigma(M) = m_{ij} \pi_\rho(N)$$

where $\rho \in S_{n-1}$ is defined by the formula

$$\rho(k) = \begin{cases} \sigma(k) & \text{if } k < i \text{ and } \sigma(k) < j, \\ \sigma(k) - 1 & \text{if } k < i \text{ and } \sigma(k) > j, \\ \sigma(k+1) & \text{if } k \geq i \text{ and } \sigma(k+1) < j, \\ \sigma(k+1) - 1 & \text{if } k \geq i \text{ and } \sigma(k+1) > j. \end{cases}$$

On the one hand, we have

$$\frac{\partial \pi_\sigma}{\partial m_{ij}}(M) = \pi_\rho(N).$$

And we also have

$$\varepsilon(\sigma) = (-1)^{i+j}\varepsilon(\rho).$$

We conclude that

$$\frac{\partial \operatorname{Det}}{\partial m_{ij}}(M) = (-1)^{i+j} \sum_{\rho \in S_{n-1}} \varepsilon(\rho)\pi_\rho(N),$$

which is the announced formula. □

3.2.1.1 A Fundamental Application

Let us choose an index $1 \le j \le n$ and a matrix $M \in \mathbf{M}_n(A)$. Given a vector $X \in A^n$, we differentiate the determinant at M, with respect to the jth column in the direction X. Because Δ is a linear function of each column, this gives

$$\Delta(\dots, M^{(j-1)}, X, M^{(j+1)}, \dots) = \sum_{i=1}^{n} x_i \hat{m}_{ij}. \tag{3.4}$$

Applying (3.4) to $X = M^{(j)}$, we obtain

$$\det M = \sum_{i=1}^{n} m_{ij} \hat{m}_{ij}.$$

Choosing $X = M^{(k)}$ with $k \ne j$ instead, and recalling Proposition 3.2, we find

$$0 = \sum_{i=1}^{n} m_{ik} \hat{m}_{ij}.$$

Assembling both these equalities, we have proved $(\operatorname{adj} M)M = (\det M)I_n$.

Let us apply this identity to M^T instead. Remarking that $\operatorname{adj}(M^T) = (\operatorname{adj} M)^T$, and recalling (3.3), we have

$$(\operatorname{adj} M)^T M^T = (\det M)I_n.$$

Transposing the latter identity, there remains $M(\operatorname{adj} M) = (\det M)I_n$. We summarize our results.

Proposition 3.3 *If $M \in \mathbf{M}_n(A)$, one has*

$$M(\operatorname{adj} M) = (\operatorname{adj} M)M = \det M \cdot I_n. \tag{3.5}$$

Proposition 3.3 contains the well-known and important expansion formula for the determinant with respect to either a row or a column. The expansion with respect to the ith row is written

$$\det M = m_{i1}\hat{m}_{i1} + \cdots + m_{in}\hat{m}_{in},$$

and the expansion with respect to the ith column is

$$\det M = m_{1i}\hat{m}_{1i} + \cdots + m_{ni}\hat{m}_{ni}.$$

3.2.2 The Cauchy–Binet Formula

In the sequel, we also use the following result.

Proposition 3.4 *Let $B \in \mathbf{M}_{n\times m}(A)$, $C \in \mathbf{M}_{m\times l}(A)$, and an integer $p \leq n, l$ be given. Let $1 \leq i_1 < \cdots < i_p \leq n$ and $1 \leq k_1 < \cdots < k_p \leq l$ be indices. Then the corresponding minor in the product BC is given by the formula*

$$(BC)\begin{pmatrix} i_1 & i_2 & \cdots & i_p \\ k_1 & k_2 & \cdots & k_p \end{pmatrix} = \sum_{1 \leq j_1 < j_2 < \cdots < j_p \leq m} B\begin{pmatrix} i_1 & i_2 & \cdots & i_p \\ j_1 & j_2 & \cdots & j_p \end{pmatrix} \cdot C\begin{pmatrix} j_1 & j_2 & \cdots & j_p \\ k_1 & k_2 & \cdots & k_p \end{pmatrix}.$$

Corollary 3.2 *Let $b, c \in A$. If b divides every minor of order p of B and if c divides every minor of order p of C, then bc divides every minor of order p of BC.*

The particular case $l = m = n = p$ is fundamental.

Theorem 3.1 *If $B, C \in \mathbf{M}_n(A)$, then $\det(BC) = \det B \cdot \det C$.*

In other words, the determinant is a *multiplicative homomorphism* from $\mathbf{M}_n(A)$ to A.

Proof. The corollaries are trivial. We only prove the Cauchy–Binet formula. The calculation of the ith row (respectively, the jth column) of BC involves only the ith row of B (respectively, the jth column of C), thus one may assume that $p = n = l$. The minor to be evaluated is then $\det BC$. If $m < n$, there is nothing to prove, because the rank of BC is less than or equal to m, thus $\det BC$ is zero by Corollary 3.1, and on the other hand the left-hand side sum in the formula is empty.

There remains the case $m \geq n$. Let us write the determinant of a matrix P as that of its columns P_j and let us use the multilinearity of the determinant:

$$\det BC = \det\left(\sum_{j_1=1}^{n} c_{j_1 1} B_{j_1}, (BC)_2, \ldots, (BC)_n\right)$$

$$= \sum_{j_1=1}^{n} c_{j_1 1} \det\left(B_{j_1}, \sum_{j_2=1}^{n} c_{j_2 2} B_{j_2}, (BC)_3, \ldots, (BC)_n\right)$$

$$= \cdots = \sum_{1 \leq j_1, \ldots, j_n \leq n} c_{j_1 1} \cdots c_{j_n n} \det(B_{j_1}, \ldots, B_{j_n}).$$

In the sum the determinant is zero as soon as $f \mapsto j_f$ is not injective, because then there are two identical columns. When j is injective, this determinant is a minor of B, up to the sign. This sign is that of the permutation that puts $j_1, \ldots, j_p$ in increasing order. Grouping in the sum the terms corresponding to the same minor, we obtain

$$\det BC = \sum_{1 \leq k_1 < \cdots < k_n \leq m, \, \sigma \in S_n} \varepsilon(\sigma) c_{k_1 \sigma(1)} \cdots c_{k_n \sigma(n)} B \begin{pmatrix} 1 & 2 & \cdots & n \\ k_1 & k_2 & \cdots & k_n \end{pmatrix},$$

which is the required formula. □

3.2.3 Irreducibility of the Determinant

Theorem 3.2 *The polynomial* Det *is irreducible in* $A[x_{11}, \ldots, x_{nn}]$.

Proof. We proceed by induction on the size n. If $n = 1$, there is nothing to prove. Thus let us assume that $n \geq 2$. We denote by D the ring of polynomials in the x_{ij} with $(i, j) \neq (1, 1)$, so that $A[x_{11}, \ldots, x_{nn}] = D[x_{11}]$. Expanding with respect to the first row, we see that Det $= x_{11} P + Q$, with $P, Q \in D$. Because Det is of degree one as a polynomial in x_{11}, any factorization must be of the form $(x_{11} R + S) T$, with $R, S, T \in D$. In particular, $RT = P$.

By induction, and inasmuch as P is the polynomial Det of $(n-1) \times (n-1)$ matrices, it is irreducible in E, the ring of polynomials in the x_{ij}s with $i, j > 1$. Therefore, it is also irreducible in D, because D is the polynomial ring

$$E[x_{12}, \ldots, x_{1n}, x_{21}, \ldots, x_{n1}].$$

Therefore, we may assume that either R or T equals 1.

If the factorization is nontrivial, then $R = 1$ and $T = P$. It follows that P divides Det. An expansion with respect to various rows shows similarly that every minor of size $n - 1$, considered as an element of $A[x_{11}, \ldots, x_{nn}]$, divides Det. However, each such minor is irreducible, and they are pairwise distinct, because they do not depend on the same set of x_{ij}s. We conclude that the product of all minors of size $n - 1$ divides Det. In particular, the degree n of Det is greater than or equal to the degree $n^2(n-1)$ of this product, an obvious contradiction. □

3.3 Invertibility

Let us recall $\mathbf{M}_n(A)$ is not an integral domain. Thus the notion of invertible elements of $\mathbf{M}_n(A)$ needs an auxiliary result, presented below.

Proposition 3.5 *Given $M \in \mathbf{M}_n(A)$, the following assertions are equivalent.*

1. *There exists $N \in \mathbf{M}_n(A)$ such that $MN = I_n$.*
2. *There exists $N' \in \mathbf{M}_n(A)$ such that $N'M = I_n$.*
3. $\det M$ *is invertible.*

If M satisfies one of these equivalent conditions, then the matrices N, N' are unique and one has $N = N'$.

Definition 3.2 *One then says that M is* invertible, *or that M is* nonsingular. *One calls the matrix N = N' the* inverse *of M. It is denoted by* M^{-1}. *If M is not invertible, one says that M is* singular.

Proof. Let us show that (1) is equivalent to (3). If $MN = I_n$, then $\det M \cdot \det N = 1$; hence $\det M \in A^*$. Conversely, if $\det M$ is invertible, $(\det M)^{-1}\hat{M}^T$ is an inverse of M by (3.5). Analogously, (2) is equivalent to (3). The three assertions are thus equivalent.

If $MN = N'M = I_n$, one has $N = (N'M)N = N'(MN) = N'$. This equality between the left and right inverses shows that these are unique. $\quad\square$

The set of the invertible elements of $\mathbf{M}_n(A)$ is denoted by $\mathbf{GL}_n(A)$ (for "general linear group"). It is a multiplicative group.

Proposition 3.6 *Let M, N be nonsingular n × n matrices. Then we have*

$$(MN)^{-1} = N^{-1}M^{-1}, \quad (M^k)^{-1} = (M^{-1})^k, \quad (M^T)^{-1} = (M^{-1})^T.$$

Proof. We calculate

$$(N^{-1}M^{-1})(MN) = N^{-1}(M^{-1}M)N = N^{-1}I_n N = N^{-1}N = I_n,$$

whence the first identity. The second one is standard. Finally

$$(M^{-1})^T M^T = (MM^{-1})^T = I_n^T = I_n$$

gives the last one. $\quad\square$

The matrix $(M^T)^{-1}$ is also written M^{-T}. If $k \in \mathbb{N}$, one writes $M^{-k} = (M^k)^{-1}$ and one has $M^j M^k = M^{j+k}$ for every $j, k \in \mathbb{Z}$.

The set of the matrices of determinant one is a normal subgroup of $\mathbf{GL}_n(A)$, because it is the kernel of the homomorphism $M \mapsto \det M$. It is called the *special linear group* and is denoted by $\mathbf{SL}_n(A)$.

Recall that the orthogonal matrices are invertible, and they satisfy the relation $M^{-1} = M^T$. In particular, orthogonality is equivalent to $MM^T = I_n$. The set of orthogonal matrices with entries in a field K is obviously a multiplicative group, and is denoted by $\mathbf{O}_n(K)$. It is called the *orthogonal group*. The determinant of an orthogonal matrix equals ± 1, because

$$1 = \det M \cdot \det M^T = (\det M)^2.$$

The set $\mathbf{SO}_n(K)$ of orthogonal matrices with determinant equal to 1 is obviously a normal subgroup of the orthogonal group. It is called the *special orthogonal group*, the intersection of $\mathbf{O}_n(K)$ with $\mathbf{SL}_n(K)$.

Proposition 3.7 *A triangular matrix is invertible if and only if its diagonal entries are invertible; its inverse is then triangular of the same type, upper or lower.*

Proof. Let us write the proof for upper-triangular matrices. Because of Propositions 3.1 and 3.5, invertibility of the matrix amounts to that of its diagonal entries.

We thus assume that each m_{ii} is nonzero. Let T be the inverse of M and denote its columns by $T^{(1)}, \ldots, T^{(n)}$. Because of $MT^{(1)} = e^1$, we have

$$m_{nn}t_{n1} = 0,$$
$$m_{n-1,n-1}t_{n-1,1} + m_{n-1,n}t_{n1} = 0,$$
$$\vdots$$
$$m_{22}t_{21} + \cdots + m_{2n}t_{n1} = 0.$$

This gives inductively $t_{n1} = 0, \ldots, t_{21} = 0$.

We next write that $MT^{(2)}$ is colinear to e^1 and e^2, obtaining $t_{n2} = \cdots = t_{32} = 0$. By induction on the columns, we obtain $t_{ij} = 0$ whenever $i > j$. $\square$

The proposition below is an immediate application of theorem 3.1.

Proposition 3.8 *If $M, M' \in \mathbf{M}_n(A)$ are similar (i.e., $M' = P^{-1}MP$ with $P \in \mathbf{GL}_n(A)$), then*

$$\det M' = \det M.$$

3.3.1 Inverting Blockwise

The determinant of a 2×2 matrix

$$M = \begin{pmatrix} a & b \\ c & d \end{pmatrix}$$

equals $ad - bc$. We might wonder whether something similar happens when a matrix of size $n \times n$ is given blockwise:

$$M = \begin{pmatrix} A & B \\ C & D \end{pmatrix}.$$

For instance, can we say that $\det M$ equals $\det(AD - BC)$? The answer is obviously No, because the expressions AD and BC might not make sense. Even when they do, the matrices AD and BC might not be square. The only case where the answer needs a little investigation is when the four blocks have the same size $n/2 \times n/2$ (and thus n is even). Even in the latter situation, we might hesitate between AD versus DA, and BC versus CB. This makes four candidates, from $\det(AD - BC)$ to $\det(DA - CB)$. Actually *none of them equals the determinant of M in general.* The correct answer is given by the following proposition.

Proposition 3.9 (Schur complement formula.) *Let $M \in \mathbf{M}_n(K)$ read blockwise*

$$M = \begin{pmatrix} A & B \\ C & D \end{pmatrix},$$

where the diagonal blocks are square and A is invertible. Then

$$\det M = \det A \det(D - CA^{-1}B). \tag{3.6}$$

Definition 3.3 *The matrix $D - CA^{-1}B$ is called the* Schur complement *of A in M.*

We emphasize that in (3.6), we do not need the blocks to be of equal size. In particular, B and C need not be square, although they have to be of transpose sizes. The Schur complement formula turns out to be an effective tool in practical calculations.

Proof. We use a trick that is developed in Chapter 11. Because A is invertible, M factorizes as a product LU of block-triangular matrices, with

$$L = \begin{pmatrix} I & 0 \\ CA^{-1} & I \end{pmatrix}, \quad U = \begin{pmatrix} A & B \\ 0 & D - CA^{-1}B \end{pmatrix}.$$

Then $\det M = \det L \det U$ furnishes the expected formula. $\square$

Inverting L and U, we see that as soon as $M \in \mathbf{GL}_n(k)$ and $A \in \mathbf{GL}_p(k)$ (even if $p \neq n/2$), then

$$M^{-1} = U^{-1}L^{-1} = \begin{pmatrix} A^{-1} & * \\ 0 & (D - CA^{-1}B)^{-1} \end{pmatrix} \begin{pmatrix} I & 0 \\ -CA^{-1} & I \end{pmatrix} = \begin{pmatrix} \cdot & \cdot \\ \cdot & (D - CA^{-1}B)^{-1} \end{pmatrix}.$$

More generally, we have the following corollary.

Corollary 3.3 *Let $M \in \mathbf{GL}_n(k)$, with $n = 2m$, read blockwise*

$$M = \begin{pmatrix} A & B \\ C & D \end{pmatrix}, \quad A, B, C, D \in \mathbf{GL}_m(k).$$

Then

$$M^{-1} = \begin{pmatrix} (A - BD^{-1}C)^{-1} & (C - DB^{-1}A)^{-1} \\ (B - AC^{-1}D)^{-1} & (D - CA^{-1}B)^{-1} \end{pmatrix}.$$

Proof. We can verify the formula by multiplying by M. The only point to show is that the inverses are meaningful; that is, $A - BD^{-1}C, \ldots$ are invertible. Because of the symmetry of the formulæ, it is enough to check it for a single term, namely $D - CA^{-1}B$. However, $\det(D - CA^{-1}B) = \det M / \det A$ is well defined and nonzero by assumption. $\square$

3.3.2 Cramer's Formulæ

When a matrix $M \in \mathbf{M}_n(A)$ is invertible, the linear system

$$Mx = b, \tag{3.7}$$

where $b \in A^n$ is a datum and $x \in A^n$ the unknown, admits a unique solution, given by $x = M^{-1}b$. The identity

$$x = \frac{1}{\det M}(\text{adj}M)b$$

gives an expression of the solution, which we want to make more explicit. This is the role of *Cramer's formulae*:

Proposition 3.10 *If M is invertible, the coordinates of the solution of equation* (3.7) *are given by*

$$x_i = \frac{\det M(i;b)}{\det M}, \tag{3.8}$$

where $M(i;b)$ is the matrix formed by replacing in M the ith column by the vector b.

Proof. Let us denote by $X_i(b)$ the expression in the right-hand side of (3.8), and $X(b)$ the vector whose coordinates are $X_i(b)$ for $i = 1, \ldots, n$. The map $b \mapsto X(b)$ is linear, and thus corresponds to a matrix $N \in \mathbf{M}_n(A)$.

When $b := M^{(j)}$ is the jth column of M, the determinant of $M(i; M^{(j)})$ vanishes for every $i \neq j$ because this matrix has two identical columns. Because $M(j; M^{(j)}) = M$, we deduce $X_i(M^{(j)}) = \delta_i^j$, that is $X(M^{(j)}) = \mathbf{e}^j$.

Because $M\mathbf{e}^j = M^{(j)}$, we infer that N coincides with M^{-1} over the set of columns of M. The matrix being invertible, its columns span A^n : every vector b is a linear combination of $M^{(1)}, \ldots, M^{(n)}$. Just write

$$b = \sum_j b_j \mathbf{e}^j = \sum_j b_j M^{-1} M^{(j)} = \sum_i \beta_i M^{(i)}, \qquad \beta_i := \sum_j (M^{-1})_{ij} b_j.$$

Therefore $N = M^{-1}$, which means that $X(b) = M^{-1}b$ for every $b \in A^n$. □

3.4 Eigenvalues and Eigenvectors

We repeat the analysis made for an endomorphism. If $M \in \mathbf{M}_n(K)$, then $n = \dim \ker M + \text{rk} M$. Therefore M is bijective over K^n if and only if it is either injective or surjective. Applying this to $\lambda I_n - M$ where $\lambda \in K$, we therefore have the alternative

- Either there exists a nonzero vector $X \in K^n$ such that $Mx = \lambda X$. One then says that λ is an *eigenvalue* of M in K, and that X is an *eigenvector* associated with λ.
- Or, for every $b \in K^n$, the following equation admits a unique solution $Y \in K^n$,

$$MY - \lambda Y = b.$$

The set of the eigenvalues of M in K is called the *spectrum* of M and is denoted by $\mathrm{Sp}_K(M)$.

The matrix $\lambda I_n - M$ acts bijectively over K^n if and only if it is nonsingular, therefore we find that the eigenvalues are the solutions of the polynomial equation

$$\det(\lambda I_n - M) = 0. \tag{3.9}$$

A matrix in $\mathbf{M}_n(K)$ may have no eigenvalues in K, as the following example demonstrates, with $K = \mathbb{R}$:

$$\begin{pmatrix} 0 & 1 \\ -1 & 0 \end{pmatrix}.$$

To understand in detail the structure of a square matrix $M \in \mathbf{M}_n(K)$, one is thus led to consider M as a matrix with entries in the algebraic closure $\overline{K}$. One then writes $\mathrm{Sp}(M)$ instead of $\mathrm{Sp}_{\overline{K}}(M)$, and one has $\mathrm{Sp}_K(M) = K \cap \mathrm{Sp}(M)$, because the eigenvalues are characterized by (3.9), and this equality has the same meaning in $\overline{K}$ as in K when $\lambda \in K$.

3.5 The Characteristic Polynomial

Given $M \in \mathbf{M}_n(K)$, the equation (3.9) for the eigenvalues suggests defining a polynomial

$$P_M(X) := \det(XI_n - M).$$

Let us observe in passing that if X is an indeterminate, then $XI_n - M \in \mathbf{M}_n(K(X))$. Its determinant P_M is thus well-defined, because $K(X)$ is a commutative integral domain with a unit element. One calls P_M the *characteristic polynomial* of M. Substituting 0 for X, one sees that the constant term in P_M is simply $(-1)^n \det M$. The term corresponding to the permutation $\sigma = \mathrm{id}$ in the computation of the determinant is of degree n (it is $\prod_i (X - m_{ii})$) and the products corresponding to the other permutations are of degree less than or equal to $n - 2$, therefore one sees that P_M is of degree n, with

$$P_M(X) = X^n - \left(\sum_{i=1}^{n} m_{ii} \right) X^{n-1} + \cdots + (-1)^n \det M. \tag{3.10}$$

The coefficient

$$\sum_{i=1}^{n} m_{ii}$$

is called the *trace* of M and is denoted by $\mathrm{Tr}\, M$. One has the trivial formula that if $N \in \mathbf{M}_{n \times m}(K)$ and $P \in \mathbf{M}_{m \times n}(K)$, then

$$\mathrm{Tr}(NP) = \mathrm{Tr}(PN).$$

For square matrices, this identity also becomes

$$\text{Tr}[N,P] = 0.$$

Because P_M possesses n roots in $\overline{K}$, counting multiplicities, a square matrix always has at least one eigenvalue that, however, does not necessarily belong to K. The multiplicity of λ as a root of P_M is called the *algebraic multiplicity* of the eigenvalue λ. The *geometric multiplicity* of λ is the dimension of $\ker(\lambda I_n - M)$ in K^n. The sum of the algebraic multiplicities of the eigenvalues of M (considered in $\overline{K}$) is $\deg P_M = n$, the size of the matrix. An eigenvalue of algebraic multiplicity one (i.e., a simple root of P_M) is called a *simple eigenvalue*. An eigenvalue is *geometrically simple* if its geometric multiplicity equals one.

The characteristic polynomial is a *similarity invariant*, in the following sense.

Proposition 3.11 *If M and M' are similar, then $P_M = P_{M'}$. In particular, $\det M = \det M'$ and $\text{Tr}\, M = \text{Tr}\, M'$.*

The proof is immediate. One deduces that the eigenvalues and their algebraic multiplicities are similarity invariants. This is also true for the geometric multiplicities, by a direct comparison of the kernel of $\lambda I_n - M$ and of $\lambda I_n - M'$. Furthermore, the expression (3.10) provides the following result.

Proposition 3.12 *The product of the eigenvalues of M (considered in $\overline{K}$), counted with their algebraic multiplicities, is $\det M$. Their sum is $\text{Tr}\, M$.*

The characteristic polynomials of M and M^T are equal. Thus, M and M^T have the same eigenvalues. We show a much deeper result in Chapter 9, namely that M and M^T are similar.

3.5.1 Eigenvalues of the Transpose Matrix

We recall (Proposition 2.8) that $\dim \ker M^T = \dim \ker M$. Applying this statement to $M - \lambda I_n$, we find that λ is an eigenvalue of M if and only if it is an eigenvalue of M^T, and that the geometric multiplicities coincide. In addition, because of (3.3), we have $P_{M^T} = P_M$ and the algebraic multiplicities coincide too. Finally we have the following.

Proposition 3.13 *A square matrix and its transpose have the same eigenvalues, with the same geometric multiplicities, and the same algebraic multiplicities.*

3.5.2 The Theorem of Cayley–Hamilton

Theorem 3.3 (Cayley–Hamilton) *Let $M \in \mathbf{M}_n(K)$. Let*

$$P_M(X) = X^n + a_1 X^{n-1} + \cdots + a_n$$

be its characteristic polynomial. Then the matrix

$$M^n + a_1 M^{n-1} + \cdots + a_n I_n$$

equals 0_n.

One also writes $P_M(M) = 0$. Although this formula looks trivial (the equality $\det(MI_n - M) = 0$ is obvious) at first glance, it is not. Actually, it must be understood in the following way. Let us consider the expression $XI_n - M$ as a matrix with entries in $K[X]$. When one substitutes a matrix N for the indeterminate X in $XI_n - M$, one obtains a matrix of $\mathbf{M}_n(A)$, where A is the subring of $\mathbf{M}_n(K)$ spanned by I_n and N (denoted above by $K(N)$). The ring A is commutative (but is not an integral domain in general), because it is the set of the $q(N)$ for $q \in K[X]$. Therefore,

$$P_M(N) = \begin{pmatrix} N - m_{11}I_n & & \\ & \ddots & -m_{ij}I_n \\ & & N - m_{nn}I_n \end{pmatrix}.$$

The Cayley–Hamilton theorem expresses that the determinant (which is an element of $\mathbf{M}_n(K)$, rather than a scalar) of this matrix is zero.

Proof. Let $R \in \mathbf{M}_n(K(X))$ be the matrix $XI_n - M$, and let S be the adjugate of R. Each s_{ij} is a polynomial of degree less than or equal to $n-1$, because the products arising in the calculation of the cofactors involve $n-1$ linear or constant terms. Thus we may write

$$S = S_0 X^{n-1} + \cdots + S_{n-1},$$

where $S_j \in \mathbf{M}_n(K)$. Let us now write $RS = (\det R)I_n = P_M(X)I_n$:

$$(XI_n - M)(S_0 X^{n-1} + \cdots + S_{n-1}) = (X^n + a_1 X^{n-1} + \cdots + a_n)I_n.$$

Identifying the powers of X, we obtain

$$S_0 = I_n,$$
$$S_1 - MS_0 = a_1 I_n,$$
$$\vdots$$
$$S_j - MS_{j-1} = a_j I_n,$$
$$\vdots$$
$$S_{n-1} - MS_{n-2} = a_{n-1} I_n,$$
$$-MS_{n-1} = a_n I_n.$$

Let us multiply these lines by the powers of M, beginning with M^n and ending with $M^0 = I_n$. Summing all these equalities, we obtain the expected formula. □

For example, every 2×2 matrix satisfies the identity

$$M^2 - (\operatorname{Tr} M)M + (\det M)I_2 = 0.$$

3.5.3 The Minimal Polynomial

For a square matrix $M \in \mathbf{M}_n(K)$, let us denote by J_M the set of polynomials $Q \in K[X]$ such that $Q(M) = 0$. It is clearly an ideal of $K[X]$. Because $K[X]$ is Euclidean, hence principal (see Sections 9.1.1 and 9.1.2 for these notions), there exists a polynomial π_M such that $J_M = K[X]\pi_M$ is the set of polynomials divisible by π_M: if $Q(M) = 0$ and $Q \in K[X]$ then $\pi_M | Q$. Theorem 3.3 shows that the ideal J_M does not reduce to $\{0\}$, because it contains the characteristic polynomial. Hence, $\pi_M \neq 0$ and one may choose it monic. This choice determines π_M in a unique way, and one calls it the *minimal polynomial* of M. It divides the characteristic polynomial P_M.

In Section 9.3.2, we show that if L is a field containing K, then the minimal polynomials of M in $K[X]$ and $L[X]$ are the same. Therefore the minimal polynomial is independent of the choice of a field containing the entries of M.

Two similar matrices obviously have the same minimal polynomial, inasmuch as

$$Q(P^{-1}MP) = P^{-1}Q(M)P.$$

If λ is an eigenvalue of M, associated with an eigenvector X, and if $q \in K[X]$, then $q(\lambda)X = q(M)X$. Applied to the minimal polynomial, this equality shows that the minimal polynomial is divisible by $X - \lambda$. Hence, if P_M splits over $\bar{K}$ in the form

$$\prod_{j=1}^{r}(X - \lambda_j)^{n_j},$$

the λ_j all being distinct, then the minimal polynomial can be written as

$$\prod_{j=1}^{r}(X - \lambda_j)^{m_j},$$

for some $1 \leq m_j \leq n_j$. In particular, if every eigenvalue of M is simple, the minimal polynomial and the characteristic polynomial are equal.

An eigenvalue is called *semisimple* if it is a simple root of the minimal polynomial.

3.5.4 Semisimplicity

Let μ be the geometric multiplicity of an eigenvalue λ of M. Let us choose a basis γ of $\ker(\lambda I_n - M)$, and then a basis of β of K^n that completes γ. Using the change-of-basis matrix from the canonical basis to β, one sees that M is similar to a matrix $M' = P^{-1}MP$ of the form

$$\begin{pmatrix} \lambda I_\mu & R \\ 0_{n-\mu,\mu} & S \end{pmatrix},$$

where $S \in \mathbf{M}_{n-\mu}(K)$. Because of the block-triangular form, we have $P_{M'}(X) = (X - \lambda)^\mu P_S(X)$. Because $P_{M'} = P_M$ by similarity, we thus have the following.

Proposition 3.14 *The geometric multiplicity of an eigenvalue is less than or equal to its algebraic multiplicity.*

Definition 3.4 *We say that an eigenvalue of $M \in \mathbf{M}_n(K)$ is semisimple if its algebraic and geometric multiplicities coincide.*

For instance, a simple eigenvalue is semisimple. With the calculation above, λ is semisimple if and only if it is not an eigenvalue of the block S; that is, $S - \lambda I_{n-\mu}$ is nonsingular.

Theorem 3.4 *An eigenvalue $\lambda \in K$ of M is semisimple if and only if*

$$K^n = R(M - \lambda I_n) \oplus \ker(M - \lambda I_n).$$

Proof. We may assume that M is in block-triangular form as above. We decompose the vectors blockwise accordingly:

$$x = \begin{pmatrix} x_+ \\ x_- \end{pmatrix}.$$

The eigenspace associated with λ is that spanned by $\mathbf{e}^1, \ldots, \mathbf{e}^\mu$. Therefore $x \in \ker(M - \lambda I_n)$ if and only if $x_- = 0$.

If λ is semisimple, then $S - \lambda I_{n-\mu}$ is nonsingular. Let $x \in R(M - \lambda I_n) \cap \ker(M - \lambda I_n)$ be given. There exists a y such that $x = (M - \lambda I_n)y$. We get $(S - \lambda I_{n-\mu})y_- = x_- = 0$, which implies $y_- = 0$. Therefore $y \in \ker(M - \lambda I_n)$; that is, $x = 0$.

If instead λ is not semisimple, we may choose a nonzero vector y_- in the kernel of $S - \lambda I_{n-\mu}$. Choosing $y_+ = 0$ and defining $x := (M - \lambda I_n)y$, we have $x_- = 0$; that is, $x \in \ker(M - \lambda I_n)$. However y is not itself in the eigenspace, and therefore x is a nonzero element of $R(M - \lambda I_n) \cap \ker(M - \lambda I_n)$. $\square$

We now investigate in greater detail the case of a geometrically simple eigenvalue λ. According to Proposition 3.13, λ is also a geometrically simple eigenvalue of M^T. Let X be a generator of $\ker(M - \lambda I_n)$ and Y a generator of $\ker(M^T - \lambda I_n)$. Thanks to Corollary 2.1, the equation $(M - \lambda I_n)x = X$ is solvable if and only if $Y^T X = 0$. This solvability amounts to saying that $R(M - \lambda I_n) \cap \ker(M - \lambda I_n)$ is nontrivial. With theorem 3.4, this gives the following proposition.

Proposition 3.15 *Let* $\lambda \in K$ *be a geometrically simple eigenvalue of* $M \in \mathbf{M}_n(K)$. *Let* X *and* Y *be eigenvectors of* M *and* M^T, *respectively, associated with* λ.
Then λ *is a semisimple (thus simple) eigenvalue if and only if* $Y^T X \neq 0$.

When λ is simple, we may normalize X or Y above in such a way that $Y^T X = 1$.

3.6 Diagonalization

If $\lambda \in K$ is an eigenvalue of M, the linear subspace $E_K(\lambda) = \ker(M - \lambda I_n)$ in K^n is nontrivial and is called the *eigenspace* associated with λ. It is formed of eigenvectors associated with λ and of the zero vector.

The following statement uses the notion of tensor product, here in the restricted situation of extension of scalars, presented in Chapter 4.

Proposition 3.16 *Let* $\lambda \in K$ *be given. If* L *is a field containing* K *with* $[L : K] < \infty$ *(an "extension" of* K *of finite degree), then* $E_L(\lambda) = E_K(\lambda) \otimes_K L$ *and thus* $\dim_K E_K(\lambda) = \dim_L E_L(\lambda)$.

Proof. Let $a_1, \ldots, a_\ell$ be a basis of L over K. If $z \in E_K(\lambda) \otimes_K L$, then $z = a_1 x^1 + \cdots + a_\ell x^\ell$ where $x^1, \ldots, x^\ell$ belong to $E_K(\lambda)$. Then $Mz = a_1 M x^1 + \cdots + a_\ell M x^\ell = \lambda z$ and thus $z \in E_L(\lambda)$.
Conversely, if $z \in E_L(\lambda)$, let us write $z = a_1 x^1 + \cdots + a_\ell x^\ell$ with $x^j \in K^n$. Then

$$0 = (M - \lambda)z = a_1(M - \lambda)x_1 + \cdots + a_\ell(M - \lambda)x_\ell.$$

The coefficient of a_j must vanish, which means $x^j \in E_K(\lambda)$. This gives us $z \in E_K(\lambda) \otimes_K L$.   □

If $\lambda_1, \ldots, \lambda_r$ are distinct eigenvalues, the corresponding eigenspaces arc in direct sum. That is,

$$(x_1 \in E_K(\lambda_1), \ldots, x_r \in E_K(\lambda_r), x_1 + \cdots + x_r = 0) \Longrightarrow (x_1 = \cdots = x_r = 0).$$

As a matter of fact, if there existed a relation $x_1 + \cdots + x_s = 0$ where $x_1, \ldots, x_s$ did not vanish simultaneously, one could choose such a relation of minimal length r. One then would have $r \geq 2$. Multiplying this relation by $M - \lambda_r I_n$, one would obtain

$$(\lambda_1 - \lambda_r)x_1 + \cdots + (\lambda_{r-1} - \lambda_r)x_{r-1} = 0,$$

which is a nontrivial relation of length $r - 1$ for the vectors $(\lambda_j - \lambda_r)x_j \in E_K(\lambda_j)$. This contradicts the minimality of r.

If all the eigenvalues of M are in K (we say that the characteristic polynomial P_M *splits* in K), the sum of the dimensions of the eigenspaces equals the sum of geometric multiplicities. If in addition the algebraic and geometric multiplicities coincide for each eigenvalue of M, this sum is n and the subspace

$$\oplus_{j=1}^{s} E(\lambda_j)$$

has dimension n. It is thus equal to K^n:

$$K^n = E(\lambda_1) \oplus \cdots \oplus E(\lambda_r).$$

Thus one may choose a basis of K^n formed of eigenvectors. If P is the change-of-basis matrix from the canonical basis to the new one, then $M' = P^{-1}MP$ is diagonal, and its diagonal entries are the eigenvalues, repeated with their multiplicities. One says that M is *diagonalizable* in K. A particular case is that in which the eigenvalues of M are in K and are simple.

Proposition 3.17 *Let* $M \in \mathbf{M}_n(K)$ *be given, such that* P_M *splits in* K *and has simple roots. Then* M *is diagonalizable in* $\mathbf{M}_n(K)$.

Conversely, if M is similar in $\mathbf{M}_n(K)$ to a diagonal matrix $D = P^{-1}MP$, then P is a change-of-basis matrix from the canonical basis to an *eigenbasis* (i.e., a basis composed of eigenvectors) of M, hence the following.

Proposition 3.18 *A square matrix M is diagonalizable in* $\mathbf{M}_n(K)$ *if and only if* K^n *admits a basis of eigenvectors of M, or in other words:*

- P_M *splits in* K.
- *The algebraic and geometric multiplicities of each eigenvalue coincide.*

Two obstacles could prevent M from being diagonalizable in K. The first one is that an eigenvalue of M does not belong to K. One can always overcome this difficulty by extending our field of scalars to $\overline{K}$ or simply to a suitable finite extension L over K. Thus many matrices are not diagonalizable over K, but are so over an extension L. The second cause is more serious: even if an eigenvalue λ is in K, its geometric multiplicity can be strictly less than its algebraic multiplicity, and this remains true if we replace K by an extension. For instance, a triangular matrix whose diagonal vanishes has only one eigenvalue, zero, of algebraic multiplicity n. Such a matrix is nilpotent. However, it is diagonalizable only if it is 0_n, because if $M = PDP^{-1}$ and D is diagonal, then $D = 0_n$ and this implies $M = 0_n$. For instance,

$$\begin{pmatrix} 0 & 1 \\ 0 & 0 \end{pmatrix}$$

is not diagonalizable.

3.7 Trigonalization

Definition 3.5 *A square matrix is* trigonalizable *in* $\mathbf{M}_n(K)$ *if it is similar to a triangular matrix.*

Because the characteristic polynomial is invariant under conjugation and that of a triangular matrix T is the product of linear factors $X - t_{jj}$, a necessary condition for trigonalizability of M is that P_M split over K. This turns out to be sufficient.

Theorem 3.5 *A square matrix M is trigonalizable over $\mathbf{M}_n(K)$ if and only if its characteristic polynomial splits over K.*

Proof. We proceed by induction over n. The sufficiency is obvious for $n = 1$.

Let us assume that $n \geq 2$ and that P_M splits. Thus M has an eigenvalue λ. Let x be a corresponding eigenvector. We form a basis $\mathscr{B}$, whose first element is x. Let P be the matrix of the change of basis from the canonical one to $\mathscr{B}$. Then $MPe^1 = Mx = \lambda x = \lambda Pe^1$, or equivalently $P^{-1}MPe^1 = \lambda e^1$. This means that

$$P^{-1}MP = \begin{pmatrix} \lambda & \cdots \\ 0 & M' \end{pmatrix}$$

is block-triangular.

Because of the triangular form, one has $P_M = (X - \lambda)P_{M'}$. Thus $P_{M'}$ splits over K too. By induction hypothesis, M' is similar to an upper-triangular matrix T': $Q^{-1}M'Q = T'$. Let us form $Q_0 := \mathrm{diag}(1, Q)$, which is nonsingular. Finally, define $R := PQ_0$. Then

$$R^{-1}MR = \begin{pmatrix} \lambda & \cdots \\ 0 & T' \end{pmatrix}$$

is triangular.   □

Corollary 3.4 *Every square matrix $M \in \mathbf{M}_n(K)$ is trigonalizable over a suitable extension of K.*

We now give a more accurate reduction of matrices under the same assumption as above. We begin with an application of the Cayley–Hamilton theorem.

Proposition 3.19 *Let $M \in \mathbf{M}_n(K)$ and let P_M be its characteristic polynomial. If $P_M = QR$ with coprime factors $Q, R \in K[X]$, then $K^n = E \oplus F$, where E, F are the ranges of $Q(M)$ and $R(M)$, respectively. Moreover, one has $E = \ker R(M)$, $F = \ker Q(M)$.*

More generally, if $P_M = R_1 \cdots R_s$, where the R_s are coprime, one has $K^n = E_1 \oplus \cdots \oplus E_s$ with $E_j = \ker R_j(M)$. The subspace E_j is also the range of

$$\left(\frac{P_M}{R_j} \right)(M).$$

Proof. It is sufficient to prove the first assertion and then to work by induction over the number of factors s.

From Bézout's theorem, there exist $T, S \in K[X]$ such that $RT + QS = 1$. Hence, every $x \in K^n$ can be written as a sum $y + z$ with $y = Q(M)(S(M)x) \in E$ and $z = R(M)(T(M)x) \in F$. Hence $K^n = E + F$ and

$$n \leq \dim E + \dim F. \tag{3.11}$$

For every $y \in E$, the Cayley–Hamilton theorem says that $R(M)y = 0$. This means that $E \subset \ker R(M)$. Likewise, $F \subset \ker Q(M)$. We deduce

$$\dim E \leq \dim \ker R(M), \qquad \dim F \leq \dim \ker Q(M). \qquad (3.12)$$

If $x \in \ker Q(M) \cap \ker R(M)$, one has

$$x = T(M)(R(M)x) + S(M)(Q(M)x) = T(M)0 + S(M)0 = 0,$$

whence $\ker Q(M) \cap \ker R(M) = \{0\}$. This tells us that

$$\dim(\ker Q(M) + \ker R(M)) = \dim \ker Q(M) + \dim \ker R(M). \qquad (3.13)$$

Assembling (3.11-3.13), we obtain $n \leq \dim(\ker Q(M) + \ker R(M))$; that is, $n = \dim(\ker Q(M) + \ker R(M))$. We infer that the equalities hold in (3.11) and (3.12):

$$n = \dim E + \dim F, \qquad \dim E = \dim \ker R(M), \qquad \dim F = \dim \ker Q(M).$$

We conclude that

$$K^n = E \oplus F, \qquad E = \ker R(M), \qquad \dim F = \ker Q(M).$$

$\square$

Let us now assume that P_M splits over K:

$$P_M(X) = \prod_{\lambda \in \mathrm{Sp}(M)} (X - \lambda)^{n_\lambda}.$$

Proposition 3.19 tells us that $K^n = \oplus_\lambda E_\lambda$, where $E_\lambda = \ker(M - \lambda I)^{n_\lambda}$ is called a *generalized eigenspace*. Choosing a basis in each E_λ, we obtain a new basis $\mathscr{B}$ of K^n. If P is the matrix of the linear transformation from the canonical basis to $\mathscr{B}$, the matrix PMP^{-1} is block-diagonal, because each E_λ is stable under the action of M:

$$PMP^{-1} = \mathrm{diag}(\ldots, M_\lambda, \ldots).$$

The matrix M_λ is that of the restriction of M to E_λ. Because $E_\lambda = \ker(M - \lambda I)^{n_\lambda}$, one has $(M_\lambda - \lambda I)^{n_\lambda} = 0$, so that λ is the unique eigenvalue of M_λ. Let us define $N_\lambda = M_\lambda - \lambda I_{n_\lambda}$, which is nilpotent. Let us also write

$$D' = \mathrm{diag}(\ldots, \lambda I_{n_\lambda}, \ldots),$$
$$N' = \mathrm{diag}(\ldots, N_\lambda, \ldots),$$

and then $D = P^{-1}D'P$, $N = P^{-1}N'P$. The matrices D', N' are, respectively, diagonal and nilpotent. Moreover, they commute to each other: $D'N' = N'D'$. One deduces the existence part in the following result.

Proposition 3.20 *Let us assume that the characteristic polynomial of $M \in \mathbf{M}_n(K)$ splits over K. Then M decomposes as a sum $M = D + N$, where D is diagonalizable, N is nilpotent, $DN = ND$, and $\mathrm{Sp}(D) = \mathrm{Sp}(M)$.*

In addition, D and N are unique, and are polynomials in M. The formula $M = D + N$ is the Dunford decomposition *of M.*

Proof. There remains to prove the uniqueness and polynomiality.

Polynomiality.   It is sufficient to prove that D' and N' are polynomials in $D' + N'$. This is done if we find *one* polynomial R, such that $\lambda I_{n_\lambda} = R(\lambda I_{n_\lambda} + N_\lambda)$ for *every* λ, because then we have $D = R(M)$ and $N = (X - R)(M)$. Such a polynomial needs only to satisfy the following properties. For each eigenvalue λ, $R(\lambda) = \lambda$, and the multiplicity of λ as a root of $R - \lambda$ is n_λ. This is an interpolation problem which does have a solution.

Uniqueness.   Let $D + N$ be the Dunford decomposition of M constructed above. Up to conjugation, we may assume that D is already diagonal, of the form

$$D = \mathrm{diag}(a_1 I_{n_1}, \dots, a_r I_{n_r})$$

where the a_js are pairwise distinct.

If $D' + N'$ is another Dunford decomposition, then $D + N = D' + N'$. In addition, D' (respectively, N, N') commutes with M, thus with every polynomial in M, in particular with D. Writing D' blockwise, we obtain $(a_k - a_j)D'_{jk} = 0$, whence D' is block-diagonal:

$$D = \mathrm{diag}(D'_1, \dots, D'_r).$$

Likewise, N' and N are block-diagonal, and the identity $D' + N' = M$ reduces to the list $a_j I_{n_j} + N_j = D_j + N'_j$.

We are led to prove that if $a I_m + N = D' + N'$ where D' is diagonal and N, N' are nilpotent, these matrices commuting pairwise, then $D' = a I_m$ and $N' = N$. Replacing $D' - a I_m$ by D', we may assume that $a = 0$. Let λ be an eigenvalue of D': $D'x = \lambda x$. Then $(D' + N' - \lambda)x = N'x$. Because of commutation, this implies $(D' + N' - \lambda)^k x = N'^k x$. Because N' is nilpotent, this yields $(\lambda - D' - N')^m x = 0$, thus $\det(\lambda - D' - N')^m = 0$, meaning that λ is an eigenvalue of $D' + N'$. But $D' + N' = N$ is nilpotent, therefore $\lambda = 0$. Finally D', being diagonalizable with the only eigenvalue 0, equals 0_m. At last, $N' = N - D' = N$.
□

3.8 Rank-One Perturbations

In a rank-one matrix $M \in \mathbf{M}_{n \times m}(K)$, the columns are colinear to a given vector $x \in K^n$. Writing that $M^{(j)} = y_j x$, we obtain $M = xy^T$, where $y \in K^m$. Conversely, every matrix of the form xy^T with $x, y \neq 0$ has rank one.

We now focus on the case of a square matrix. The spectrum of a rank-one matrix xy^T is easily described. First of all, its kernel $y^\perp$ has dimension $n - 1$. Because $y^\perp =$

$\ker M$, the geometric multiplicity of the eigenvalue 0 is $n-1$. Thus the spectrum consists in $(0,\ldots,0,\mu)$. We identify μ by computing the trace of xy^T, recalling that the trace equals the sum of the eigenvalues:

$$x_1y_1 + \cdots + x_ny_n = 0 + \cdots + 0 + \mu.$$

Finally $\mu = x \cdot y$, using the canonical scalar product in K^n.

Lemma 2. *Given vectors $x,y \in K^n$, the spectrum of xy^T is $(0,\ldots,0,x\cdot y)$.*

Notice that 0 might have algebraic multiplicity n, if $x \cdot y = 0$.

We apply this lemma to the calculation of $\det(I_n + xy^T)$. The spectrum of $I_n + xy^T$ is just shifted from that of xy^T and equals $(1,\ldots,1,1+x\cdot y^T)$. The determinant is the product of the eigenvalues, therefore we have

$$\det(I_n + xy^T) = 1 + x \cdot y. \qquad (3.14)$$

We now derive

Proposition 3.21 *Given a matrix $M \in \mathbf{M}_n(K)$ and vectors $x,y \in K^n$, we have*

$$\det(M + xy^T) = \det M + x^T \hat{M} y, \qquad (3.15)$$

where $\hat{M}$ is the cofactor matrix.

If in addition M and $M + xy^T$ are nonsingular, then we have the Sherman–Morrison *formula*

$$(M + xy^T)^{-1} = M^{-1} - \frac{1}{1 + y^T M^{-1} x} M^{-1} xy^T M^{-1}. \qquad (3.16)$$

We notice that the function

$$t \mapsto \det(M + txy^T)$$

is affine over K. We say that the determinant is *rank-one affine*. As a consequence, every minor is rank-one affine. The following alternative holds true on a line ℓ whose direction is a rank-one matrix: either every element of ℓ is a singular matrix, or at most one is singular. In the case where none is singular, M is nonsingular and $x^T \hat{M} y = 0$, with the notations of the proposition above.

Proof. Let us begin with the case where M is nonsingular. Then $M + xy^T = M(I_n + M^{-1}xy^T)$ gives

$$\det(M + xy^T) = (\det M)\det(I_n + M^{-1}xy^T) = (1 + y^T M^{-1}x)\det M,$$

thanks to (3.14). We obtain (3.15) by using the formula $(\det M)M^{-1} = \hat{M}^T$.

To extend this formula, we begin by remarking that there exists a polynomial $\Pi \in \mathbb{Z}[X_1,\ldots,X_{n^2+2n}]$ such that

$$\Pi(M,x,y) \equiv \det(M + xy^T) - \det M - x^T \hat{M} y.$$

We have seen that $\Pi(M,x,y) = 0$ whenever M is nonsingular and for every x and y. Let us focus on complex data: Π vanishes on a dense subset of $\mathbf{M}_n(\mathbb{C}) \times \mathbb{C}^n \times \mathbb{C}^n$, namely $\mathbf{GL}_n(\mathbb{C}) \times \mathbb{C}^n \times \mathbb{C}^n$. Being continuous, it vanishes everywhere. This means that, as a polynomial, $\Pi = 0$. Thus formula (3.15) is valid for every M,x,y and every field[1] K.

To prove (3.16), we multiply the right-hand side at the left by $M + xy^T$. We obtain the expression $I_n + \beta xy^T M^{-1}$ where we check immediately that $\beta = 0$. $\Box$

3.9 Alternate Matrices and the Pfaffian

The very simple structure of alternate bilinear forms is described in the following statement.

Proposition 3.22 *Let B be an alternate bilinear form on a vector space E, of dimension n. Then there exists a basis $\{x_1,y_1,\ldots,x_k,y_k,z_1,\ldots,z_{n-2k}\}$ such that the matrix of B in this basis is block-diagonal, equal to $\mathrm{diag}(J,\ldots,J,0,\ldots,0)$, with k blocks J defined by*

$$J = \begin{pmatrix} 0 & 1 \\ -1 & 0 \end{pmatrix}.$$

Proof. We proceed by induction on the dimension n. If $B = 0$, there is nothing to prove. If B is nonzero, there exist two vectors x_1,y_1 such that $B(x_1,y_1) \neq 0$. Multiplying one of them by $B(x_1,y_1)^{-1}$, one may assume that $B(x_1,y_1) = 1$. Because B is alternate, $\{x_1,y_1\}$ is free. Let N be the plane spanned by x_1,y_1. The set of vectors x satisfying $B(x,v) = 0$ (or equivalently $B(v,x) = 0$, inasmuch as B must be skew-symmetric) for every v in N is denoted by $N^\perp$. The formulæ

$$B(ax_1 + by_1, x_1) = -b, \quad B(ax_1 + by_1, y_1) = a$$

show that $N \cap N^\perp = \{0\}$. In addition, every vector $x \in E$ can be written as $x = y + n$, where $n \in N$ and $y \in N^\perp$ are given by

$$n = B(x,y_1)x_1 - B(x,x_1)y_1, \quad y := x - n.$$

Therefore, $E = N \oplus N^\perp$. We now consider the restriction of B to the subspace $N^\perp$ and apply the induction hypothesis. There exists a basis $\{x_2,y_2,\ldots,x_k,y_k,z_1,\ldots,z_{n-2k}\}$ such that the matrix of the restriction of B in this basis is block-diagonal, equal to $\mathrm{diag}(J,\ldots,J,0,\ldots,0)$, with $k-1$ blocks J, which means that $B(x_j,y_j) = 1 = -B(y_j,x_j)$ and $B(u,v) = 0$ for every other choice of u,v in the basis. Obviously, this property extends to the form B itself and the basis $\{x_1,y_1,\ldots,x_k,y_k,z_1,\ldots,z_{n-2k}\}$. $\Box$

We now choose an alternate matrix $M \in M_n(K)$ and apply Proposition 3.22 to the form defined by M. In view of Section 1.3, we have the following.

---

[1] This argument is known as the *extension of polynomial identities*.

Corollary 3.5 *Given an alternate matrix* $M \in M_n(K)$, *there exists a matrix* $Q \in$ $\mathbf{GL}_n(K)$ *such that*

$$M = Q^T \operatorname{diag}(J, \dots, J, 0, \dots, 0) Q. \tag{3.17}$$

Obviously, the rank of M, being the same as that of the block-diagonal matrix, equals twice the number of J blocks. Finally, because $\det J = 1$, we have $\det M = \varepsilon (\det Q)^2$, where $\varepsilon = 0$ if there is a zero diagonal block in the decomposition, and $\varepsilon = 1$ otherwise. Thus we have proved the following result.

Proposition 3.23 *The rank of an alternate matrix* M *is even. The number of* J *blocks in the identity* (3.17) *is the half of that rank. In particular, it does not depend on the decomposition. Finally, the determinant of an alternate matrix is a square in* K.

A very important application of Proposition 3.23 concerns the *Pfaffian*, whose crude definition is a polynomial whose square is the determinant of the general alternate matrix. First of all, because the rank of an alternate matrix is even, $\det M = 0$ whenever n is odd. Therefore, we restrict our attention from now on to the even-dimensional case $n = 2m$. Let us consider the field $F = \mathbb{Q}(x_{ij})$ of rational functions with rational coefficients, in $n(n-1)/2$ indeterminates x_{ij}, $i < j$. We apply the proposition to the alternate matrix X whose (i,i)-entry is 0 and (i,j)-entry (respectively, (j,i)-entry) is x_{ij} (respectively, $-x_{ij}$): its determinant, a polynomial in $\mathbb{Z}[x_{ij}]$, is the square of some irreducible rational function f/g, where f and g belong to $\mathbb{Z}[x_{ij}]$. From $g^2 \det X = f^2$, we see that g divides f in $\mathbb{Z}[x_{ij}]$. But because f and g are coprime, one finds that g is invertible; in other words $g = \pm 1$. Thus

$$\det X = f^2. \tag{3.18}$$

Now let k be a field and let $M \in M_n(k)$ be alternate. There exists a unique homomorphism from $\mathbb{Z}[x_{ij}]$ into k sending x_{ij} to m_{ij}. From equation (3.18) we obtain

$$\det M = (f(m_{12}, \dots, m_{n-1,n}))^2. \tag{3.19}$$

In particular, if $k = \mathbb{Q}$ and $M = \operatorname{diag}(J, \dots, J)$, one has $f(M)^2 = 1$. Up to multiplication by ± 1, which leaves unchanged the identity (3.18), we may assume that $f(M) = 1$ for this special case. This determination of the polynomial f is called the Pfaffian and is denoted by Pf. It may be viewed as a polynomial function on the vector space of alternate matrices with entries in a given field k. Equation (3.19) now reads

$$\det M = (\operatorname{Pf}(M))^2. \tag{3.20}$$

Given an alternate matrix $M \in M_n(k)$ and a matrix $Q \in M_n(k)$, we consider the Pfaffian of the alternate matrix $Q^T M Q$. We first treat the case of the field of fractions $\mathbb{Q}(x_{ij}, y_{ij})$ in the $n^2 + n(n-1)/2$ indeterminates x_{ij} $(1 \le i < j \le n)$ and y_{ij} $(1 \le i, j \le n)$. Let Y be the matrix whose (i,j)-entry is y_{ij}. Then, with X as above,

$$(\operatorname{Pf}(Y^T X Y))^2 = \det Y^T X Y = (\det Y)^2 \det X = (\operatorname{Pf}(X) \det Y)^2.$$

Because $\mathbb{Z}[x_{ij}, y_{ij}]$ is an integral domain, we have the polynomial identity

$$\mathrm{Pf}\left(Y^T X Y\right) = \varepsilon\,\mathrm{Pf}(X)\det Y, \quad \varepsilon = \pm 1.$$

As above, one infers that $\mathrm{Pf}(Q^T M Q) = \pm\mathrm{Pf}(M)\det Q$ for every field k, matrix $Q \in M_n(k)$, and alternate matrix $M \in M_n(k)$. Inspection of the particular case $Q = I_n$ yields $\varepsilon = 1$. We summarize these results now.

Theorem 3.6 *Let $n = 2m$ be an even integer. There exists a unique polynomial* Pf *in the indeterminates x_{ij} $(1 \leq i < j \leq n)$ with rational integer coefficients such that:*

- *For every field k and every alternate matrix $M \in M_n(k)$, one has $\det M = \mathrm{Pf}(M)^2$.*
- *If $M = \mathrm{diag}(J,\ldots,J)$, then $\mathrm{Pf}(M) = 1$.*

Moreover, if $Q \in M_n(k)$ is given, then

$$\mathrm{Pf}\left(Q^T M Q\right) = \mathrm{Pf}(M)\det Q. \tag{3.21}$$

We warn the reader that if $m > 1$, there does not exist a matrix $Z \in \mathbb{Q}[x_{ij}]$ such that $X = Z^T \mathrm{diag}(J,\ldots,J)Z$. The factorization of the polynomial $\det X$ does not correspond to a similar factorization of X itself. In other words, the decomposition $X = Q^T \mathrm{diag}(J,\ldots,J)Q$ in $M_n(\mathbb{Q}(x_{ij}))$ cannot be written within $M_n(\mathbb{Q}[x_{ij}])$.

The Pfaffian is computed easily for small values of n. For instance, $\mathrm{Pf}(X) = x_{12}$ if $n = 2$, and $\mathrm{Pf}(X) = x_{12}x_{34} - x_{13}x_{24} + x_{14}x_{23}$ if $n = 4$.

3.10 Calculating the Characteristic Polynomial

The algorithm of Leverrier is a method for computing the characteristic polynomial of a square matrix. It applies to matrices with entries in any field of characteristic 0. It is not used for the approximation of eigenvalues in the complex or real setting, because the practical calculation of the roots of $P \in \mathbb{C}[X]$ is best done by ... calculating accurately the eigenvalues of the so-called companion matrix! However, it has historical and combinatorial interests. The latter has been reinforced recently, after an improvement found by Preparata and Sarwate [31].

3.10.1 The Algorithm of Leverrier

Let K be a field of characteristic 0 and $M \in \mathbf{M}_n(K)$ be given. Let us denote by $\lambda_1,\ldots,\lambda_n$ the eigenvalues of M, counted with multiplicity. We define two lists of n numbers.

Elementary symmetric polynomials

$$\sigma_1 := \lambda_1 + \cdots + \lambda_n = \operatorname{Tr} M,$$
$$\sigma_2 := \sum_{j<k} \lambda_j \lambda_k,$$
$$\vdots$$
$$\sigma_r := \sum_{j_1 < \cdots < j_r} \lambda_{j_1} \cdots \lambda_{j_r},$$
$$\vdots$$
$$\sigma_n := \prod_j \lambda_j = \det M.$$

Newton sums

$$s_m := \sum_j \lambda_j^m, \quad 1 \le m \le n.$$

Because of $P_M(X) = \prod_j (X - \lambda_j)$, the numbers $(-1)^j \sigma_j$ are the coefficients of the characteristic polynomial of M:

$$P_M(X) = X^n - \sigma_1 X^{n-1} + \sigma_2 X^{n-2} - \cdots + (-1)^n \sigma_n.$$

The s_m are the traces of the powers M^m. One can obtain them by computing $M^2, \ldots, M^n$. Each of these matrices is obtained in $O(n^3)$ elementary operations.[2] In all, the computation of $s_1, \ldots, s_n$ is done $O(n^4)$ operations when n is large.

The passage from Newton sums to elementary symmetric polynomials is done through Newton's formulæ. If $\Sigma_j = (-1)^j \sigma_j$ and $\Sigma_0 := 1$, we have

$$m\Sigma_m + \sum_{k=1}^{m} s_k \Sigma_{m-k} = 0, \quad 1 \le m \le n.$$

One uses these formulæ in increasing order, beginning with $\Sigma_1 = -s_1$. When $\Sigma_1, \ldots, \Sigma_{m-1}$ are known, one computes

$$\Sigma_m = -\frac{1}{m}(s_1 \Sigma_{m-1} + \cdots + s_m \Sigma_0). \qquad (3.22)$$

This computation, which needs only $O(n^2)$ operations, has a negligible cost, compared to the $O(n^4)$ above.

In conclusion, the algorithm reads as follows.

- Compute $M^2, \ldots, M^n$,
- Compute the traces of $M, M^2, \ldots, M^n$, whence $s_1, \ldots, s_n$,
- Derive the $\sigma_1, \ldots, \sigma_n$,
- Form P_M.

---

[2] An elementary operation is a sum or a product in K.

Besides the high cost (in n^4) of this method, its instability is unfortunate when $k = \mathbb{R}$ or $k = \mathbb{C}$: when n is large, s_k has to be computed for large values of k. But then s_k is dominated by the powers λ^k of a few eigenvalues only, those of largest modulus. Thus a lot of information, corresponding to the smaller eigenvalues, is lost in the roundoff. This is even worse because of the large number of operations, which multiplies the roundoff errors.

This is the reason why, on the analytical side, the Leverrier's algorithm is rarely used.

When the field is of nonzero characteristic p, Leverrier method may be employed only if $n < p$. Because $s_p = \sigma_1^p$, the computation of the s_ms for $m \geq p$ does not bring any new information about the σ_js.

3.10.1.1 A Characterization of Nilpotent Matrices

Newton's formulæ have the following interesting consequence.

Proposition 3.24 *If K is of characteristic 0 and $A \in \mathbf{M}_n(K)$, then A is nilpotent; that is, $A^n = 0_n$ if and only if*

$$\mathrm{Tr}(A^k) = 0, \qquad \forall 1 \leq k \leq n.$$

Remark

The proposition is still valid if we replace K by an Abelian ring in which $mx = 0$ implies $x = 0$ whenever m is a positive integer.

Proof. If A is nilpotent, its only eigenvalue is 0, as well as for A^k, whence $\mathrm{Tr}(A^k) = 0$. Conversely, if all these traces vanish, then $s_1 = \cdots = s_n = 0$ and (3.22) yields $\Sigma_1 = \cdots \Sigma_n = 0$, whence $P_A = X^n$. We conclude with the Cayley–Hamilton theorem. $\square$

3.10.2 The Improvement by Preparata and Sarwate

The cost $O(n^4)$ can be reduced significantly, to $O(n^{3.5})$, thanks to the following trick. In order to compute the traces of $M, \ldots, M^n$, we do not really need to calculate all the powers $M^2, \ldots, M^n$. Let us start with the observation that if $A, B \in \mathbf{M}_n(K)$ are given, the trace of the product

$$\mathrm{Tr}(AB) = \sum_{i,j=1}^{n} a_{ij} b_{ji}$$

needs only $2n^2 - 1$ operations.

Let r be the least integer larger than or equal to $\sqrt{n}$. Assuming that we have computed $M^2, \ldots, M^r$, we then also compute powers of M^r: $M^{2r}, \ldots, M^{(r-1)r}$. Doing so, we have performed only $2r - 3$ matrix products.

Given an integer k between 1 and n, we make the Euclidean division of k by r: $k = ar + b$ with $0 \leq b \leq r - 1$. Writing that $M^k = M^{ar}M^b$, we see that the calculation of $\mathrm{Tr}(M^k)$ needs only $2n^2 - 1$ operations.

Let us evaluate the complexity of the algorithm with this improvement:

- The calculation of $M^2, \ldots, M^r$ needs $O(rn^3)$ operations.
- The calculation of $M^{2r}, \ldots, M^{(r-1)r}$ needs $O(rn^3)$ operations.
- The calculation of $\mathrm{Tr}\,M, \ldots, \mathrm{Tr}\,M^n$ needs $O(n^3)$ operations (which is significantly more than the $O(n^2)$ in the original method).
- The calculation of $\sigma_2, \ldots, \sigma_n$ needs $O(n^2)$ operations.

Inasmuch as $r \approx \sqrt{n}$, this amounts to an $O(n^{3.5})$ operations.

3.11 Irreducible Matrices

Let us consider the most classical problems in matrix theory:

- Solve a linear system
$$Ax = b, \qquad (b \text{ given}). \qquad (3.23)$$

- Find the eigenvalues of A.

When a matrix is block-triangular,

$$A = \begin{pmatrix} B & C \\ 0_{p,n-p} & D \end{pmatrix}, \qquad (3.24)$$

these problems become easier. As a matter of fact, the spectrum of A is the union of those of B and D, adding the algebraic multiplicities, because $P_A(X) = P_B(X)P_D(X)$. As far as system (3.23) is concerned, it decouples when decomposing the data b and the unknown x blockwise:

$$b = \begin{pmatrix} b_- \\ b_+ \end{pmatrix}, \qquad x = \begin{pmatrix} x_- \\ x_+ \end{pmatrix},$$

because it is enough to solve the smaller system $Dx_+ = b_+$ first, and then solve the other one $Bx_- = b_- - Cx_+$.

More generally, it may happen that A is blockwise-triangular up to a relabeling of the components of the vectors, that is, after a conjugation by a permutation matrix. We say then that A is *reducible*. Reducibility means that there exists a nontrivial partition $\{1, \ldots, n\} = I \cup J$ such that $(i, j) \in I \times J$ implies $a_{ij} = 0$. It is *irreducible* otherwise. We show in Exercise 12 a characterization of irreducible matrices in terms of graphs.

The inverse of the permutation matrix

$$P := \begin{pmatrix} 0_{p \times q} & I_p \\ I_q & 0_{q \times p} \end{pmatrix}$$

is

$$P^{-1} = \begin{pmatrix} 0_{q \times p} & I_q \\ I_p & 0_{p \times q} \end{pmatrix}.$$

It follows that a lower block-triangular matrix is permutationally similar to an upper block-triangular one:

$$\begin{pmatrix} A & 0_{p \times q} \\ B & C \end{pmatrix} = P \begin{pmatrix} C & B \\ 0_{p \times q} & A \end{pmatrix} P^{-1}.$$

In particular, we infer the next proposition

Proposition 3.25 *The notion of reducibility is invariant under transposition: M^T is reducible if and only if M is reducible.*

3.11.1 Hessenberg Irreducible Matrices

The notions of (ir-)reducibility are of practical interest because reducibility is usually easy to catch by human eyes. A useful consequence of irreducibility concerns *Hessenberg matrices*, defined as the square matrices M such that $i \geq j + 2$ implies $m_{ij} = 0$.

Proposition 3.26 *Let $M \in \mathbf{M}_n(K)$ be an irreducible Hessenberg matrix. Then the eigenvalues of M are geometrically simple.*

Proof. The hypothesis implies that all entries $m_{j+1,j}$ are nonzero. If λ is an eigenvalue, let us consider the matrix $N \in \mathbf{M}_{n-1}(\bar{K})$, obtained from $M - \lambda I_n$ by deleting the first row and the last column. It is a triangular matrix, whose diagonal terms are nonzero. It is thus invertible, which implies (Proposition 2.2) $\mathrm{rk}(M - \lambda I_n) = n - 1$. Hence $\ker(M - \lambda I_n)$ is of dimension one. $\quad\square$

Exercises

1. Verify that the product of two triangular matrices of the same type (upper or lower) is triangular, of the same type.
2. Prove in full detail that the determinant of a triangular matrix (respectively, a block-triangular one) equals the product of its diagonal terms (respectively, the product of the determinants of its diagonal blocks).
3. Find matrices $M, N \in \mathbf{M}_2(K)$ such that $MN = 0_2$ and $NM \neq 0_2$. Such an example shows that MN and NM are not necessarily similar, although they would be if either M or N was invertible.

4. Characterize the square matrices that are simultaneously orthogonal and triangular.

5. One calls any square matrix M satisfying $M^2 = M$ a *projection matrix*, or *projector*.

   a. Let $P \in \mathbf{M}_n(K)$ be a projector, and let $E = \ker P$, $F = \ker(I_n - P)$. Show that $K^n = E \oplus F$.

   b. Let P, Q be two projectors. Prove the identity

   $$(P - Q)^2 + (I_n - P - Q)^2 = I_n.$$

6. If $A, B, C, D \in \mathbf{M}_m(K)$ and if $AC = CA$, show that the determinant of

$$M = \begin{pmatrix} A & B \\ C & D \end{pmatrix}$$

equals $\det(AD - CB)$. Begin with the case where A is invertible, by computing the product

$$\begin{pmatrix} I_m & 0_m \\ -C & A \end{pmatrix} M.$$

Then apply this intermediate result to the matrix $A - zI_n$, with $z \in \bar{K}$ a suitable scalar.

7. Let $A \in \mathbf{M}_n(K)$ be given. One says that a list $(a_{1\sigma(1)}, \ldots, a_{n\sigma(n)})$ is a *diagonal* of A if σ is a permutation (in that case, the diagonal given by the identity is the *main* diagonal). Show the equivalence of the following properties.

   • Every diagonal of A contains a zero element.
   • There exists a null matrix extracted from A of size $k \times l$ with $k + l > n$.

8. Compute the number of elements in the group $\mathbf{GL}_2(\mathbb{Z}/2\mathbb{Z})$. Show that it is not commutative. Show that it is isomorphic to the symmetric group S_m, for a suitable integer m.

9. If $(a_0, \ldots, a_{n-1}) \in \mathbb{C}^n$ is given, the *circulant matrix* $\mathrm{circ}(a_0, \ldots, a_{n-1}) \in \mathbf{M}_n(\mathbb{C})$ is

$$\mathrm{circ}(a_0, \ldots, a_{n-1}) := \begin{pmatrix} a_0 & a_1 & \cdots & a_{n-1} \\ a_{n-1} & a_0 & \ddots & \vdots \\ \vdots & \ddots & \ddots & a_1 \\ a_1 & \cdots & a_{n-1} & a_0 \end{pmatrix}.$$

We denote by $\mathscr{C}_n$ the set of circulant matrices. The matrix $\mathrm{circ}(0, 1, 0, \ldots, 0)$ is denoted by π.

   a. Show that $\mathscr{C}_n$ is a subalgebra of $\mathbf{M}_n(\mathbb{C})$, equal to $\mathbb{C}[\pi]$. Deduce that it is isomorphic to the quotient ring $\mathbb{C}[X]/(X^n - 1)$.

   b. Let C be a circulant matrix. Show that C^*, as well as $P(C)$, is circulant for every polynomial P. If C is nonsingular, show that C^{-1} is circulant.

   c. Show that the elements of $\mathscr{C}_n$ are diagonalizable in a common eigenbasis.

d. Replace $\mathbb{C}$ by any field K. If K contains a primitive nth root ω of unity (that is, $\omega^n = 1$, and $\omega^m = 1$ implies $m \in n\mathbb{Z}$), show that the elements of $\mathscr{C}_n$ are diagonalizable.
   Note: A thorough presentation of circulant matrices and applications is given in Davis's book [12].

e. One assumes that the charc$(K)|n$. Show that $\mathscr{C}_n$ contains matrices that are not diagonalizable.

10. Show that the Pfaffian is linear with respect to any row or column of an alternate matrix. Deduce that the Pfaffian is an irreducible polynomial in $\mathbb{Z}[x_{ij}]$.

11. (Schur's lemma).
    Let k be an algebraically closed field and S a subset of $\mathbf{M}_n(k)$. Assume that the only linear subspaces of k^n that are stable under every element of S are $\{0\}$ and k^n itself. Let $A \in \mathbf{M}_n(k)$ be a matrix that commutes with every element of S. Show that A is of the form cI_n.

12. a. Show that $A \in \mathbf{M}_n(K)$ is irreducible if and only if for every pair (j,k) with $1 \le j,k \le n$, there exists a finite sequence of indices $j = l_1, \ldots, l_r = k$ such that $a_{l_p, l_{p+1}} \ne 0$.

    b. Show that a tridiagonal matrix $A \in \mathbf{M}_n(K)$, for which none of the $a_{j,j+1}$s and $a_{j+1,j}$s vanish, is irreducible.

13. Let $A \in \mathbf{M}_n(k)$ ($k = \mathbb{R}$ or $\mathbb{C}$) be given, with minimal polynomial π. If $x \in k^n$, the set
$$I_x := \{p \in k[X] \mid p(A)x = 0\}$$
is an ideal of $k[X]$, which is therefore principal.

    a. Show that $I_x \ne (0)$ and that its monic generator, denoted by p_x, divides π.

    b. One writes r_j instead of p_x when $x = \mathbf{e}^j$. Show that π is the least common multiple of $r_1, \ldots, r_n$.

    c. If $p \in k[X]$, show that the set
$$V_p := \{x \in k^n \mid p_x \in (p)\}$$
    (the vectors x such that p divides p_x) is open.

    d. Let $x \in k^n$ be an element for which p_x is of maximal degree. Show that $p_x = \pi$. **Note**: In fact, the existence of an element x such that p_x equals the minimal polynomial holds true for every field k.

14. Let k be a field and $A \in \mathbf{M}_{n \times m}(k)$, $B \in \mathbf{M}_{m \times n}(k)$ be given.

    a. Let us define
$$M = \begin{pmatrix} XI_n & A \\ B & XI_m \end{pmatrix}.$$
    Show that $X^m \det M = X^n \det(X^2 I_m - BA)$ (find a lower-triangular matrix M' such that $M'M$ is upper-triangular).

    b. Find an analogous relation between $\det(X^2 I_n - AB)$ and $\det M$. Deduce that $X^n P_{BA}(X) = X^m P_{AB}(X)$.

    c. What do you deduce about the eigenvalues of A and of B ?

15. Let k be a field and $\theta : \mathbf{M}_n(k) \to k$ a linear form satisfying $\theta(AB) = \theta(BA)$ for every $A, B \in \mathbf{M}_n(k)$.

    a. Show that there exists $\alpha \in k$ such that for all $X, Y \in k^n$, one has $\theta(XY^T) = \alpha \sum_j x_j y_j$.

    b. Deduce that $\theta = \alpha \operatorname{Tr}$.

16. Let A_n be the ring $K[X_1, \ldots, X_n]$ of polynomials in n variables. Consider the matrix $M \in \mathbf{M}_n(A_n)$ defined by

$$M = \begin{pmatrix} 1 & \cdots & 1 \\ X_1 & \cdots & X_n \\ X_1^2 & \cdots & X_n^2 \\ \vdots & & \vdots \\ X_1^{n-1} & \cdots & X_n^{n-1} \end{pmatrix}.$$

Let us denote by $\Delta(X_1, \ldots, X_n)$ the determinant of M.

    a. Show that for every $i \neq j$, the polynomial $X_j - X_i$ divides Δ.

    b. Deduce that

$$\Delta = a \prod_{i<j} (X_j - X_i),$$

    where $a \in K$.

    c. Determine the value of a by considering the monomial

$$\prod_{j=1}^{n} X_j^{j-1}.$$

    d. Redo this analysis for the matrix

$$\begin{pmatrix} X_1^{p_1} & \cdots & X_n^{p_1} \\ \vdots & & \vdots \\ X_1^{p_n} & \cdots & X_n^{p_n} \end{pmatrix},$$

    where $p_1, \ldots, p_n$ are nonnegative integers.

17. Deduce from the previous exercise that the determinant of the *Vandermonde* matrix

$$\begin{pmatrix} 1 & \cdots & 1 \\ a_1 & \cdots & a_n \\ a_1^2 & \cdots & a_n^2 \\ \vdots & & \vdots \\ a_1^{n-1} & \cdots & a_n^{n-1} \end{pmatrix}, \quad a_1, \ldots, a_n \in K,$$

vanishes if and only if at least two of the a_js coincide.

18. Multiplying a Vandermonde matrix by its transpose, show that

$$
\det \begin{pmatrix} n & s_1 & \cdots & s_{n-1} \\ s_1 & s_2 & \ddots & \vdots \\ \vdots & & \ddots & \vdots \\ s_{n-1} & \cdots & \cdots & s_{2n-2} \end{pmatrix} = \prod_{i<j}(a_j - a_i)^2,
$$

where $s_q := a_1^q + \cdots + a_n^q$.

19. The *discriminant* of a matrix $A \in M_n(k)$ is the number

$$
d(A) := \prod_{i<j}(\lambda_j - \lambda_i)^2,
$$

where $\lambda_1, \ldots, \lambda_n$ are the eigenvalues of A, counted with multiplicity.

a. Verify that the polynomial

$$
\Delta(X_1, \ldots, X_n) := \prod_{i<j}(X_j - X_i)^2
$$

is symmetric. Therefore, there exists a unique polynomial $Q \in \mathbb{Z}[Y_1, \ldots, Y_n]$ such that

$$
\Delta = Q(\sigma_1, \ldots, \sigma_n),
$$

where the σ_js are the elementary symmetric polynomials

$$
\sigma_1 = X_1 + \cdots + X_n, \ldots, \sigma_n = X_1 \cdots X_n.
$$

b. Deduce that there exists a polynomial $D \in \mathbb{Z}[x_{ij}]$ in the indeterminates x_{ij}, $1 \leq i, j \leq n$, such that for every k and every square matrix A,

$$
d(A) = D(a_{11}, a_{12}, \ldots, a_{nn}).
$$

c. Consider the restriction D_S of the discriminant to symmetric matrices, where x_{ji} is replaced by x_{ij} whenever $i < j$. Prove that D_S takes only non-negative values on $\mathbb{R}^{n(n+1)/2}$. Show, however, that D_S is not the square of a polynomial if $n \geq 2$ (consider first the case $n = 2$).

20. (Formanek [14].) This exercise is reminiscent of theorem 4.4. Let k be a field of characteristic 0.

a. Show that for every $A, B, C \in \mathbf{M}_2(k)$,

$$
[[A,B]^2, C] = 0.
$$

Hint: use the Cayley–Hamilton theorem.

b. Show that for every $M, N \in \mathbf{M}_2(k)$,

$$MN + NM - \mathrm{Tr}(M)N - \mathrm{Tr}(N)M + (\mathrm{Tr}(M)\,\mathrm{Tr}(N) - \mathrm{Tr}(MN))I_2 = 0.$$

Hint: One may begin with the case $M = N$ and recognize a classical theorem, then "bilinearize" the formula.

c. If $\pi \in S_r$ (S_r is the symmetric group over $\{1,\dots,r\}$), one defines a map $T_\pi : \mathbf{M}_2(k)^r \to k$ in the following way. One decomposes π as a product of disjoint cycles, including the cycles of order one, which are the fixed points of π:

$$\pi = (a_1,\dots,a_{k_1})(b_1,\dots,b_{k_2})\cdots.$$

One then sets

$$T_\pi(N_1,\dots,N_r) = \mathrm{Tr}(N_{a_1}\cdots N_{a_{k_1}})\,\mathrm{Tr}(N_{b_1}\cdots N_{b_{k_2}})\cdots$$

(note that the right-hand side depends neither on the order of the cycles in the product nor on the choice of the first index inside each cycle, because of the formula $\mathrm{Tr}(AB) = \mathrm{Tr}(BA)$). Show that for every $N_1, N_2, N_3 \in \mathbf{M}_2(k)$, one has

$$\sum_{\pi \in S_3} \varepsilon(\pi) T_\pi(N_1, N_2, N_3) = 0.$$

d. Generalize this result to $\mathbf{M}_n(k)$: for every $N_1,\dots,N_{n+1} \in \mathbf{M}_n(k)$, one has

$$\sum_{\pi \in S_{n+1}} \varepsilon(\pi) T_\pi(N_1,\dots,N_{n+1}) = 0.$$

21. Let k be a field and let $A \in \mathbf{M}_n(k)$ be given. For every set $J \subset \{1,\dots,n\}$, denote by A_J the matrix extracted from A by keeping only the indices $i, j \in J$. Hence, $A_J \in \mathbf{M}_p(k)$ for $p = \mathrm{card}\, J$. Let λ be a scalar.

   a. Assume that for every J whose cardinality is greater than or equal to $n - p$, λ is an eigenvalue of A_J. Show that λ is an eigenvalue of A, of algebraic multiplicity greater than or equal to $p + 1$ (express the derivatives of the characteristic polynomial).

   b. Conversely, let q be the geometric multiplicity of λ as an eigenvalue of A. Show that if $\mathrm{card}\, J > n - q$, then λ is an eigenvalue of A_J.

22. Let $A \in \mathbf{M}_n(k)$ and $l \in \mathbb{N}$ be given. Show that there exists a polynomial $q_\ell \in k[X]$, of degree at most $n - 1$, such that $A^\ell = q_\ell(A)$. If A is invertible, show that there exists $r_\ell \in k[X]$, of degree at most $n - 1$, such that $A^{-l} = r_\ell(A)$.

23. Let k be a field and $A, B \in \mathbf{M}_n(k)$. Assume that $\mathrm{Sp}\, A \cap \mathrm{Sp}\, B = \emptyset$.

   a. Show, using the Cayley–Hamilton theorem, that the linear map $M \mapsto AM - MB$ is one-to-one over $\mathbf{M}_n(k)$. **Hint:** The spectrum of $P_A(B)$ is the image of that of B under P_A. This is proved in Proposition 5.7 in the complex case, but is valid for every field k.

   b. Deduce that given $C \in \mathbf{M}_n(k)$, the *Sylvester equation*

$$AX - XB = C$$

admits a unique solution $X \in \mathbf{M}_n(k)$.

24. Let k be a field and $(M_{jk})_{1 \leq j,k \leq n}$ a set of matrices of $\mathbf{M}_n(k)$, at least one of which is nonzero, such that $M_{ij}M_{kl} = \delta_j^k M_{il}$ for all $1 \leq i,j,k,l \leq n$.

   a. Show that none of the matrices M_{jk} vanishes.
   b. Verify that each M_{ii} is a projector. Denote its range by E_i.
   c. Show that $E_1, \ldots, E_n$ are in direct sum. Deduce that each E_j is a line.
   d. Show that there exist generators e_j of each E_j such that $M_{jk}e_\ell = \delta_k^\ell e_j$.
   e. Deduce that every algebra automorphism of $\mathbf{M}_n(k)$ is interior: For every $\sigma \in \mathrm{Aut}(\mathbf{M}_n(k))$, there exists $P \in \mathbf{GL}_n(k)$ such that $\sigma(M) = P^{-1}MP$ for every $M \in \mathbf{M}_n(k)$.

25. Set $n = 2m$.

   a. Show the following formula for the Pfaffian, as an element of $\mathbb{Z}[x_{ij}; 1 \leq i < j \leq n]$,
   $$\mathrm{Pf}(X) = \sum (-1)^\sigma x_{i_1 i_2} \cdots x_{i_{2m-1} i_{2m}}.$$
   Hereabove, the sum runs over all the possible ways the set $\{1, \ldots, n\}$ can be partitioned in pairs,
   $$\{1, \ldots, n\} = \{i_1, i_2\} \cup \cdots \cup \{i_{2m-1} i_{2m}\}.$$
   To avoid redundancy in the list of partitions, one normalized by
   $$i_{2k-1} < i_{2k}, \quad 1 \leq k \leq m,$$
   and $i_1 < i_3 < \cdots < i_{2m-1}$ (in particular, $i_1 = 1$ and $i_{2m} = 2m$). At last, σ is the signature of the permutation $(i_1, i_2, \cdots, i_{2m-1}, i_{2m})$.
   Compute the number of monomials in the Pfaffian.
   b. Deduce an "expansion formula with respect to the ith row" for the Pfaffian: if i is given, then
   $$\mathrm{Pf}(X) = \sum_{j(\neq i)} \alpha(i, j)(-1)^{i+j+1} x_{ij} \mathrm{Pf}(X^{ij}),$$
   where $X^{ij} \in \mathbf{M}_{n-2}(k)$ denotes the alternate matrix obtained from X by removing the ith and the jth rows and columns, and $\alpha(i, j)$ is $+1$ if $i < j$ and is -1 if $j < i$.

26. Let k be a field and n an even integer. If $x, y \in k^n$, denote by $x \wedge y$ the alternate matrix $xy^T - yx^T$. Show the formula
   $$\mathrm{Pf}(A + x \wedge y) = (1 + y^T A^{-1} x) \mathrm{Pf} A$$
   for every nonsingular alternate $n \times n$ matrix A.
   Hint: Use proposition 3.21.

27. Check the easy formula valid whenever the inverses concern nonsingular $n \times n$ matrices:

$$(I_n + A^{-1})^{-1} + (I_n + A)^{-1} = I_n.$$

Deduce Hua's identity

$$(B + BA^{-1}B)^{-1} + (A + B)^{-1} = B^{-1}.$$

Hint: Transport the algebra structure of $\mathbf{M}_n(k)$ by the linear map $M \mapsto BM$. This procedure is called *isotopy*; remark that the multiplicative identity in the new structure is B. Then apply the easy formula.

28. Prove the determinantal identity (Cauchy's *double alternant*)

$$\left\| \frac{1}{a_i + b_j} \right\|_{1 \le i, j \le n} = \frac{\prod_{i<j}(a_j - a_i) \prod_{k<l}(b_k - b_\ell)}{\prod_{i,k}(a_i + b_k)}.$$

Hint: One may assume that $a_1, \ldots, a_n, b_1, \ldots, b_n$ are indeterminate, and then work in the field $\mathbb{Q}(a_1, \ldots, a_n, b_1, \ldots, b_n)$. This determinant is a homogeneous rational function whose denominator is quite trivial. Some specializations make it vanishing; this gives accurate information about the numerator. There remains to find a scalar factor. That can be done by induction on n, with an expansion with respect to the last row and column.

29. Prove Schur's Pfaffian identity

$$\mathrm{Pf}\left(\left(\frac{a_j - a_i}{a_i + a_j} \right) \right)_{1 \le i, j \le 2n} = \prod_{i<j} \frac{a_j - a_i}{a_j + a_i}.$$

See the previous exercise for a hint.

30. We denote

$$X := \begin{pmatrix} 1 & 1 \\ 0 & 1 \end{pmatrix}, \qquad Y := \begin{pmatrix} 1 & 0 \\ 1 & 1 \end{pmatrix}.$$

   a. Let $M \in \mathbf{SL}_2(\mathbb{N})$ be given. If $M \ne I_2$, show that the columns of N are ordered: $(m_{11} - m_{12})(m_{21} - m_{22}) \ge 0$.
   b. Under the same assumption, deduce that there exists a matrix $M' \in \mathbf{SL}_2(\mathbb{N})$ such that either $M = M'X$ or $M = M'Y$. Check that $\mathrm{Tr}\,M' \le \mathrm{Tr}\,M$. Under which circumstances do we have $\mathrm{Tr}\,M' < \mathrm{Tr}\,M$?
   c. Let $M \in \mathbf{SL}_2(\mathbb{N})$ be given. Arguing by induction, show that there exists a word w_0 in two letters, and a triangular matrix $T \in \mathbf{SL}_2(\mathbb{N})$, such that $M = Tw_0(X, Y) \in \mathbf{SL}_2(\mathbb{N})$.
   d. Conclude that for every $M \in \mathbf{SL}_2(\mathbb{N})$, there exists a word w in two letters, such that $M = w(X, Y)$.

Comment. One can show that every element of $\mathbf{SL}_2(\mathbb{Z})$, whose trace is larger than 2, is conjugated in $\mathbf{SL}_2(\mathbb{Z})$ to a word in X and Y. This word is not unique in general, because if $M \sim w_2(X, Y)w_1(X, Y)$, then $M \sim w_1(X, Y)w_2(X, Y)$ too.

31. Let $A \in \mathbf{M}_n(k)$, $B \in \mathbf{M}_m(k)$, and $M \in \mathbf{M}_{n \times m}(k)$ be such that $AM = MB$. It is well-known that if $n = m$ and M is nonsingular, then the characteristic polynomials of A and B are equal: $P_A = P_B$. If $\mathrm{rk}M = r$, prove that the degree of $\gcd\{P_A, P_B\}$ is larger than or equal to r. **Hint:** Reduce to the case where M is quasidiagonal, that is, $m_{ij} = 0$ whenever $i \neq j$.

32. Let k be a field. Given two vectors X, Y in k^3, we define the vector product as usual:

$$X \times Y := \begin{pmatrix} x_2 y_3 - x_3 y_2 \\ x_3 y_1 - x_1 y_3 \\ x_1 y_2 - x_2 y_1 \end{pmatrix}.$$

Prove the following identity in $\mathbf{M}_3(k)$:

$$X(Y \times Z)^T + Y(Z \times X)^T + Z(X \times Y)^T = \det(X, Y, Z) I_3, \qquad \forall X, Y, Z \in k^3.$$

33. Among the class of Hessenberg matrices, we distinguish the *unit* ones, which have 1s below the diagonal:

$$M = \begin{pmatrix} * & \cdots & & \cdots & * \\ 1 & \ddots & & & \vdots \\ 0 & \ddots & & & \\ \vdots & \ddots & \ddots & \ddots & \vdots \\ 0 & \cdots & 0 & 1 & * \end{pmatrix}.$$

a. Let $M \in \mathbf{M}_n(k)$ be a unit Hessenberg matrix. We denote by M_k the submatrix obtained by retaining the first k rows and columns. For instance, $M_n = M$. We set P_k the characteristic polynomial of M_k.
   Show that

$$P_n(X) = (X - m_{nn}) P_{n-1} - m_{n-1,n} P_{n-2} - \cdots - m_{1n}.$$

b. Let $Q_1, \ldots, Q_n \in k[X]$ be monic polynomials, with $\deg Q_k = k$. Show that there exists one and only one unit Hessenberg matrix M such that, for every $k = 1, \ldots, n$, the characteristic polynomial of M_k equals Q_k. **Hint:** Argue by induction over n.

Chapter 4
Tensor and Exterior Products

4.1 Tensor Product of Vector Spaces

4.1.1 Construction of the Tensor Product

Let E and F be K-vector spaces whose dimensions are finite. We construct their *tensor product* $E \otimes_K F$ as follows.

We start with their Cartesian product. We warn the reader that we do not equip $E \times F$ with the usual addition. Thus we do not think of it as a vector space. The first step is to consider the set G of formal linear combinations of elements of $E \times F$

$$\sum_{j=1}^{r} \lambda_j (x_j, y_j),$$

where r is an arbitrary natural integer, λ_j are scalars, and $(x_j, y_j) \in E \times F$.

The set G has a natural structure of K-vector space, where $E \times F$ is a basis. The zero element is the empty sum ($r = 0$). We warn the reader that $(x + x', y) - (x, y) - (x', y)$ and $\lambda(x, y) - (\lambda x, y)$ cannot be simplified, and $(0_E, 0_F)$ is not equal to 0_G. Actually, G is infinite-dimensional whenever K is infinite!

We now consider the subspace G_0 generated by all elements of the form $(x + x', y) - (x, y) - (x', y)$, $(x, y + y') - (x, y) - (x, y')$, $\lambda(x, y) - (\lambda x, y)$ or $\lambda(x, y) - (x, \lambda y)$. The quotient space G/G_0 is what we call the tensor product of E and F (in this order) and denote by $E \otimes_K F$. When there is no ambiguity about the scalars, we simply write $E \otimes F$.

By construction, $E \otimes_K F$ is a vector space. The class of an elementary pair (x, y) is denoted $x \otimes y$. The elements of G_0 can be viewed as the *simplification rules* in $E \otimes_K F$:

$$(\lambda x + x') \otimes y = \lambda(x \otimes y) + x' \otimes y, \qquad x \otimes (\lambda y + y') = \lambda(x \otimes y) + x \otimes y'.$$

In particular, we have $x \otimes 0_F = 0_E \otimes y = 0$ for every x and y.

Theorem 4.1 *Let* E, F, H *be K-vector spaces. Then* $\textbf{Bil}(E \times F; H)$ *is isomorphic to* $\mathscr{L}(E \otimes F; H)$ *through the formula*

$$b(x, y) = u(x \otimes y).$$

Proof. If $u \in \mathscr{L}(E \otimes F; H)$ is given, then $b(x, y) := u(x \otimes y)$ clearly defines a bilinear map.

Conversely, let $b \in \textbf{Bil}(E \times F; H)$ be given. Then $\beta : G \mapsto H$, defined by

$$\beta \left(\sum_{j=1}^{r} \lambda_j(x_j, y_j) \right) := \sum_{j=1}^{r} \lambda_j b(x_j, y_j),$$

is linear. Because of the bilinearity, β vanishes identically over G_0, thus passes to the quotient as a linear map u.   □

Corollary 4.1 *Let* $x \in E$ *and* $y \in F$ *be given vectors. If* $x \neq 0$ *and* $y \neq 0$, *then* $x \otimes y \neq 0$.

Proof. There exist linear forms $\ell \in E'$ and $m \in F'$ such that $\ell(x) = m(y) = 1$. Then the map

$$(w, z) \mapsto b(w, z) := \ell(w) m(z)$$

is bilinear over $E \times F$ and is such that $b(x, y) = 1$. Theorem 4.1 provides a linear form u over $E \otimes F$ such that $u(x \otimes y) = 1$. Thus $x \otimes y \neq 0$.   □

We say that an element $x \otimes y$ is *rank one* if $x \neq 0$ and $y \neq 0$. More generally, the *rank* of an element of $E \otimes_K F$ is the minimal length r among all of its representations of the form

$$x_1 \otimes y_1 + \cdots + x_r \otimes y_r.$$

Given a basis $\mathscr{B}_E$ of E and a basis $\mathscr{B}_F$ of F, one may form a basis of $E \otimes_K F$ by taking all the products $\mathbf{e}^i \otimes \mathbf{f}^j$ where $\mathbf{e}^i \in \mathscr{B}_F$ and $\mathbf{f}^j \in \mathscr{B}_F$. This is obviously a generating family. To see that it is free, let us consider an identity

$$\sum_{i,j} \lambda_{ij} \mathbf{e}^i \otimes \mathbf{f}^j = 0.$$

Let ℓ_p and m_q be the elements of the dual bases such that $\ell_p(\mathbf{e}^i) = \delta_p^i$ and $m_q(\mathbf{f}^j) = \delta_q^j$. The bilinear map $b(x, y) := \ell_p(x) m_q(y)$ extends as a linear map u over $E \otimes F$, according to Theorem 4.1. We have

$$0 = u \left(\sum_{i,j} \lambda_{ij} \mathbf{e}^i \otimes \mathbf{f}^j \right) = \sum_{i,j} \lambda_{ij} \ell_p(\mathbf{e}^i) m_q(\mathbf{f}^j) = \lambda_{pq}.$$

The dimension of $E \otimes_K F$ is thus equal to the product of $\dim E$ and $\dim F$:

$$\dim E \otimes_K F = \dim E \cdot \dim F. \tag{4.1}$$

This contrasts with the formula

$$\dim E \times F = \dim E + \dim F.$$

4.1.2 Linearity versus Bilinearity

4.1.2.1 The Dual of a Tensor Product

If $\ell \in E'$ and $m \in F'$ are linear forms, then the pair (m, ℓ) defines a bilinear form over $E \times F$, by

$$(x, y) \mapsto \ell(x) \cdot m(y).$$

By Theorem 4.1, there corresponds a linear form $T_{(m,\ell)}$ over $E \otimes F$. We notice that the map $(m, \ell) \mapsto T_{(m,\ell)}$ is bilinear too. Invoking the theorem again, we infer a linear map

$$T : F' \otimes E' \to (E \otimes F)',$$

defined by

$$[T(m \otimes \ell)](x \otimes y) = \ell(x) \cdot m(y).$$

We verify easily that T is an isomorphism, a canonical one.

4.1.2.2 $\mathscr{L}(E; F)$ as a Tensor Product

Given a linear map $f : E \mapsto F$, we may construct a bilinear form over $E \times F'$ by

$$(x, m) \mapsto m(f(x)).$$

By Theorem 4.1, it extends as a linear form over $E \otimes F'$, satisfying

$$x \otimes m \mapsto m(f(x)).$$

By the previous paragraph, f can be identified as an element of $(E \otimes F')' = (F'') \otimes E' = F \otimes E'$. This provides a homomorphism from $\mathscr{L}(E; F)$ into $F \otimes E'$. This morphism is obviously one-to-one: if $m(f(x)) \equiv 0$ for every m and x, then $f(x) = 0$ for every x; that is, $f = 0$. It is also onto: an element

$$\sum_{j=1}^{r} v^j \otimes m^j$$

is the image of the linear map

$$x \mapsto f(x) := \sum_j m^j(x) v^j.$$

Therefore $\mathscr{L}(E;F)$ identifies canonically with $F \otimes E'$.

4.1.3 Iterating the Tensor Product

Given three vector spaces E, F, and G over K, with bases $(\mathbf{e}^i)_{1 \le i \le m}$, $(\mathbf{f}^j)_{1 \le j \le n}$, $(\mathbf{g}^k)_{1 \le k \le p}$, the spaces $(E \otimes F) \otimes G$ and $E \otimes (F \otimes G)$ are isomorphic through $(\mathbf{e}^i \otimes \mathbf{f}^j) \otimes \mathbf{g}^k \longleftrightarrow \mathbf{e}^i \otimes (\mathbf{f}^j \otimes \mathbf{g}^k)$. We thus identify both spaces, and denote them simply by $E \otimes F \otimes G$.

This rule allows us to define the tensor product of an arbitrary finite number of vector spaces $E^1, \ldots, E^r$, denoted as $E^1 \otimes \cdots \otimes E^r$. The following generalization of Theorem 4.1 is immediate.

Theorem 4.2 *Let $E^1, \ldots, E^r, F$ be K-vector spaces. Then the vector space of r-linear maps from $E^1 \times \cdots \times E^r$ into F is isomorphic to $\mathscr{L}(E^1 \otimes \cdots \otimes E^r; F)$ through the formula*

$$\psi(x^1, \ldots, x^r) = u(x^1 \otimes \cdots \otimes x^r).$$

4.2 Exterior Calculus

4.2.1 Tensors of Degree Two

We assume temporarily that the characteristic of the field K is not 2, even though in general we do not need this hypothesis. Let E be a finite-dimensional K-vector space, with basis $\{\mathbf{e}^1, \ldots, \mathbf{e}^n\}$. The linear map over G

$$\sum_i \lambda_i(x_i, y_i) \mapsto \sum_i \lambda_i(y_i, x_i)$$

sends G_0 into G_0. It thus passes to the quotient, defining a linear map $\sigma : x \otimes y \mapsto y \otimes x$. Because σ is an involution, the group $\mathbb{Z}/2\mathbb{Z} \sim \{1, \sigma\}$ operates over $E \otimes E$ by

$$1 \cdot (x \otimes y) = (x \otimes y), \qquad \sigma \cdot (x \otimes y) = (y \otimes x).$$

$$\mathbf{Sym}^2(E) := \{w \in E \otimes E \mid \sigma(w) = w\}$$

the set of *symmetric* tensors, and

$$\Lambda^2(E) := \{w \in E \otimes E \mid \sigma(w) = -w\}$$

the set of *skew-symmetric* tensors. These are subspaces, with obviously $\mathbf{Sym}^2(E) \cap \Lambda^2(E) = \{0\}$. If $w \in E \otimes E$, we have

$$\frac{1}{2}(w + \sigma(w)) \in \mathbf{Sym}^2(E), \qquad \frac{1}{2}(w - \sigma(w)) \in \Lambda^2(E),$$

and their sum equals w. We deduce

$$E \otimes E = \mathbf{Sym}^2(E) \oplus \Lambda^2(E).$$

Thus we can view as well $\Lambda^2(E)$ as the quotient of $E \otimes E$ by the subspace spanned by the tensors of the form $x \otimes x$ (which is nothing but $\mathbf{Sym}^2(E)$).

Proposition 4.1 *A basis of* $\mathbf{Sym}^2(E)$ *is provided by the tensors*

$$\mathbf{e}^j \mathbf{e}^k := \frac{1}{2}(\mathbf{e}^j \otimes \mathbf{e}^k + \mathbf{e}^k \otimes \mathbf{e}^j), \qquad 1 \leq j \leq k \leq n,$$

and a basis of $\Lambda^2(E)$ *is provided by the tensors*

$$\mathbf{e}^j \wedge \mathbf{e}^k := \frac{1}{2}(\mathbf{e}^j \otimes \mathbf{e}^k - \mathbf{e}^k \otimes \mathbf{e}^j), \qquad 1 \leq j < k \leq n.$$

Proof. Clearly, the elements $\mathbf{e}^j \mathbf{e}^k$ span a subspace S of $\mathbf{Sym}^2(E)$, whereas the elements $\mathbf{e}^j \wedge \mathbf{e}^k$ span a subspace A of $\Lambda^2(E)$. We thus have $S \cap A = \{0\}$. Also, these elements together span $E \otimes E$ because

$$\mathbf{e}^j \otimes \mathbf{e}^k = \mathbf{e}^j \mathbf{e}^k + \mathbf{e}^j \wedge \mathbf{e}^k,$$

and $\mathbf{e}^j \otimes \mathbf{e}^k$ form a basis of $E \otimes E$. Therefore $E \otimes E = S \oplus A$, which implies $S = \mathbf{Sym}^2(E)$ and $A = \Lambda^2(E)$. Because the family made of elements $\mathbf{e}^j \mathbf{e}^k$ and $\mathbf{e}^j \wedge \mathbf{e}^k$ has cardinal n^2 and is generating, it is a basis of $E \otimes E$. Therefore the elements $\mathbf{e}^j \mathbf{e}^k$ form a basis of $\mathbf{Sym}^2(E)$ and the elements $\mathbf{e}^j \wedge \mathbf{e}^k$ form a basis of $\Lambda^2(E)$. $\square$

As a by-product, we have

$$\dim \mathbf{Sym}^2(E) = \frac{n(n+1)}{2}, \qquad \dim \Lambda^2(E) = \frac{n(n-1)}{2}.$$

We point out that

$$\mathbf{e}^j \mathbf{e}^k = \mathbf{e}^k \mathbf{e}^j, \qquad \mathbf{e}^j \wedge \mathbf{e}^k = -\mathbf{e}^k \wedge \mathbf{e}^j.$$

4.2.2 Exterior Products

Let $k \geq 1$ be an integer. We denote $T^k(E) = E \otimes \cdots \otimes E$ the tensor product of k copies of E. We also write $T^k(E) = V^{\otimes k}$. When $k = 1$, $T^1(E)$ is nothing but E. By convention, we set $T^0(E) = K$.

We extend the definition of $\Lambda^2(E)$ to other integers $k \geq 0$ as follows. We first consider the subspace L_k of $T^k(E)$ spanned by the elementary products $x^1 \otimes \cdots \otimes x^k$ in which at least two vectors x^i and x^j are equal. Notice that $L_0 = \{0\}$ and $L_1 = \{0\}$. Then we define the quotient space

$$\Lambda^k(E) := T^k(E)/L_k.$$

In particular, $\Lambda^0(E) = K$ and $\Lambda^1(E) = E$. The class of $w^1 \otimes \cdots \otimes w^k$ is denoted $w^1 \wedge \cdots \wedge w^k$. The operation $\wedge$ is called the *exterior product* of tensors. It is extended by multilinearity.

By definition, and because

$$x \otimes y + y \otimes x = (x+y) \otimes (x+y) - x \otimes x - y \otimes y \in L_k,$$

the exchange of two consecutive factors w^s and w^{s+1} just flips the sign:

$$\cdots \wedge w^{s+1} \wedge w^s \wedge \cdots = -(\cdots \wedge w^s \wedge w^{s+1} \wedge \cdots).$$

By induction, we infer the next lemma.

Lemma 3.

$$w^{\sigma(1)} \wedge \cdots \wedge w^{\sigma(k)} = \varepsilon(\sigma) w^1 \wedge \cdots \wedge w^k, \qquad \forall \sigma \in S_k.$$

Theorem 4.3 *If* $\dim E = n$, *then*

$$\dim \Lambda^k(E) = \binom{n}{k}.$$

A basis of $\Lambda^k(E)$ *is given by the set of tensors* $\mathbf{e}^{i_1} \wedge \cdots \wedge \mathbf{e}^{i_k}$ *with* $i_1 < \cdots < i_k$.
In particular, $\Lambda^k(E) = \{0\}$ *if* $k > n$.

Proof. Because $T^k(E)$ is spanned by the vectors of the form $\mathbf{e}^{i_1} \otimes \cdots \otimes \mathbf{e}^{i_k}$, $\Lambda^k(E)$ is spanned by elements of the form $\mathbf{e}^{i_1} \wedge \cdots \wedge \mathbf{e}^{i_k}$. However, this vector vanishes if two indices i_r are equal, by construction. Thus $\Lambda^k(E)$ is spanned by those for which the indices $i_1, \ldots, i_k$ are pairwise distinct. And because of Lemma 3, those that satisfy $i_1 < \cdots < i_k$ form a generating family. In particular, $\Lambda^k(E) = \{0\}$ if $k > n$.

There remains to prove that this generating family is free. It suffices to treat the case $k \leq n$.

Case $k = n$.   We may assume that $E = K^n$. With $\mathbf{w} = (w^1, \ldots, w^n) \in V^n$, we associate the matrix W whose jth column is w^j for each j. The map $\mathbf{w} \mapsto \det W$ is n-linear and thus corresponds to a linear form D over $T^n(E)$, according to Theorem 4.2. Obviously, D vanishes over L_n, and thus defines a linear form Δ over the quotient $\Lambda^n(E)$. This form is nontrivial, because $\Delta(\mathbf{e}^1, \ldots, \mathbf{e}^n) = \det I_n = 1 \neq 0$. This shows that $\dim \Lambda^n(E) \geq 1$. This space is spanned by at most one element $\mathbf{e}^1 \wedge \cdots \wedge \mathbf{e}^n$, therefore we deduce $\dim \Lambda^n(E) = 1$, and $\mathbf{e}^1 \wedge \cdots \wedge \mathbf{e}^n \neq 0$ too.
Case $k < n$.   Let us assume that

$$\sum_J \mu_J \mathbf{e}^{j_1} \wedge \cdots \wedge \mathbf{e}^{j_k} = 0, \tag{4.2}$$

where the sum runs over increasing lists $J = (j_1 < \cdots < j_k)$ and the μ_J are scalars. Let us choose an increasing list I of length k. We denote by I^c the complement of I in $\{1, \ldots, n\}$, arranged in increasing order. To be specific,

$$I = (i_1, \ldots, i_k), \qquad I^c = (i_{k+1}, \ldots, i_n).$$

We define a linear map $S : T^k(E) \to T^n(E)$ by

$$S(\mathbf{e}^{j_1} \otimes \cdots \otimes \mathbf{e}^{j_n}) := \mathbf{e}^{j_1} \otimes \cdots \otimes \mathbf{e}^{j_n} \otimes \mathbf{e}^{i_{k+1}} \otimes \cdots \otimes \mathbf{e}^{i_n}.$$

Obviously, we have $S(L_k) \subset L_n$ and thus S passes to the quotient. This yields a linear map $s : \Lambda^k(E) \to \Lambda^n(E)$ satisfying

$$s(\mathbf{e}^{j_1} \wedge \cdots \wedge \mathbf{e}^{j_n}) := \mathbf{e}^{j_1} \wedge \cdots \wedge \mathbf{e}^{j_n} \wedge \mathbf{e}^{i_{k+1}} \wedge \cdots \wedge \mathbf{e}^{i_n}.$$

Applying s to (4.2), and remembering that $\mathbf{e}^{j_1} \wedge \cdots \wedge \mathbf{e}^{j_n} \wedge \mathbf{e}^{i_{k+1}} \wedge \cdots \wedge \mathbf{e}^{i_n}$ vanishes if an index occurs twice, we obtain

$$\mu_I \mathbf{e}^1 \wedge \cdots \wedge \mathbf{e}^n = 0.$$

Because $\mathbf{e}^1 \wedge \cdots \wedge \mathbf{e}^n \neq 0$ (from the case $k = n$ above), we deduce $\mu_I = 0$. This proves that our generating set is free.
□

4.2.3 The Tensor and Exterior Algebras

The spaces $T^k(E)$ may be summed up so as to form the *tensor algebra* of E, denoted $T(E)$:

$$T(E) = K \oplus E \oplus T^2(E) \oplus \cdots \oplus T^k(E) \oplus \cdots.$$

We recall that an element of $T(E)$ is a sequence $(x^0, x^1, \ldots, x^k, \ldots)$ whose kth element is in $T^k(E)$ and only finitely many of them are nonzero. It is thus a *finite* sum $x^0 \oplus \cdots \oplus x^k + \cdots$.

The word *algebra* is justified by the following bilinear operation, defined from $T^k(E) \times T^\ell(E)$ into $T^{k+\ell}(E)$ and then extended to $T(E)$ by linearity. It makes $T(E)$ a *graded algebra*. We need only to define a product over elements of the bases:

$$(\mathbf{e}^{i_1} \otimes \cdots \otimes \mathbf{e}^{i_k}) \cdot (\mathbf{e}^{j_1} \otimes \cdots \otimes \mathbf{e}^{j_\ell}) = \mathbf{e}^{i_1} \otimes \cdots \otimes \mathbf{e}^{i_k} \otimes \mathbf{e}^{j_1} \otimes \cdots \otimes \mathbf{e}^{j_\ell}.$$

When $k = 0$, we simply have

$$\lambda \cdot (\mathbf{e}^{j_1} \otimes \cdots \otimes \mathbf{e}^{j_\ell}) = (\lambda \mathbf{e}^{j_1}) \otimes \mathbf{e}^{j_2} \otimes \cdots \otimes \mathbf{e}^{j_\ell}.$$

Because of the identities above, it is natural to denote this product with the same tensor symbol $\otimes$. We thus have

$$(\mathbf{e}^{i_1} \otimes \cdots \otimes \mathbf{e}^{i_k}) \otimes (\mathbf{e}^{j_1} \otimes \cdots \otimes \mathbf{e}^{j_\ell}) = \mathbf{e}^{i_1} \otimes \cdots \otimes \mathbf{e}^{i_k} \otimes \mathbf{e}^{j_1} \otimes \cdots \otimes \mathbf{e}^{j_\ell}.$$

If $E \neq \{0\}$, the vector space $T(E)$ is infinite dimensional. As an algebra, it is associative, inasmuch as associativity holds true for elements of the canonical basis.

The bilinear map $B : (x,y) \mapsto x \otimes y$, from $T^k(E) \times T^\ell(E)$ into $T^{k+\ell}(E)$, obviously satisfies

$$B(L_k \times T^\ell(E)) \subset L_{k+\ell}, \qquad B(T^k(E) \times L_\ell) \subset L_{k+\ell}.$$

It therefore corresponds to a bilinear map $b : \Lambda^k(E) \times \Lambda^\ell(E) \to \Lambda^{k+\ell}(E)$ verifying

$$b(x^1 \wedge \cdots \wedge x^k, x^{k+1} \wedge \cdots \wedge x^{k+\ell}) = x^1 \wedge \cdots \wedge x^{k+\ell}.$$

This operation is called the *exterior product*. From its definition, it is natural to denote it again with the same wedge symbol $\wedge$. We thus have

$$(x^1 \wedge \cdots \wedge x^k) \wedge (x^{k+1} \wedge \cdots \wedge x^{k+\ell}) = x^1 \wedge \cdots \wedge x^{k+\ell}.$$

Again, the exterior product allows us to define the *graded algebra*

$$\Lambda(E) = K \oplus E \oplus \Lambda^2(E) \oplus \cdots \oplus \Lambda^n(E).$$

We point out that because of Theorem 4.3, this sum involves only $n+1$ terms. Again, the *exterior algebra* $\Lambda(E)$ is associative. Its dimension equals 2^n, thanks to the identity

$$\sum_{k=0}^{n} \binom{n}{k} = 2^n.$$

4.2.3.1 Rules

Let $x, y \in E$ be given. Because $x \otimes y + y \otimes x = (x+y) \otimes (x+y) - x \otimes x - y \otimes y \in L_2$, we have

$$y \wedge x = -x \wedge y, \qquad \forall x, y \in E. \tag{4.3}$$

When dealing with the exterior product of higher order, the situation is slightly different. For instance, if x, y, z belong to E, then (4.3) together with associativity give

$$x \wedge (y \wedge z) = (x \wedge y) \wedge z = -(y \wedge x) \wedge z = -y \wedge (x \wedge z) = y \wedge (z \wedge x) = (y \wedge z) \wedge x.$$

By linearity, we deduce that if $x \in E$ and $w \in \Lambda^2(E)$, then $x \wedge w = w \wedge x$. More generally, we prove the following by induction over k and ℓ.

Proposition 4.2 *If $w \in \Lambda^k(E)$ and $z \in \Lambda^\ell(E)$, then*

$$z \wedge w = (-1)^{k\ell} w \wedge z.$$

4.2.3.2 A Commutative Subalgebra

The sum

$$\Lambda_{\text{even}}(E) = K \oplus \Lambda^2(E) \oplus \Lambda^4(E) \oplus \cdots$$

is obviously a subalgebra of $\Lambda(E)$, of dimension 2^{n-1} because of

$$\sum_{0 \le k \le n/2} \binom{n}{2k} = 2^{n-1}.$$

Because of Proposition 4.2, it is actually commutative.

Corollary 4.2 *If $w, z \in \Lambda_{\text{even}}(E)$, then $w \wedge z = z \wedge w \in \Lambda_{\text{even}}(E)$.*

4.3 Tensorization of Linear Maps

4.3.1 Tensor Product of Linear Maps

Let E_0, E_1, F_0, F_1 be vector spaces over K. If $u_j \in \mathscr{L}(E_j; F_j)$, Theorem 4.1 allows us to define a linear map $u_0 \otimes u_1 \in \mathscr{L}(E_0 \otimes E_1; F_0 \otimes F_1)$, satisfying

$$(u_0 \otimes u_1)(x \otimes y) = u_0(x) \otimes u_1(y).$$

A similar construction is available with an arbitrary number of linear maps $u_j : E_j \to F_j$.

Let us choose bases $\{\mathbf{e}^{01}, \ldots, \mathbf{e}^{0m}\}$ of E_0, $\{\mathbf{e}^{11}, \ldots, \mathbf{e}^{1q}\}$ of E_1, $\{\mathbf{f}^{01}, \ldots, \mathbf{f}^{0n}\}$ of F_0, $\{\mathbf{f}^{11}, \ldots, \mathbf{f}^{1p}\}$ of F_1. Let A and B be the respective matrices of u_0 and u_1 in these bases. Then

$$(u_0 \otimes u_1)(\mathbf{e}^{0i} \otimes \mathbf{e}^{1j}) = \left(\sum_k a_{ki} \mathbf{f}^{0k}\right) \otimes \left(\sum_\ell b_{\ell j} \mathbf{f}^{1\ell}\right)$$
$$= \sum_{k,\ell} a_{ki} b_{\ell j} \mathbf{f}^{0k} \otimes \mathbf{f}^{1\ell}.$$

This shows that the $((k, l), (i, j))$-entry of the matrix of $u_0 \otimes u_1$ in the tensor bases is the product $a_{ki} b_{\ell j}$. If we arrange the bases $(\mathbf{e}^{0i} \otimes \mathbf{e}^{1j})_{i,j}$ and $(\mathbf{f}^{0k} \otimes \mathbf{f}^{1\ell})_{k,\ell}$ in lexicographic order, then this matrix reads blockwise

$$\begin{pmatrix} a_{11}B & a_{12}B & \ldots & a_{1m}B \\ a_{21}B & \ddots & & \vdots \\ \vdots & & & \\ a_{n1}B & \ldots & & a_{nm}B \end{pmatrix}.$$

This matrix is called the *tensor product* of A and B, denoted $A \otimes B$.

4.3.2 Exterior Power of an Endomorphism

If $u \in \mathbf{End}(E)$, then $u \otimes u \otimes \cdots \otimes u =: u^{\otimes k}$ has the property that $u(L_k) \subset L_k$. Therefore there exists a unique linear map $\Lambda^k(u)$ such that

$$u(x^1 \wedge \cdots \wedge x^k) = u(x^1) \wedge \cdots \wedge u(x^k).$$

Proposition 4.3 *If A is the matrix of $u \in \mathbf{End}(E)$ in a basis $\{\mathbf{e}^1, \ldots, \mathbf{e}^n\}$, then the entries of the matrix $A^{(k)}$ of $\Lambda^k(u)$ in the basis of vectors $\mathbf{e}^{j_1} \wedge \cdots \wedge \mathbf{e}^{j_n}$ are the $k \times k$ minors of A.*

Proof. This is essentially the same line as in the proof of the Cauchy–Binet formula (Proposition 3.4). $\square$

Corollary 4.3 *If $\dim E = n$ and $u \in \mathbf{End}(E)$, then $\Lambda^n(u)$ is multiplication by $\det u$.*

4.4 A Polynomial Identity in $\mathbf{M_n(K)}$

We already know a polynomial identity in $\mathbf{M}_n(K)$, namely the Cayley–Hamilton theorem: $P_A(A) = 0_n$. However, it is a bit complicated because the matrix is involved both as the argument of the polynomial and in its coefficients. We prove here a remarkable result, where a multilinear application vanishes identically when the arguments are arbitrary $n \times n$ matrices. To begin with, we introduce some special polynomials in *noncommutative* indeterminates.

4.4.1 The Standard Noncommutative Polynomial

Noncommutative polynomials in indeterminates $X_1, \ldots, X_\ell$ are linear combinations of *words* written in the alphabet $\{X_1, \ldots, X_\ell\}$. The important rule is that in a word, you are not allowed to permute two distinct letters: $X_i X_j \neq X_j X_i$ if $j \neq i$, contrary to what occurs in ordinary polynomials.

The *standard polynomial* $\mathscr{S}_\ell$ in noncommutative indeterminates $X_1, \ldots, X_\ell$ is defined by

$$\mathscr{S}_\ell(X_1, \ldots, X_\ell) := \sum \varepsilon(\sigma) X_{\sigma(1)} \cdots X_{\sigma(\ell)}.$$

Hereabove, the sum runs over the permutations of $\{1, \ldots, \ell\}$, and $\varepsilon(\sigma)$ denotes the signature of σ. For instance, $\mathscr{S}_2(X, Y) = XY - YX = [X, Y]$. The standard polynomial is thus a tool for measuring the defect of commutativity in a ring or an algebra: a ring R is Abelian if $\mathscr{S}_2$ vanishes identically over $R \times R$.

The following formula is obvious.

Lemma 4. *Let $A_1, \ldots, A_r \in \mathbf{M}_n(K)$ be given. We form the matrix $A \in \mathbf{M}_n(\Lambda(K^r)) \sim \mathbf{M}_n(K) \otimes_K \Lambda(K^r)$ by*

$$A = A_1 \mathbf{e}^1 + \cdots + A_r \mathbf{e}^r.$$

We emphasize that A has entries in a noncommutative ring.

Then we have

$$A^\ell = \sum_{i_1 < \cdots < i_\ell} \mathscr{S}_\ell(A_{i_1}, \ldots, A_{i_\ell}) \, \mathbf{e}^{i_1} \wedge \cdots \wedge \mathbf{e}^{i_\ell}.$$

In particular, we have

$$A^r = \mathscr{S}_r(A_1, \ldots, A_r) \, \mathbf{e}^1 \wedge \cdots \wedge \mathbf{e}^r. \tag{4.4}$$

The other important formula generalizes the well-known identity $\mathrm{Tr}[A, B] = 0$. To begin with, we easily have

$$\mathscr{S}_\ell(X_{\sigma(1)}, \ldots, X_{\sigma(\ell)}) = \varepsilon(\sigma) \mathscr{S}_\ell(X_1, \ldots, X_\ell), \qquad \forall \sigma \in \mathbf{S}_\ell.$$

Applying this to a cycle, we deduce

$$\mathscr{S}_\ell(X_2, \ldots, X_\ell, X_1) = (-1)^{\ell-1} \mathscr{S}_\ell(X_1, \ldots, X_\ell).$$

Because $\mathrm{Tr}(A_2 \cdots A_\ell A_1) = \mathrm{Tr}(A_1 \cdots A_\ell)$, we infer the following.

Lemma 5. *If ℓ is even and $A_1, \ldots, A_\ell \in \mathbf{M}_n(R)$ (R a commutative ring), then*

$$\mathrm{Tr}\, \mathscr{S}_\ell(A_1, \ldots, A_\ell) = 0.$$

Proof. If ℓ is even, we have

$$\mathrm{Tr}\, \mathscr{S}_\ell(A_1, \ldots, A_\ell) = -\mathrm{Tr}\, \mathscr{S}_\ell(A_2, \ldots, A_\ell, A_1) = -\mathrm{Tr}\, \mathscr{S}_\ell(A_1, \ldots, A_\ell),$$

the first equality because this is true even before taking the trace, and the last equality because of $\mathrm{Tr}(AB) = \mathrm{Tr}(BA)$. If $2x = 0$ implies $x = 0$ in R, we deduce

$$\mathrm{Tr}\, \mathscr{S}_\ell(A_1, \ldots, A_\ell) = 0.$$

For instance, this is true if $R = \mathbb{C}$. Because $\mathrm{Tr}\, \mathscr{S}_\ell(\cdots)$ belongs to $\mathbb{Z}[Y_1, \ldots, Y_{\ell n^2}]$, it must vanish as a polynomial. Thus the identity is valid in every commutative ring R. $\square$

4.4.2 The Theorem of Amitsur and Levitzki

A beautiful as well as surprising fact is that $\mathbf{M}_n(K)$ does have some amount of commutativity, although it seems at first glance to be a paradigm for noncommutative algebras.

Theorem 4.4 (A. Amitsur and J. Levitzki) *The standard polynomial in $2n$ noncommutative indeterminates vanishes over $\mathbf{M}_n(K)$: for every $A_1,\ldots,A_{2n} \in \mathbf{M}_n(K)$, we have*

$$\mathscr{S}_{2n}(A_1,\ldots,A_{2n}) = 0_n. \tag{4.5}$$

Comments

- This result is accurate in the sense that $\mathscr{S}_k$ does not vanish identically over $n \times n$ matrices. For instance, $\mathscr{S}_k$ does not vanish over the $(2n-1)$-uplet of matrices $\mathbf{e}^n \otimes \mathbf{e}^1, \mathbf{e}^{n-1} \otimes \mathbf{e}^1, \ldots, \mathbf{e}^1 \otimes \mathbf{e}^1, \mathbf{e}^1 \otimes \mathbf{e}^2, \ldots, \mathbf{e}^1 \otimes \mathbf{e}^n$.
- Actually, it is known that $\mathscr{S}_{2n} \equiv 0_n$ is the only polynomial identity of degree less than or equal to $2n$ over $\mathbf{M}_n(\mathbb{C})$.
- Theorem 4.4 has long remained mysterious, with a complicated proof, until Rosset published a simple proof in 1976. This is the presentation that we give below.
- We notice that inasmuch as $\mathscr{S}_{2n}$ may be viewed as a list of n^2 elements of $\mathbb{Z}[Y_1,\ldots,Y_{2n^3}]$, the identity (4.5) must be valid for matrices whose entries are independent *commuting* indeterminates m_{ij}^α with $1 \le \alpha \le 2n$ and $1 \le i,j \le n$.

The theorem is thus a list of n^2 identities in $\mathbb{Z}\left[m_{ij}^\alpha \mid 1 \le \alpha \le 2n, 1 \le i,j \le n\right]$.

Proof. (Taken from Rosset [32].)

We thus assume that $K = \mathbb{C}$. As above, we form the matrix $A \in \mathbf{M}_n(\Lambda(\mathbb{C}^{2n}))$ by

$$A = A_1\mathbf{e}^1 + \cdots A_{2n}\mathbf{e}^{2n}.$$

Because of (4.4), what we have to prove is that $A^{2n} = 0_n$.

The matrix A has the flaw of having noncommutative entries. However, Corollary 4.2 tells us that the entries of A^2 belong to the abelian ring $\Lambda_{\text{even}}(\mathbb{C}^{2n})$. We thus may apply Proposition 3.24 (recall that it is valid for matrices with entries in an Abelian ring R in which $mx = 0$ implies $x = 0$ whenever m is a positive integer): in order to prove that $A^{2n} = (A^2)^n = 0_n$, it is sufficient to prove the identities

$$\text{Tr}(A^{2k}) = 0, \qquad 1 \le k \le n.$$

The latter follow immediately from Lemmas 4 and 5. □

Exercises

1. Show that the rank of an element of $E \otimes F$ is bounded by $\min\{\dim E, \dim F\}$.
2. Let $u \in \mathbf{End}(E)$ be a diagonalizable map, with eigenvalues $\lambda_1,\ldots,\lambda_n$, counting multiplicities. Show that $u^{\otimes k}$ and $\Lambda^k(u)$ are diagonalizable and identify their eigenvalues.
3. Let $u \in \mathbf{End}(E)$ be given. Show that the following formula defines an endomorphism over $T^k(E)$, which we denote $u^{\oplus k}$,

$$x^1 \otimes \cdots \otimes x^k \mapsto (u(x^1) \otimes \cdots \otimes x^k) + (x^1 \otimes u(x^2) \otimes \cdots \otimes x^k)$$
$$+ \cdots + (x^1 \otimes \cdots \otimes u(x^k)).$$

a. If u is diagonalizable, show that $u^{\oplus k}$ is so and compute its eigenvalues.

b. Show that $u^{\oplus k}$ passes to the quotient, yielding an endomorphism over $\Lambda^k(E)$, which we denote $u^{\wedge k}$:

$$x^1 \wedge \cdots \wedge x^k \mapsto (u(x^1) \wedge \cdots \wedge x^k) + (x^1 \wedge u(x^2) \wedge \cdots \wedge x^k)$$
$$+ \cdots + (x^1 \wedge \cdots \wedge u(x^k)).$$

c. Again, show that $u^{\wedge k}$ is diagonalizable if u is so, and compute its eigenvalues.

4. Complete the proof of Proposition 4.3.

5. Prove that $\mathscr{S}_{2n-1}$, when applied to the matrices E^{ij} for either $i = 1$ or $j = 1$ (this makes a list of $2n - 1$ matrices), where

$$(E^{ij})_{k\ell} = \delta_i^k \delta_j^\ell, \qquad 1 \leq i, j, k, \ell \leq n,$$

gives a nonzero matrix.

6. If $A \in \mathbf{M}_n(k)$ is alternate, we define $\mathbb{A} \in \mathbf{M}_n(\Lambda(k^n))$ by

$$\mathbb{A} := \sum_{i<j} a_{ij} \mathbf{e}^i \wedge \mathbf{e}^j.$$

We assume that $n = 2m$. Prove that

$$\mathbb{A}^m = \mathrm{Pf}(A)\mathbf{e}^1 \wedge \cdots \wedge \mathbf{e}^n.$$

Hint: Use the expansion formula established in Exercise 25 of Chapter 3.

Chapter 5
Matrices with Real or Complex Entries

A matrix $M \in \mathbf{M}_{n \times m}(K)$ is an element of a vector space of finite dimension n^2. When $K = \mathbb{R}$ or $K = \mathbb{C}$, this space has a natural topology, that of K^{nm}. Therefore we may manipulate such notions as open and closed sets, and continuous and differentiable functions.

5.1 Special Matrices

5.1.1 Hermitian Adjoint

When considering matrices with complex entries, a useful operation is complex conjugation $z \mapsto \bar{z}$. One denotes by $\bar{M}$ the matrix obtained from M by conjugating the entries. We then define the *Hermitian adjoint* matrix of M by

$$M^* := (\bar{M})^T = \overline{M^T}.$$

One has $m_{ij}^* = \overline{m_{ji}}$ and $\det M^* = \overline{\det M}$. The map $M \mapsto M^*$ is an *antiisomorphism*, which means that it is antilinear (meaning that $(\lambda M)^* = \bar{\lambda} M^*$) and bijective. In addition, we have the product formula

$$(MN)^* = N^* M^*.$$

If M is nonsingular, this implies $(M^*)^{-1} = (M^{-1})^*$; this matrix is sometimes denoted M^{-*}.

The interpretation of the Hermitian adjoint is that if we endow $\mathbb{C}^n$ with the canonical scalar product

$$\langle x, y \rangle = \bar{x}_1 y_1 + \cdots + \bar{x}_n y_n,$$

and with the canonical basis, then M^* is the matrix of the adjoint $(u_M)^*$; that is,

$$\langle Mx, y \rangle = \langle x, M^* y \rangle, \qquad \forall x, y \in \mathbb{C}^n.$$

5.1.2 Normal Matrices

Definition 5.1 *A matrix* $M \in \mathbf{M}_n(C)$ *is* normal *if* M *and* M^* *commute:* $M^*M = MM^*$.

If M has real entries, this amounts to having $MM^T = M^T M$.

Because a square matrix M always commutes with M, $-M$, or M^{-1} (assuming that the latter exists), we can define sub-classes of normal matrices. The following statement serves also as a definition of such classes.

Proposition 5.1 *The following matrices $M \in \mathbf{M}_n(\mathbb{C})$ are normal.*

- Hermitian *matrices, meaning that $M^* = M$*
- Skew-Hermitian *matrices, meaning that $M^* = -M$*
- Unitary *matrices, meaning that $M^* = M^{-1}$*

The Hermitian, skew-Hermitian, and unitary matrices are thus normal. One verifies easily that H is Hermitian (respectively, skew-Hermitian) if and only if x^*Hx is real (respectively, pure imaginary) for every $x \in \mathbb{C}^n$.

For real-valued matrices, we have instead

Definition 5.2 *A square matrix $M \in \mathbf{M}_n(\mathbb{R})$ is*

- Symmetric *if $M^T = M$*
- Skew-symmetric *if $M^T = -M$*
- Orthogonal *if $M^T = M^{-1}$*

We denote by $\mathbf{H}_n$ the set of Hermitian matrices in $\mathbf{M}_n(\mathbb{C})$. It is an $\mathbb{R}$-linear subspace of $\mathbf{M}_n(\mathbb{C})$, but not a $\mathbb{C}$-linear subspace, becausee iM is skew-Hermitian when M is Hermitian. If $M \in \mathbf{M}_{n \times m}(\mathbb{C})$, the matrices $M + M^*$, $i(M^* - M)$, MM^*, and M^*M are Hermitian. One sometimes calls $\frac{1}{2}(M + M^*)$ the *real part* of M and denotes it $\Re M$. Likewise, $\frac{1}{2i}(M - M^*)$ is the *imaginary part* of M and is denoted $\Im M$. Both are Hermitian and we have

$$M = \Re M + i\Im M.$$

This terminology anticipates Chapter 10.

A matrix M is unitary if u_M is an isometry, that is $\langle Mx, My \rangle \equiv \langle x, y \rangle$. This is equivalent to saying that $\|Mx\| \equiv \|x\|$. The set of unitary matrices in $\mathbf{M}_n(\mathbb{C})$ forms a multiplicative group, denoted by $\mathbf{U}_n$. Unitary matrices satisfy $|\det M| = 1$, because $\det M^*M = |\det M|^2$ for every matrix M and $M^*M = I_n$ when M is unitary. The set of unitary matrices whose determinant equals 1, denoted by $\mathbf{SU}_n$ is obviously a normal subgroup of $\mathbf{U}_n$.

A matrix with *real* entries is orthogonal (respectively, symmetric, skew-symmetric) if and only if it is unitary, Hermitian, or skew-Hermitian.

5.1.3 Matrices and Sesquilinear Forms

Given a matrix $M \in \mathbf{M}_n(\mathbb{C})$, the map

$$(x,y) \mapsto \langle x,y \rangle_M := \sum_{j,k} m_{jk} \overline{x_j} y_k = x^* M y,$$

defined on $\mathbb{C}^n \times \mathbb{C}^n$, is a sesquilinear form. When $M = I_n$, this is nothing but the scalar product. It is Hermitian if and only if M is Hermitian. It follows that $M \mapsto \langle \cdot, \cdot \rangle_M$ is an isomorphism between $\mathbf{H}_n$ and the set of Hermitian forms over $\mathbb{C}^n$. We say that a Hermitian matrix H is *degenerate* (respectively, *nondegenerate*) if the form $\langle \cdot, \cdot \rangle_H$ is so. Nondegeneracy amounts to saying that $x \mapsto Hx$ is one-to-one. In other words, we have the following.

Proposition 5.2 *A Hermitian matrix H is degenerate (respectively, nondegenerate) if and only if $\det H = 0$ (respectively, $\neq 0$).*

We say that the Hermitian matrix H is *positive-definite* if $\langle \cdot, \cdot \rangle_H$ is so. Then $\sqrt{\langle \cdot, \cdot \rangle_H}$ is a norm over $\mathbb{C}^n$. If $-H$ is positive-definite, we say that H is *negative-definite*. We denote by $\mathbf{HPD}_n$ the set of positive-definite Hermitian matrices. If H and K are positive-definite, and if λ is a positive real number, then $\lambda H + K$ is positive-definite. Therefore $\mathbf{HPD}_n$ is a convex cone in $\mathbf{H}_n$. This cone turns out to be open. The Hermitian matrices H for which $\langle \cdot, \cdot \rangle_H$ is a positive-semidefinite Hermitian form over $\mathbb{C}^n$ are called *positive-semidefinite* Hermitian matrices. They also form a convex cone $\mathbf{H}_n^+$. If $H \in \mathbf{H}_n^+$ and ε is a positive real number, then $H + \varepsilon I_n$ is positive-definite. Because $H + \varepsilon I_n$ tends to H as $\varepsilon \to 0^+$, we see that the closure of $\mathbf{HPD}_n$ is $\mathbf{H}_n^+$.

One defines similarly, among the real symmetric matrices, the positive-definite, respectively, positive-semidefinite, ones. Again, the positive-definite real symmetric matrices form an open cone in $\mathbf{Sym}_n(\mathbb{R})$, denoted by $\mathbf{SPD}_n$, whose closure $\mathbf{Sym}_n^+$ is made of positive-semidefinite ones.

The cone $\mathbf{HPD}_n$ defines an order over $\mathbf{H}_n$: we write $K > H$ when $K - H \in \mathbf{HPD}_n$, and more generally $K \geq H$ if $K - H$ is positive-semidefinite. The fact that

$$(K \geq H \geq K) \Longrightarrow (K = H)$$

follows from the next lemma.

Lemma 6. *Let H be Hermitian. If $x^* H x = 0$ for every $x \in \mathbb{C}^n$, then $H = 0_n$.*

Proof. Using (1.1), we have $y^* H x = 0$ for every $x, y \in \mathbb{C}^n$. Therefore $Hx = 0$ for every x, which means $H = 0_n$. $\square$

We likewise define an ordering on real-valued symmetric matrices, referring to the ordering on real-valued quadratic forms.[1]

If U is unitary, the matrix $U^* M U$ is similar to M, and we say that they are *unitary similar*. If M is normal, Hermitian, skew-Hermitian, or unitary, and if U is unitary, then $U^* M U$ is still normal, Hermitian, skew-Hermitian, or unitary. When

---

[1] We warn the reader that another order, a completely different one, although still denoted by the same symbol $\geq$, is defined in Chapter 8. The latter concerns general $n \times m$ real-valued matrices, whereas the present one deals only with symmetric matrices. In practice, the context is never ambiguous.

$O \in \mathbf{O}_n(\mathbb{R})$ and $M \in \mathbf{M}_n(\mathbb{R})$, we again say that $O^T M O$ and M are *orthogonally similar*.

We notice that Lemma 6 implies the following stronger result.

Proposition 5.3 *Let $M \in \mathbf{M}_n(\mathbb{C})$ be given. If $x^* M x = 0$ for every $x \in \mathbb{C}^n$, then $M = 0_n$.*

Proof. We decompose $M = H + iK$ into its real and imaginary parts. Recall that H, K are Hermitian. Then

$$x^* M x = x^* H x + i x^* K x$$

is the decomposition of a complex number into real and imaginary parts. From the assumption, we therefore have $x^* H x = 0$ and $x^* K x = 0$ for every x. Then Lemma 6 tells us that $H = K = 0_n$. □

5.2 Eigenvalues of Real- and Complex-Valued Matrices

Let us recall that $\mathbb{C}$ is algebraically closed. Therefore the characteristic polynomial of a complex-valued square matrix has roots if $n \geq 1$. Therefore every endomorphism of a nontrivial $\mathbb{C}$-vector space possesses eigenvalues. A real-valued square matrix may have no eigenvalues in $\mathbb{R}$, but it has at least one in $\mathbb{C}$. If n is odd, $M \in \mathbf{M}_n(\mathbb{R})$ has at least one real eigenvalue, because P_M is real of odd degree.

5.2.1 Unitary Trigonalization

If $K = \mathbb{C}$, one sharpens Theorem 3.5.

Theorem 5.1 (Schur) *If $M \in \mathbf{M}_n(\mathbb{C})$, there exists a unitary matrix U such that $U^* M U$ is upper-triangular.*

One also says that every matrix with complex entries is *unitarily trigonalizable*.

Proof. We proceed by induction over the size n of the matrices. The statement is trivial if $n = 1$. Let us assume that it is true in $\mathbf{M}_{n-1}(\mathbb{C})$, with $n \geq 2$. Let $M \in \mathbf{M}_n(\mathbb{C})$ be a matrix. Because $\mathbb{C}$ is algebraically closed, M has at least one eigenvalue λ. Let X be an eigenvector associated with λ. By dividing X by $\|X\|$, we can assume that X is a unit vector. One can then find a unitary basis $\{X^1, X^2, \ldots, X^n\}$ of $\mathbb{C}^n$ whose first element is X. Let us consider the matrix $V := (X^1 = X | X^2 | \cdots | X^n)$, which is unitary, and let us form the matrix $M' := V^* M V$. Because

$$V M' \mathbf{e}^1 = M V \mathbf{e}^1 = M X = \lambda X = \lambda V \mathbf{e}^1,$$

one obtains $M' \mathbf{e}^1 = \lambda \mathbf{e}^1$. In other words, M' has the block-triangular form:

$$M' = \begin{pmatrix} \lambda & \cdots \\ 0_{n-1} & N \end{pmatrix},$$

where $N \in \mathbf{M}_{n-1}(\mathbb{C})$. Applying the induction hypothesis, there exists $W \in \mathbf{U}_{n-1}$ such that W^*NW is upper-triangular. Let us denote by $\hat{W}$ the (block-diagonal) matrix $\text{diag}(1, W) \in \mathbf{U}_n$. Then $\hat{W}^*M'\hat{W}$ is upper-triangular. Hence, $U = V\hat{W}$ satisfies the conditions of the theorem. $\square$

A useful consequence of Theorem 5.1 is the following.

Corollary 5.1 *The set of diagonalizable matrices is a dense subset in* $\mathbf{M}_n(\mathbb{C})$.

Remark

The set of real matrices diagonalizable within $\mathbf{M}_n(\mathbb{R})$ is not dense in $\mathbf{M}_n(\mathbb{R})$. For instance, the matrix

$$\begin{pmatrix} 0 & 1 \\ -1 & 0 \end{pmatrix},$$

whose eigenvalues $\pm i$ are nonreal, is interior to the set of nondiagonalizable matrices; this is a consequence of Theorem 5.2 of continuity of the spectrum. The set of real matrices diagonalizable within $\mathbf{M}_n(\mathbb{C})$ is dense in $\mathbf{M}_n(\mathbb{R})$, but the proof is more involved.

Proof. The triangular matrices with pairwise distinct diagonal entries are diagonalizable, because of Proposition 3.17, and form a dense subset of the triangular matrices. Conjugation preserves diagonalizability and is a continuous operation. Thus the closure of the diagonalizable matrices contains the matrices conjugated to a triangular matrix, that is, all of $\mathbf{M}_n(\mathbb{C})$. $\square$

5.2.2 The Spectrum of Special Matrices

Proposition 5.4 *The eigenvalues of Hermitian matrices, as well as those of real symmetric matrices, are real.*

Proof. Let $M \in \mathbf{M}_n(\mathbb{C})$ be an Hermitian matrix and let λ be one of its eigenvalues. Let us choose an eigenvector X: $MX = \lambda X$. Taking the Hermitian adjoint, we obtain $X^*M = \bar{\lambda}X^*$. Hence,

$$\lambda X^*X = X^*(MX) = (X^*M)X = \bar{\lambda}X^*X,$$

or

$$(\lambda - \bar{\lambda})X^*X = 0.$$

However, $X^*X = \sum_j |x_j|^2 > 0$. Therefore, we are left with $\bar{\lambda} - \lambda = 0$. Hence λ is real. $\square$

We leave it to the reader to show, as an exercise, that the eigenvalues of skew-Hermitian matrices are purely imaginary.

Proposition 5.5 *The eigenvalues of the unitary matrices, as well as those of real orthogonal matrices, are complex numbers of modulus one.*

Proof. As before, if X is an eigenvector associated with λ, one has

$$|\lambda|^2 \|X\|^2 = (\lambda X)^*(\lambda X) = (MX)^* MX = X^* M^* MX = X^* X = \|X\|^2,$$

and therefore $|\lambda|^2 = 1$. $\square$

5.2.3 Continuity of Eigenvalues

We study the continuity of the spectrum as a function of the matrix. The spectrum is an n-uplet $(\lambda_1, \ldots, \lambda_n)$ of complex numbers. Mind that each eigenvalue is repeated according to its algebraic multiplicity. At first glance, $\mathrm{Sp}(M)$ seems to be a well-defined element of $\mathbb{C}^n$, but this is incorrect because there is no way to define a natural ordering between the eigenvalues; thus another n-uplet $(\lambda_{\sigma(1)}, \ldots, \lambda_{\sigma(n)})$ describes the same spectrum for every permutation σ. For this reason, the spectrum of M must be viewed as an element of the quotient space $A_n := \mathbb{C}^n / \mathscr{R}$, where $a\mathscr{R}b$ is true if and only if there exists a permutation σ such that $b_j = a_{\sigma(j)}$ for all j. There is a natural topology on A_n, given by the distance

$$d(\dot{a}, \dot{b}) := \min_{\sigma \in S_n} \max_{1 \leq j \leq n} |b_j - a_{\sigma(j)}|.$$

The metric space (A_n, d) is complete.

The way to study the continuity of

$$M \mapsto \mathrm{Sp}(M)$$
$$\mathbf{M}_n(\mathbb{C}) \to A_n$$

is to split this map into

$$M \mapsto \quad P_M \quad \mapsto \mathrm{Sp}(M) = \mathrm{root}(P_M),$$
$$\mathbf{M}_n(\mathbb{C}) \to \mathbf{Unit}_n \to A_n,$$

where $\mathbf{Unit}_n$ is the affine space of monic polynomials of degree n, and the map *root* associates with every element of $\mathbf{Unit}_n$ its set of roots, counted with multiplicities. The first arrow is continuous, because every coefficient of P_M is a combination of minors, thus is polynomial in the entries of M. There remains to study the continuity of *root*. For the sake of completeness, we prove the following.

Lemma 7. *The map* root : $\mathbf{Unit}_n \to A_n$ *is continuous.*

Proof. Let $P \in \mathbf{Unit}_n$ be given. Let $a_1, \ldots, a_r$ be the distinct roots of P, m_j their multiplicities, and ρ the minimum distance between them. We denote by D_j the open disk with center a_j and radius $\rho/2$, and C_j its boundary. The union of the C_js is a compact set on which P does not vanish. The number

$$\eta := \inf \left\{ |P(z)| \, ; z \in \bigcup_j C_j \right\}$$

is thus strictly positive.

The affine space $\mathbf{Unit}_n$ is equipped with the distance $d(Q,R) := \|Q - R\|$ deriving from one of the (all equivalent) norms of $\mathbb{C}_{n-1}[X]$. For instance, we can take

$$\|q\| := \sup \left\{ |q(z)| \, ; z \in \bigcup_j C_j \right\}, \qquad q \in \mathbb{C}_{n-1}[X].$$

If $d(P,Q) < \eta$, then we have

$$|P(z) - Q(z)| < |P(z)|, \qquad \forall z \in C_j, \forall j = 1, \ldots, r.$$

Rouché's theorem asserts that if two holomorphic functions f and g on a disk D, continuous over $\overline{D}$, satisfy $|f(z) - g(z)| < |f(z)|$ on the boundary of D, then they have the same number of zeroes in D, counting with multiplicities. In our case, we deduce that Q has exactly m_j roots in D_j. This sums up to $m_1 + \cdots + m_r = n$ roots in the (disjoint) union of the D_js. Because its degree is n, Q has no other roots. Therefore $d(\mathrm{root}(P), \mathrm{root}(Q)) < \rho$.

This proves the continuity of $Q \mapsto \mathrm{root}(Q)$ at P. $\quad\square$

As a corollary, we have the following fundamental theorem.

Theorem 5.2 *The map* $\mathrm{Sp} : \mathbf{M}_n(\mathbb{C}) \to A_n$ *is continuous.*

One often invokes this theorem by saying that *the eigenvalues of a matrix are continuous functions of its entries.*

5.2.4 Regularity of Simple Eigenvalues

The continuity result in Theorem 5.2 cannot be improved without further assumptions. For instance, the eigenvalues of

$$\begin{pmatrix} 0 & 1 \\ s & 0 \end{pmatrix}$$

are $\pm\sqrt{s}$, and thus are not differentiable functions, at least at the origin. It turns out that the obstacle to the regularity of eigenvalues is the crossing of two or more eigenvalues. But simple eigenvalues are analytic (thus $\mathscr{C}^\infty$) functions of the entries of the matrix.

Theorem 5.3 *Let λ_0 be an algebraically simple eigenvalue of a matrix $M_0 \in \mathbf{M}_n(\mathbb{C})$. Then there exists an open neighbourhood $\mathscr{M}$ of M_0 in $\mathbf{M}_n(\mathbb{C})$, and two analytic functions*

$$M \mapsto \Lambda(M), \qquad M \mapsto X(M)$$

over $\mathscr{M}$, such that

- $\Lambda(M)$ *is an eigenvalue of M.*
- $X(M)$ *is an eigenvector, associated with $\Lambda(M)$.*
- $\Lambda(M_0) = \lambda_0$.

Remarks

- From Theorem 5.2, if M is close to M_0, there is exactly one eigenvalue of M close to λ_0. Theorem 5.3 is a statement about this eigenvalue.
- The theorem is valid as well in $\mathbf{M}_n(\mathbb{R})$, with the same proof.

Proof. Let X_0 be an eigenvector of M_0 associated with λ_0. We know that λ_0 is also a simple eigenvalue of M_0^T. Thanks to Proposition 3.15, an eigenvector Y_0 of M_0^T (associated with λ_0) satisfies $Y_0^T X_0 \neq 0$. We normalize Y_0 in such a way that $Y_0^T X_0 = 1$.

Let us define a polynomial function F over $\mathbf{M}_n(\mathbb{C}) \times \mathbb{C} \times \mathbb{C}^n$, with values in $\mathbb{C} \times \mathbb{C}^n$, by

$$F(M, \lambda, x) := (Y_0^T x - 1, Mx - \lambda x).$$

We have $F(M_0, \lambda_0, X_0) = (0, 0)$.

The differential of F with respect to (λ, x), at the base point (M_0, λ_0, X_0), is the linear map

$$(\mu, y) \overset{\delta}{\mapsto} (Y_0^T y, (M_0 - \lambda_0)y - \mu X_0).$$

Let us show that δ is one-to-one. Let (μ, y) be such that $\delta(\mu, y) = (0, 0)$. Then $\mu = \mu Y_0^T X_0 = Y_0^T (M_0 - \lambda_0)y = 0^T y = 0$. After that, there remains $(M_0 - \lambda_0)y = 0$. Inasmuch as λ_0 is simple, this means that y is colinear to X_0; now the fact that $Y_0^T y = 0$ yields $y = 0$.

Because δ is a one-to-one endomorphism of $\mathbb{C} \times \mathbb{C}^n$, it is an isomorphism. We may then apply the implicit function theorem to F: there exist neighborhoods $\mathscr{M}$, $\mathscr{V}$, and $\mathscr{W}$ and analytic functions $(\Lambda, X) : \mathscr{M} \to \mathscr{V}$ such that

$$\begin{pmatrix} (M, \lambda, x) \in \mathscr{W} \\ F(M, \lambda, x) = (0, 0) \end{pmatrix} \Longleftrightarrow \begin{pmatrix} M \in \mathscr{M} \\ (\lambda, x) = (\Lambda(M), X(M)) \end{pmatrix}.$$

Notice that $F = 0$ implies that (λ, x) is an eigenpair of M. Therefore the theorem is proved. □

5.3 Spectral Decomposition of Normal Matrices

We recall that a matrix M is *normal* if M^* commutes with M. For real matrices, this amounts to saying that M^T commutes with M. Because it is equivalent for an Hermitian matrix H to be zero or to satisfy $x^*Hx = 0$ for every vector x, we see that M is normal if and only if $\|Mx\|_2 = \|M^*x\|_2$ for every vector, where $\|x\|_2$ denotes the standard Hermitian (Euclidean) norm (take $H = MM^* - M^*M$).

Theorem 5.4 *In* $\mathbf{M}_n(\mathbb{C})$, *a matrix is normal if and only if it is unitarily diagonalizable:*

$$(M^*M = MM^*) \Longleftrightarrow (\exists U \in \mathbf{U}_n; \quad M = U^{-1}\operatorname{diag}(d_1,\ldots,d_n)U).$$

This theorem contains the following properties.

Corollary 5.2 *Unitary, Hermitian, and skew-Hermitian matrices are unitarily diagonalizable.*

Observe that among normal matrices one distinguishes each of the above families by the nature of their eigenvalues. Those of unitary matrices have modulus one, and those of Hermitian matrices are real. Finally, those of skew-Hermitian matrices are purely imaginary.

Proof. A diagonal matrix is obviously normal. If U is unitary, a matrix M is normal if and only if U^*MU is normal: we deduce that unitarily diagonalizable matrices are normal.

We now prove the converse. We proceed by induction on the size n of the matrix M. If $n = 0$, there is nothing to prove. Otherwise, if $n \geq 1$, there exists an eigenpair (λ, x):

$$Mx = \lambda x, \quad \|x\|_2 = 1.$$

Because M is normal, $M - \lambda I_n$ is, too. From the above, we see that $\|(M^* - \bar{\lambda})x\|_2 = \|(M - \lambda)x\|_2 = 0$, and hence $M^*x = \bar{\lambda}x$. Let V be a unitary matrix such that $Ve^1 = x$. Then the matrix $M_1 := V^*MV$ is normal and satisfies $M_1 e^1 = \lambda e^1$. Hence it satisfies $M_1^* e^1 = \bar{\lambda} e^1$. This amounts to saying that M_1 is block-diagonal, of the form $M_1 = \operatorname{diag}(\lambda, M')$. Obviously, M' inherits the normality of M_1. From the induction hypothesis, M', and therefore M_1 and M, are unitarily diagonalizable. $\square$

One observes that the same matrix U diagonalizes M^*, because $M = U^{-1}DU$ implies $M^* = U^*D^*U^{-1*} = U^{-1}D^*U$, because U is unitary.

Let us consider the case of a positive-semidefinite Hermitian matrix H. If $HX = \lambda X$, then $0 \leq X^*HX = \lambda\|X\|^2$. The eigenvalues are thus nonnegative. Let $\lambda_1,\ldots,\lambda_p$ be the nonzero eigenvalues of H. Then H is unitarily similar to

$$D := \operatorname{diag}(\lambda_1,\ldots,\lambda_p,0,\ldots,0).$$

From this, we conclude that $\operatorname{rk} H = p$. Let $U \in \mathbf{U}_n$ be such that $H = UDU^*$. Defining the vectors $X_\alpha = \sqrt{\lambda_\alpha}U_\alpha$, where the U_α are the columns of U, we obtain the following statement.

Proposition 5.6 *Let $H \in \mathbf{M}_n(\mathbb{C})$ be a positive-semidefinite Hermitian matrix. Let p be its rank. Then H has p real, positive eigenvalues, and the eigenvalue $\lambda = 0$ has multiplicity $n - p$. There exist p column vectors X_α, pairwise orthogonal, such that*

$$H = X_1 X_1^* + \cdots + X_p X_p^*.$$

Finally, H is positive-definite if and only if $p = n$ (in which case, $\lambda = 0$ is not an eigenvalue).

5.4 Normal and Symmetric Real-Valued Matrices

The situation is a bit more involved if M, a normal matrix, has real entries. Of course, one can consider M as a matrix with complex entries and diagonalize it on a unitary basis, but if M has a nonreal eigenvalue, we quit the field of real numbers when doing so. We prefer to allow orthonormal bases consisting of only *real* vectors. Some of the eigenvalues might be nonreal, thus one cannot in general diagonalize M. The statement is thus the following.

Theorem 5.5 *Let $M \in \mathbf{M}_n(\mathbb{R})$ be a normal matrix. There exists an orthogonal matrix O such that OMO^{-1} is block-diagonal, the diagonal blocks being 1×1 (those corresponding to the real eigenvalues of M) or 2×2, the latter being matrices of direct similitude:[2]*

$$\begin{pmatrix} a & b \\ -b & a \end{pmatrix} \quad (b \neq 0).$$

Likewise, $OM^T O^{-1}$ is block-diagonal, the diagonal blocks being eigenvalues or matrices of direct similitude.

Proof. One again proceeds by induction on n. When $n \geq 1$, the proof is the same as in the previous section whenever M has at least one real eigenvalue.

If this is not the case, then n is even. Let us first consider the case $n = 2$. Then

$$M = \begin{pmatrix} a & b \\ c & d \end{pmatrix}.$$

This matrix is normal, therefore we have $b^2 = c^2$ and $(a - d)(b - c) = 0$. However, $b \neq c$, because otherwise M would be symmetric, and hence would have two real eigenvalues. Hence $b = -c$ and $a = d$.

Now let us consider the general case, with $n \geq 4$. We know that M has an eigenpair (λ, z), where λ is not real. If the real and imaginary parts of z were colinear, M would have a real eigenvector, hence a real eigenvalue, a contradiction. In other words, the real and imaginary parts of z span a plane P in $\mathbb{R}^n$. As before, $Mz = \lambda z$ implies $M^T z = \bar{\lambda} z$. Hence we have $MP \subset P$ and $M^T P \subset P$. Now let V be an orthogonal

---

[2] A similitude is an endomorphism of a Euclidean space that preserves angles. It splits as aR, where R is orthogonal and a is a scalar. It is direct if its determinant is positive.

matrix that maps the plane $P_0 := \mathbb{R}e^1 \oplus \mathbb{R}e^2$ onto P. Then the matrix $M_1 := V^T M V$ is normal and satisfies

$$M_1 P_0 \subset P_0, \quad M_1^T P_0 \subset P_0.$$

This means that M_1 is block-diagonal. Of course, each diagonal block (of sizes 2×2 and $(n-2) \times (n-2)$) inherits the normality of M_1. Applying the induction hypothesis, we know that these blocks are unitarily similar to a block-diagonal matrix whose diagonal blocks are direct similitudes. Hence M_1 and M are unitarily similar to such a matrix. $\square$

Corollary 5.3 *Real symmetric matrices are diagonalizable over $\mathbb{R}$, through orthogonal conjugation. In other words, given $M \in \mathbf{Sym}_n(\mathbb{R})$, there exists an $O \in \mathbf{O}_n(\mathbb{R})$ such that OMO^{-1} is diagonal.*

In fact, because the eigenvalues of M are real, OMO^{-1} has only 1×1 blocks. We say that real symmetric matrices are *orthogonally diagonalizable*.

The interpretation of this statement in terms of quadratic forms is the following. For every quadratic form Q on $\mathbb{R}^n$, there exists an orthonormal basis $\{e_1, \ldots, e_n\}$ in which this form can be written with at most n squares:[3]

$$Q(x) = \sum_{i=1}^{n} a_i x_i^2.$$

Replacing the basis vector e_j by $|a_j|^{1/2} e_j$, one sees that there also exists an orthogonal basis in which the quadratic form can be written

$$Q(x) = \sum_{i=1}^{r} x_i^2 - \sum_{j=1}^{s} x_{j+r}^2,$$

with $r + s \leq n$. This quadratic form is nondegenerate if and only if $r + s = n$. The pair (r, s) is unique and called the *signature* or the *Sylvester index* of the quadratic form. In such a basis, the matrix associated with Q is

$$\begin{pmatrix} 1 & & & & & & & \\ & \ddots & & & & 0 & & \\ & & 1 & & & & & \\ & & & -1 & & & & \\ & & & & \ddots & & & \\ & & & & & -1 & & \\ & & & & & & 0 & \\ & 0 & & & & & & \ddots \\ & & & & & & & & 0 \end{pmatrix}.$$

---

[3] In solid mechanics, when Q is the matrix of inertia, the vectors of this basis are along the *inertia axes*, and the a_j, which then are positive, are the momenta of inertia.

5.5 Functional Calculus

Given a square matrix $A \in \mathbf{M}_n(\mathbb{C})$ and a function $f : \mathscr{U} \to \mathbb{C}$, we should like to define a matrix $f(A)$, in such a way that the maps $A \mapsto f(A)$ and $f \mapsto f(A)$ have nice properties.

The case of polynomials is easy. If

$$P(X) = a_0 X^m + a_1 X^{m-1} + \cdots + a_{m-1} X + a_m$$

has complex coefficients, then we define

$$P(A) = a_0 A^m + a_1 A^{m-1} + \cdots + a_{m-1} A + a_m I_n.$$

Remark that this definition does not need the scalar field to be that of complex numbers.

An important consequence of the Cayley–Hamilton theorem is that if two polynomials P and Q are such that the characteristic polynomial P_A divides $Q - P$, then $Q(A) = P(A)$. This shows that what really matters is the behavior of P and of a few derivatives at the eigenvalues of A. By *few derivatives*, we mean that if ℓ is the algebraic multiplicity of an eigenvalue λ, then one only needs to know $P(\lambda), \ldots, P^{(\ell-1)}(\lambda)$. Actually, ℓ can be chosen as the multiplicity of the root λ in the minimal polynomial of A. For instance, if $N \in \mathbf{M}_n(k)$ is a nilpotent matrix, then the Taylor formula yields

$$P(N) = P(0)I_n + P'(0)N + \cdots + \frac{1}{(n-1)!}P^{(n-1)}(0)N^{n-1}. \tag{5.1}$$

This suggests the following treatment when f is a holomorphic function. Naturally, we ask that its domain $\mathscr{U}$ contain $\mathrm{Sp}A$. We interpolate f at order n at every point of $\mathrm{Sp}A$, by a polynomial P:

$$P^{(r)}(\lambda) = f^{(r)}(\lambda), \qquad \forall \lambda \in \mathrm{Sp}A, \quad \forall 0 \le r \le n-1.$$

We then define

$$f(A) := P(A). \tag{5.2}$$

In order that this definition be meaningful, we verify that it does not depend upon the choice of the interpolation polynomial. This turns out to be true, because if Q is another interpolation polynomial as above, then $Q - P$ is divisible by

$$\prod_{\lambda \in \mathrm{Sp}A} (X - \lambda)^n,$$

thus by P_A, and therefore $Q(A) = P(A)$.

Proposition 5.7 *The functional calculus with holomorphic functions enjoys the following properties. Below, the domains of functions are such that the expressions make sense.*

- *If f and g match at order n over $\mathrm{Sp}A$, then $f(A) = g(A)$.*
- *Conjugation: If M is nonsingular, then $f(M^{-1}AM) = M^{-1}f(A)M$.*
- *Linearity: $(af + g)(A) = af(A) + g(A)$.*
- *Algebra homomorphism: $(fg)(A) = f(A)g(A)$.*
- *Spectral mapping: $\mathrm{Sp}f(A) = f(\mathrm{Sp}A)$.*
- *Composition: $(f \circ g)(A) = f(g(A))$.*

Proof. The first property follows directly from the definition, and the linearity is obvious. The conjugation formula is already true for polynomials.

The product formula is true for polynomials. Now, if f and g are interpolated by P and Q, respectively, at order n at every point of $\mathrm{Sp}A$, then fg is likewise interpolated by PQ. This proves the product formula for holomorphic functions.

We now prove the spectral mapping formula. For this, let (λ, x) be an eigenpair: $Ax = \lambda x$. Let P be an interpolation polynomial of f as above. Then

$$f(A)x = P(A)x = a_0A^m x + \cdots + a_m x = (a_0\lambda^m + \cdots + a_m)x = P(\lambda)x = f(\lambda)x.$$

Therefore $f(\lambda) \in \mathrm{Sp}f(A)$, which tells us $f(\mathrm{Sp}A) \subset \mathrm{Sp}f(A)$.

Conversely, let R be a polynomial vanishing at order n over $f(\mathrm{Sp}A)$. We have $R(f(A)) = (R \circ f)(A) = 0_n$, because $R \circ f$ is flat at order n over $\mathrm{Sp}A$. Replacing A by $f(A)$ and f by R above, we have $R(\mathrm{Sp}f(A)) \subset \mathrm{Sp}0_n = \{0\}$. We have proved that if R vanishes at order n over $f(\mathrm{Sp}A)$, then it vanishes at $\mathrm{Sp}f(A)$. Therefore $\mathrm{Sp}f(A) \subset f(\mathrm{Sp}A)$.

There remains to treat composition. Because of the spectral mapping formula, our assumption is that the domain of f contains $g(\mathrm{Sp}A)$. If f and g are interpolated at order n by P and Q at $g(\mathrm{Sp}A)$ and $\mathrm{Sp}A$ respectively, then $f \circ g$ is interpolated at $\mathrm{Sp}A$ at order n by $P \circ Q$. The formula is true for polynomials, thus it is true for $f \circ g$ too. $\square$

Remark

As long as we are interested in polynomial functions only, Proposition 5.7 is valid in $\mathbf{M}_n(k)$ for an arbitrary field.

5.5.1 The Dunford–Taylor Formula

An alternate definition can be given in terms of a Cauchy integral: the so-called *Dunford–Taylor integral*.

Proposition 5.8 *Let f be holomorphic over a domain $\mathscr{U}$ containing the spectrum of $A \in \mathbf{M}_n(\mathbb{C})$. Let Γ be a positively oriented contour around $\mathrm{Sp}A$, contained in $\mathscr{U}$. Then we have*

$$f(A) = \frac{1}{2i\pi} \int_{\Gamma} f(z)(zI_n - A)^{-1} dz.$$

Proof. Because of the conjugation property, and thanks to Proposition 3.20, it is enough to verify the formula when $A = \lambda I_n + N$ where N is nilpotent. By translation, it suffices to treat the nilpotent case.

So we assume that A is nilpotent. Therefore Γ is a disjoint union of Jordan curves. It is oriented in the trigonometric sense with index one around the origin. Because of nilpotence, we have

$$(zI_n - A)^{-1} = z^{-1}I_n + z^{-2}A + \cdots + z^{-n}A^{n-1}.$$

Thanks to the Cauchy formula

$$\frac{1}{2i\pi}\int_\Gamma f(z)z^{-m-1}\,dz = \frac{1}{m!}f^{(m)}(0),$$

we obtain

$$\frac{1}{2i\pi}\int_\Gamma f(z)(zI_n - A)^{-1}dz = f(0)I_n + f'(0)A + \cdots + \frac{1}{(n-1)!}f^{(n-1)}(0)A^{(n-1)}.$$

Now let P be a polynomial matching f at order n at the origin. We have

$$\frac{1}{2i\pi}\int_\Gamma f(z)(zI_n - A)^{-1}dz = P(0)I_n + P'(0)A + \cdots + \frac{1}{(n-1)!}P^{(n-1)}(0)A^{(n-1)}$$

$$= P(A) = f(A),$$

where we have used Formula (5.1), and the definition of $f(A)$. $\square$

Definition 5.3 *The factor* $(zI_n - A)^{-1}$ *appearing in the Dunford–Taylor formula is the* resolvent *of A at z. It is denoted $R(z;A)$. The domain of $z \mapsto R(z;A)$, which is the complement of $\mathrm{Sp}\,A$, is the* resolvent set.

5.5.2 Invariant Subspaces

An important situation occurs when f is an indicator function; that is, $f(z)$ takes its values in $\{0,1\}$. Because f is assumed to be holomorphic, hence continuous, the closures of the subdomains

$$\mathscr{U}_0 := \{z \mid f(z) = 0\}, \qquad \mathscr{U}_1 := \{z \mid f(z) = 1\}$$

are disjoint sets. Because of the multiplicative property, we see that $f(A)^2 = (f^2)(A) = f(A)$. This tells us that $f(A)$ is a *projector*. If E and F denote its kernel and range, we have $\mathbb{C}^n = E \oplus F$.

Again, the multiplicative property tells us that $f(A)$ commutes with A. This implies that both E and F are invariant subspaces for A:

$$A(E) \subset E, \qquad A(F) \subset F.$$

Using Proposition 3.20 and applying Proposition 5.8 blockwise, we see that passing from A to $f(A)$ amounts to keeping the diagonal blocks $\lambda_j I_{n_j} + N_j$ of the Dunford decomposition for which $f(\lambda_j) = 1$, while dropping those for which $f(\lambda_j) = 0$.

In particular, when $\mathscr{U}_1$ contains precisely one eigenvalue λ (which may have some multiplicity), $f(A)$ is called an *eigenprojector*, because it is the projection onto the characteristic subspace $E(\lambda) := \ker(A - \lambda I_n)^n$, parallel to the other characteristic subspaces.

5.5.2.1 Stable and Unstable Subspaces

The following notions are useful in the linear theory of differential equations, especially when studying asymptotic behavior as time goes to infinity.

Definition 5.4 *Let $A \in \mathbf{M}_n(\mathbb{C})$ be given. Its* stable *invariant subspace is the sum of the subspaces $E(\lambda)$ over the eigenvalues of negative real part. The* unstable *subspace is the sum over the eigenvalues of positive real part. At last, the* central *subspace is the sum over the pure imaginary eigenvalues.*

These spaces are denoted, respectively, $S(A)$, $U(A)$, and $C(A)$.

By invariance of $E(\lambda)$, the stable, unstable, and central subspaces are each invariant under A. From the above analysis, there are three corresponding eigenprojectors π_s, π_u, and π_c, given by the Dunford–Taylor formulæ. For instance, π_s is obtained by choosing $f_s \equiv 1$ over $\Re z < -\varepsilon$ and $f_s \equiv 0$ over $\Re z > \varepsilon$, for a small enough positive ε. In other words,

$$\pi_s = \frac{1}{2i\pi} \int_{\Gamma_s} (zI_n - A)^{-1} dz$$

for some large enough circle Γ_s contained in $\Re z < 0$.

Because $f_s + f_u + f_c \equiv 1$ around the spectrum of A, we have

$$\pi_s + \pi_u + \pi_c = f_s(A) + f_u(A) + f_c(A) = \mathbf{1}(A) = I_n.$$

In addition, the properties $f_s f_u \equiv 0$, $f_s f_c \equiv 0$, and $f_c f_u \equiv 0$ around the spectrum yield

$$\pi_s \pi_u = \pi_u \pi_c = \pi_c \pi_s = 0_n.$$

The identities above give, as expected,

$$\mathbb{C}^n = S(A) \oplus U(A) \oplus C(A).$$

We leave the following characterization to the reader.

Proposition 5.9 *The stable (respectively, unstable) subspace of A is the set of vectors $x \in \mathbb{C}^n$ such that the solution of the Cauchy problem*

$$\frac{dy}{dt} = Ay, \qquad y(0) = x$$

tends to zero exponentially fast as time goes to $+\infty$ (respectively, to $-\infty$).

5.5.2.2 Contractive/Expansive Invariant Subspaces

In the linear theory of discrete dynamical systems, that is, of iterated sequences

$$x^{m+1} = \phi(x^m),$$

what matters is the position of the eigenvalues with respect to the unit circle. We thus define the *contractive subspace* as the sum of $E(\lambda)$s over the eigenvalues of modulus less than 1. The sum over $|\lambda| > 1$ is called the *expansive subspace*. At last, the sum over $|\lambda| = 1$ is the *neutral subspace*. Again, $\mathbb{C}^n$ is the direct sum of these three invariant subspaces.

The link with the stable/unstable subspaces can be described in terms of the exponential of matrices, a notion developed in Chapter 10. The contractive (respectively, expansive, neutral) subspace of $\exp A$ coincides with $S(A)$ (respectively, with $U(A)$, $C(A)$).

5.6 Numerical Range

In this paragraph, we denote $\|x\|_2$ the Hermitian norm in $\mathbb{C}^n$. If $A \in \mathbf{M}_n(\mathbb{C})$ and x is a vector, the expression $r_A(x) = x^* A x$ is a complex number.

Definition 5.5 *The numerical range of A is the subset of the complex plane*

$$\mathscr{H}(A) = \{r_A(x) \mid \|x\|_2 = 1\}.$$

The numerical range is obviously compact. It is unitarily invariant:

$$\mathscr{H}(U^* A U) = \mathscr{H}(A), \qquad \forall U \in \mathbf{U}_n,$$

because $x \mapsto Ux$ is a bijection between unitary vectors. It is thus enough to evaluate the numerical range over upper-triangular matrices, thanks to Theorem 5.1.

5.6.1 The Numerical Range of a 2×2 Matrix

Let us begin with the case $n = 2$. As mentioned above, it is enough to consider triangular matrices. By adding μI_n, we shift the numerical range by a complex number μ. Doing so, we may reduce our analysis to the case where $\mathrm{Tr} A = 0$. Next, multiplying A by z has the effect of applying similitude to the numerical range, whose magnitude is $|z|$ and angle is $\mathrm{Arg}(z)$. Doing so, we reduce our analysis to either

$$A = 0_2, \text{ or } A = 2J_2 := \begin{pmatrix} 0 & 2 \\ 0 & 0 \end{pmatrix}, \text{ or } A = B_a := \begin{pmatrix} 1 & 2a \\ 0 & -1 \end{pmatrix},$$

for some $a \in \mathbb{C}$. At last, conjugating by $\text{diag}(1, e^{i\beta})$ for some real β, we may assume that a is real, nonnegative.

Clearly, $\mathscr{H}(0_2) = \{0\}$. The case of $2J_2$ is quite simple:

$$\mathscr{H}(2J_2) = \{2yz \mid |y|^2 + |z|^2 = 1\}.$$

This is a rotationally invariant set, containing the segment $[0, 1]$ and contained in the unit disk, by Cauchy–Schwarz. Thus it equals the unit disk.

Let us examine the third case in details.

$$\mathscr{H}(B_a) = \{|y|^2 - |z|^2 + 2a\bar{y}z \mid |y|^2 + |z|^2 = 1\}.$$

At fixed moduli $|y|$ and $|z|$, the number $r_{B_a}(x)$ runs over a circle whose center is $|y|^2 - |z|^2$ and radius is $2a|yz|$. Therefore $\mathscr{H}(B_a)$ is the union of the circles $C(r; \rho)$ where the center r is real, and (r, ρ) is constrained by

$$a^2 r^2 + \rho^2 = a^2.$$

Let $\mathscr{E} \in \mathbb{C} \sim \mathbb{R}^2$ be the filled ellipse with foci ± 1 and passing through $z = \sqrt{1 + a^2}$. It is defined by the inequality

$$\frac{(\Re z)^2}{1 + a^2} + \frac{(\Im z)^2}{a^2} \le 1.$$

If $z = r + \rho e^{i\theta} \in C(r; \rho)$, we have

$$\frac{(\Re z)^2}{1 + a^2} + \frac{(\Im z)^2}{a^2} = \frac{r^2}{1 + a^2} + \frac{\rho^2}{a^2} + \frac{2r\rho}{1 + a^2} \cos\theta - \frac{\rho^2}{a^2(1 + a^2)} \cos^2\theta.$$

Considering this expression as a quadratic polynomial in $\cos\theta$, its maximum is reached when the argument equals[4] ra^2/ρ and then it takes the value $r^2 + \rho^2/a^2$. The latter being less than or equal to one, we deduce that z belongs to $\mathscr{E}$. Therefore $\mathscr{H}(A) \subset \mathscr{E}$.

Conversely, let z belong to $\mathscr{E}$. The polynomial $r \mapsto g(r) := (\Im z)^2 + (\Re z - r)^2 + a^2(r^2 - 1)$ is convex and reaches its minimum at $r_0 = (\Re z)/(1 + a^2)$, which belongs to $[-1, 1]$. We have

$$g(r_0) = (\Im z)^2 + \frac{a^2}{1 + a^2} (\Re z)^2 - a^2,$$

which is nonpositive by assumption. Because $g(\pm 1) \ge 0$, we deduce the existence of an $r \in [-1, 1]$ such that $g(r) = 0$. This precisely means that $z \in C(r; \rho)$. Therefore $\mathscr{E} \subset \mathscr{H}(A)$.

Finally, $\mathscr{H}(A)$ is a filled ellipse whose foci are ± 1. Its great axis is $\sqrt{1 + a^2}$ and the small one is a. Its area is therefore $\pi a \sqrt{1 + a^2}$, which turns out to equal

---

[4] The value ra^2/ρ is not necesssarily within $[-1, 1]$, but we don't mind. When $ra^2/\rho > 1$, the circle is contained in the interior of $\mathscr{E}$, whereas if $ra^2/\rho \le 1$, it is interiorly tangent to $\mathscr{E}$.

$|\det[B_a^*, B_a]|^{1/2}$, because we have

$$[B_a^*, B_a] = \begin{pmatrix} -4a^2 & -4a \\ 4a & 4a^2 \end{pmatrix}.$$

We notice that the same formula holds true for the other cases $A = 0_2$ or $A = 2J_2$.

Going backward to a general 2×2 matrix through similitude and conjugation, and pointing out that affine transformations preserve the class of ellipse, while multiplying the area by the Jacobian determinant, we have established the following.

Lemma 8. *The numerical range of a 2×2 matrix is a filled ellipse whose foci are its eigenvalues. Its area equals*

$$\frac{\pi}{4} |\det[A^*, A]|^{1/2}.$$

5.6.2 The General Case

Let us turn towards matrices of sizes $n \geq 3$ (the case $n = 1$ being trivial). If z, z' belong to the numerical range, we have $z = r_A(x)$ and $z' = r_A(x')$ for suitable unit vectors. Applying Lemma 8 to the restriction of r_A to the plane spanned by x and x', we see that there is a filled ellipse, containing z and z', and contained in $\mathscr{H}(A)$. Therefore the segment $[z, z']$ is contained in the numerical range and $\mathscr{H}(A)$ is convex.

If x is a unitary eigenvector, then $r_A(x) = x^*(\lambda x) = \lambda$. Finally we have the following.

Theorem 5.6 (Toeplitz–Hausdorff) *The numerical range of a matrix $A \in \mathbf{M}_n(\mathbb{C})$ is a compact convex domain. It contains the eigenvalues of A.*

5.6.2.1 The Case of Normal Matrices

If A is normal, we deduce from Theorem 5.4 that its numerical range equals that of the diagonal matrix D with the same eigenvalues $a_1, \ldots, a_n$. Denoting $\theta_j = |x_j|^2$ when $x \in \mathbb{C}^n$, we see that

$$\mathscr{H}(A) = \{\theta_1 a_1 + \cdots + \theta_n a_n \,|\, \theta_1, \ldots, \theta_n \geq 0 \text{ and } \theta_1 + \cdots + \theta_n = 1\}.$$

This is precisely the convex envelope of $a_1, \ldots, a_n$.

Proposition 5.10 *The numerical range of a normal matrix is the convex hull of its eigenvalues.*

5.6.3 The Numerical Radius

Definition 5.6 *The* numerical radius *of $A \in \mathbf{M}_n(\mathbb{C})$ is the nonnegative real number*

$$w(A) := \sup\{|z| \,;\, z \in \mathscr{H}(A)\} = \sup_{x \neq 0} \frac{|x^*Ax|}{\|x\|} = \sup_{\|x\|=1} |x^*Ax|.$$

As a supremum of seminorms $A \mapsto |x^*Ax|$, it is a seminorm. Because of Proposition 5.3, $w(A) = 0$ implies $A = 0_n$. The numerical radius is thus a norm. We warn the reader that it is not a *matrix norm* (this notion is developed in Chapter 7) inasmuch as $w(AB)$ is not always less than or equal to $w(A)w(B)$. For instance

$$w(J_2^T J_2) = 1, \text{ and } w(J_2^T) = w(J_2) = \frac{1}{2}.$$

However, the numerical radius satisfies the inequality $w(A^k) \leq w(A)^k$ for every positive integer k (see Exercise 8). A norm with this property is called *superstable*.

Because $\mathbf{M}_n(\mathbb{C})$ is finite dimensional, the numerical radius is equivalent as a norm to any other norm, for instance to matrix norms. However, norm equivalence involves constant factors, which may depend dramatically on the dimension n. It is thus remarkable that the equivalence with the standard operator norm is uniform in n:

Proposition 5.11 *For every $A \in \mathbf{M}_n(\mathbb{C})$, we have*

$$w(A) \leq \|A\|_2 \leq 2w(A).$$

Proof. Cauchy–Schwarz gives

$$|x^*Ax| \leq \|x\|_2 \|Ax\|_2 \leq \|A\|_2 \|x\|_2^2,$$

which yields $w(A) \leq \|A\|_2$.

On the other hand, let us majorize $|y^*Ax|$ in terms of $w(A)$. We have

$$4y^*Ax = (x+y)^*A(x+y) - (x-y)^*A(x-y) + i(x+iy)^*A(x+iy) - i(x-iy)^*A(x-iy).$$

The triangle inequality and the definition of $w(A)$ then give

$$4|y^*Ax| \leq \left(\|x+y\|_2^2 + \|x-y\|_2^2 + \|x+iy\|_2^2 + \|x-iy\|_2^2 \right) w(A)$$
$$= 4(\|x\|_2^2 + \|y\|_2^2)w(A).$$

If x and y are unit vectors, this means $|y^*Ax| \leq 2w(A)$. If x is a unit vector, we now write

$$\|Ax\|_2 = \sup\{|y^*Ax| \,;\, \|y\|_2 = 1\} \leq 2w(A).$$

Taking the supremum over x, we conclude that $\|A\|_2 \leq w(A)$. $\quad\square$

Specific examples show that each one of the inequalities in Proposition 5.11 can be an equality.

5.7 The Gershgorin Domain

In this section, we use the norm $\| \cdot \|_\infty$ over $\mathbb{C}^n$, defined by

$$\|x\|_\infty := \max_i |x_i|.$$

Let $A \in \mathbf{M}_n(\mathbb{C})$, λ be an eigenvalue and x an associated eigenvector. Let i be an index such that $|x_i| = \|x\|_\infty$. Then $x_i \neq 0$ and the majorization

$$|a_{ii} - \lambda| = \left| \sum_{j \neq i} a_{ij} \frac{x_j}{x_i} \right| \leq \sum_{j \neq i} |a_{ij}|$$

gives the following.[5]

Proposition 5.12 (Gershgorin) *The spectrum of A is included in the* Gershgorin domain $\mathscr{G}(A)$, *defined as the union of the* Gershgorin disks

$$D_i(A) := D(a_{ii}; r_i), \qquad r_i := \sum_{j \neq i} |a_{ij}|.$$

Replacing A by its transpose, which has the same spectrum, we have likewise

$$\text{Sp}(A) \subset \mathscr{G}(A^T) = \bigcup_{j=1}^n D_j'(A), \qquad D_j'(A) := D_j(A^T) = D(a_{jj}; r_j'), \qquad r_j' := \sum_{i \neq j} |a_{ij}|.$$

One may improve this result by considering the connected components of $\mathscr{G}(A)$. Let G be one of them. It is the union of the D_ks that meet G. Let p be the number of such disks. One has $G = \cup_{i \in I} D_i(A)$ where I has cardinality p.

Theorem 5.7 *There are exactly p eigenvalues of A in G, counted with their multiplicities.*

Proof. For $r \in [0,1]$, we define a matrix $A(r)$ by the formula

$$a_{ij}(r) := \begin{cases} a_{ii}, & j = i, \\ r a_{ij}, & j \neq i. \end{cases}$$

---

[5] This result can also be deduced from Proposition 7.5: let us decompose $A = D + C$, where D is the diagonal part of A. If $\lambda \neq a_{ii}$ for every i, then $\lambda I_n - A = (\lambda I_n - D)(I_n - B)$ with $B = (\lambda I_n - D)^{-1} C$. Hence, if λ is an eigenvalue, then either λ is an a_{ii}, or $\|B\|_\infty \geq 1$.

It is clear that the Gershgorin domain $\mathscr{G}_r$ of $A(r)$ is included in $\mathscr{G}(A)$. We observe that $A(1) = A$, and that $r \mapsto A(r)$ is continuous. Let us denote by $m(r)$ the number of eigenvalues (counted with multiplicity) of $A(r)$ that belong to G.

Because G and $\mathscr{G}(A) \setminus G$ are compact, one can find a Jordan curve, oriented in the trigonometric sense, that separates G from $\mathscr{G}(A) \setminus G$. Let Γ be such a curve. Inasmuch as $\mathscr{G}_r$ is included in $\mathscr{G}(A)$, the residue formula expresses $m(r)$ in terms of the characteristic polynomial P_r of $A(r)$:

$$m(r) = \frac{1}{2i\pi} \int_\Gamma \frac{P_r'(z)}{P_r(z)} \, dz.$$

Because P_r does not vanish on Γ and $r \mapsto P_r, P_r'$ are continuous, $r \mapsto m(r)$ is continuous. Because $m(r)$ is an integer and $[0,1]$ is connected, $m(r)$ remains constant. In particular, $m(0) = m(1)$.

Finally, $m(0)$ is the number of entries a_{jj} (eigenvalues of $A(0)$) that belong to G. But a_{jj} is in G if and only if $D_j(A) \subset G$. Hence $m(0) = p$, which implies $m(1) = p$, the desired result. $\square$

An improvement of Gershgorin's theorem concerns irreducible matrices.

Proposition 5.13 *Let A be an irreducible matrix. If an eigenvalue of A does not belong to the interior of any Gershgorin disk, then it belongs to every circle $S(a_{ii}; r_i)$.*

Proof. Let λ be such an eigenvalue and x an associated eigenvector. By assumption, one has $|\lambda - a_{ii}| \geq \sum_{j \neq i} |a_{ij}|$ for every i. Let I be the set of indices for which $|x_i| = \|x\|_\infty$ and let J be its complement. If $i \in I$, then

$$\|x\|_\infty \sum_{j \neq i} |a_{ij}| \leq |\lambda - a_{ii}| \, \|x\|_\infty = \left| \sum_{j \neq i} a_{ij} x_j \right| \leq \sum_{j \neq i} |a_{ij}| \, |x_j|.$$

It follows that $\sum_{j \neq i}(\|x\|_\infty - |x_j|)|a_{ij}| \leq 0$, where all the terms in the sum are nonnegative. Each term is thus zero, so that $a_{ij} = 0$ for $j \in J$. Because A is irreducible, J is empty. One has thus $|x_j| = \|x\|_\infty$ for every j, and the previous inequalities show that λ belongs to every circle. $\square$

5.7.1 An Application

Definition 5.7 *A square matrix $A \in \mathbf{M}_n(\mathbb{C})$ is said to be*

1. Diagonally dominant if

$$|a_{ii}| \geq \sum_{j \neq i} |a_{ij}|, \quad 1 \leq i \leq n$$

2. Strongly diagonally dominant if it is diagonally dominant and in addition at least one of these n inequalities is strict

3. Strictly diagonally dominant *if the inequality above is strict for every index i*

Corollary 5.4 *Let A be a square matrix. If A is strictly diagonally dominant, or if A is irreducible and strongly diagonally dominant, then A is invertible.*

In fact, either zero does not belong to the Gershgorin domain, or it is not interior to the disks. In the latter case, A is assumed to be irreducible, and there exists a disk D_j that does not contain zero.

Exercises

1. Show that the eigenvalues of skew-Hermitian matrices, as well as those of real skew-symmetric matrices, are pure imaginary.
2. Let $P, Q \in \mathbf{M}_n(\mathbb{R})$ be given. Assume that $P + iQ \in \mathbf{GL}_n(\mathbb{C})$. Show that there exist $a, b \in \mathbb{R}$ such that $aP + bQ \in \mathbf{GL}_n(\mathbb{R})$. Deduce that if $M, N \in \mathbf{M}_n(\mathbb{R})$ are similar in $\mathbf{M}_n(\mathbb{C})$, then these matrices are similar in $\mathbf{M}_n(\mathbb{R})$.
3. Given an invertible matrix

$$M = \begin{pmatrix} a & b \\ c & d \end{pmatrix} \in \mathbf{GL}_2(\mathbb{R}),$$

   define a map h_M from $S^2 := \mathbb{C} \cup \{\infty\}$ into itself by

$$h_M(z) := \frac{az+b}{cz+d}.$$

   a. Show that h_M is a bijection.
   b. Show that $h : M \mapsto h_M$ is a group homomorphism. Compute its kernel.
   c. Let $\mathscr{H}$ be the upper half-plane, consisting of those $z \in \mathbb{C}$ with $\Im z > 0$. Compute $\Im h_M(z)$ in terms of $\Im z$ and deduce that the subgroup

$$\mathbf{GL}_2^+(\mathbb{R}) := \{M \in \mathbf{GL}_2(\mathbb{R}) \mid \det M > 0\}$$

   operates on $\mathscr{H}$.
   d. Conclude that the group $\mathbf{PSL}_2(\mathbb{R}) := \mathbf{SL}_2(\mathbb{R})/\{\pm I_2\}$, called the *modular group*, operates on $\mathscr{H}$.
   e. Let $M \in \mathbf{SL}_2(\mathbb{R})$ be given. Determine, in terms of $\operatorname{Tr} M$, the number of fixed points of h_M on $\mathscr{H}$.

4. Show that $M \in \mathbf{M}_n(\mathbb{C})$ is normal if and only if there exists a unitary matrix U such that $M^* = MU$.
5. Let $d : \mathbf{M}_n(\mathbb{R}) \to \mathbb{R}^+$ be a multiplicative function; that is,

$$d(MN) = d(M)d(N)$$

for every $M, N \in \mathbf{M}_n(\mathbb{R})$. If $\alpha \in \mathbb{R}$, define $\delta(\alpha) := d(\alpha I_n)^{1/n}$. Assume that d is not constant.

a. Show that $d(0_n) = 0$ and $d(I_n) = 1$. Deduce that $P \in \mathbf{GL}_n(\mathbb{R})$ implies $d(P) \neq 0$ and $d(P^{-1}) = 1/d(P)$. Show, finally, that if M and N are similar, then $d(M) = d(N)$.

b. Let $D \in \mathbf{M}_n(\mathbb{R})$ be a diagonal matrix. Find matrices $D_1, \ldots, D_{n-1}$, similar to D, such that $DD_1 \cdots D_{n-1} = (\det D)I_n$. Deduce that $d(D) = \delta(\det D)$.

c. Let $M \in \mathbf{M}_n(\mathbb{R})$ be a diagonalizable matrix. Show that $d(M) = \delta(\det M)$.

d. Using the fact that M^T is similar to M, show that $d(M) = \delta(\det M)$ for every $M \in \mathbf{M}_n(\mathbb{R})$.

6. Let $A \in \mathbf{M}_n(\mathbb{C})$ be given, and let $\lambda_1, \ldots, \lambda_n$ be its eigenvalues. Show, by induction on n, that A is normal if and only if

$$\sum_{i,j} |a_{ij}|^2 = \sum_{1}^{n} |\lambda_\ell|^2.$$

Hint: The left-hand side (whose square root is called *Schur's norm*) is invariant under conjugation by a unitary matrix. It is then enough to restrict attention to the case of a triangular matrix.

7. (Fiedler and Pták [13]) Given a matrix $A \in \mathbf{M}_n(\mathbb{R})$, we wish to prove the equivalence of the following properties:

P1   For every vector $x \neq 0$ there exists an index k such that $x_k (Ax)_k > 0$.

P2   For every vector $x \neq 0$ there exists a diagonal matrix D with positive diagonal elements such that the scalar product (Ax, Dx) is positive.

P3   For every vector $x \neq 0$ there exists a diagonal matrix D with nonnegative diagonal elements such that the scalar product (Ax, Dx) is positive.

P4   The real eigenvalues of all principal submatrices of A are positive.

P5   All principal minors of A are positive.

a. Prove that **Pj** implies **P(j+1)** for every $j = 1, \ldots, 4$.

b. Assume **P5**. Show that for every diagonal matrix D with nonnegative entries, one has $\det(A + D) > 0$.

c. Then prove that **P5** implies **P1**.

8. (Berger) We show here that the numerical radius satisfies the *power inequality*. In what follows, we use the *real part* of a square matrix

$$\Re M := \frac{1}{2}(M + M^*).$$

a. Show that $w(A) \leq 1$ is equivalent to the fact that $\Re(I_n - zA)$ is positive-semidefinite for every complex number z in the open unit disc.

b. We now assume that $w(A) \leq 1$. If $|z| < 1$, verify that $I_n - zA$ is nonsingular.

c. If $M \in \mathbf{GL}_n(\mathbb{C})$ has a nonnegative real part, prove that $\Re(M^{-1}) \geq 0_n$. Deduce that $\Re(I_n - zA)^{-1} \geq 0_n$ whenever $|z| < 1$.

d. Let $m \geq 1$ be an integer and ω be a primitive mth root of unity in $\mathbb{C}$. Check that the formula

$$\frac{1}{1-X^m} = \frac{1}{m} \sum_{k=0}^{m-1} \frac{1}{1-\omega^k X}$$

can be recast as a polynomial identity.
Deduce that

$$(I_n - z^m A^m)^{-1} = \frac{1}{m} \sum_{k=0}^{m-1} (I_n - \omega^k z A)^{-1},$$

whenever $|z| < 1$.

e. Deduce from above that

$$\Re(I_n - z^m A^m)^{-1} \geq 0_n,$$

whenever $|z| < 1$. Going backward, conclude that for every complex number y in the open unit disc, $\Re(I_n - yA^m) \geq 0_n$ and thus $w(A^m) \leq 1$.

f. Finally, prove the power inequality

$$w(M^m) \leq w(M)^m, \qquad \forall M \in \mathbf{M}_n(\mathbb{C}), \forall m \in \mathbb{N}.$$

Note: A norm that satisfies the power inequality is called a *superstable* norm.

9. Given a complex $n \times n$ matrix A, show that there exists a unitary matrix U such that $M := U^* A U$ has a constant diagonal:

$$m_{ii} = \frac{1}{n} \mathrm{Tr} A, \qquad \forall i = 1, ..., n.$$

Hint: Use the convexity of the numerical range.
In the Hermitian case, compare with Schur's theorem 6.7.

10. Let $B \in \mathbf{GL}_n(\mathbb{C})$. Verify that the inverse and the Hermitian adjoint of $B^{-1} B^*$ are similar. Conversely, let $A \in \mathbf{GL}_n(\mathbb{C})$ be a matrix whose inverse and the Hermitian adjoint are similar: $A^* = P A^{-1} P^{-1}$.

a. Show that there exists an invertible Hermitian matrix H such that $H = A^* H A$. **Hint:** Look for an H as a linear combination of P and of P^*.

b. Show that there exists a matrix $B \in \mathbf{GL}_n(\mathbb{C})$ such that $A = B^{-1} B^*$. Look for a B of the form $(a I_n + b A^*) H$.

11. a. Show that $|\det(I_n + A)| \geq 1$ for every skew-Hermitian matrix A, and that equality holds only if $A = 0_n$.

b. Deduce that for every $M \in \mathbf{M}_n(\mathbb{C})$ such that $H := \Re M$ is positive-definite,

$$\det H \leq |\det M|$$

by showing that $H^{-1}(M - M^*)$ is similar to a skew-Hermitian matrix. You may use the *square root* defined in Chapter 10.

12. Let $A \in \mathbf{M}_n(\mathbb{C})$ be a normal matrix. We decompose $A = L + D + U$ in strictly lower, diagonal, and strictly upper-triangular parts. Let us denote by ℓ_j the Euclidean length of the jth column of L, and by u_j that of the jth row of U.

   a. Show that

   $$\sum_{j=1}^{k} u_j^2 \leq \sum_{j=1}^{k} \ell_j^2 + \sum_{j=1}^{k} \sum_{m=1}^{j-1} u_{mj}^2, \quad k = 1, \dots, n-1.$$

   b. Deduce the inequality

   $$\|U\|_S \leq \sqrt{n-1}\|L\|_S,$$

   for the Schur–Frobenius norm

   $$\|M\|_S := \left(\sum_{i,j=1}^{n} |m_{ij}|^2 \right)^{1/2}.$$

   c. Prove also that

   $$\|U\|_S \geq \frac{1}{\sqrt{n-1}}\|L\|_S.$$

   d. Verify that each of these inequalities is optimal. **Hint:** Consider a circulant matrix.

13. For $A \in \mathbf{M}_n(\mathbb{C})$, define

   $$\varepsilon := \max_{i \neq j} |a_{ij}|, \quad \delta := \min_{i \neq j} |a_{ii} - a_{jj}|.$$

   We assume in this exercise that $\delta > 0$ and $\varepsilon \leq \delta/4n$.

   a. Show that each Gershgorin disk $D_j(A)$ contains exactly one eigenvalue of A.
   b. Let $\rho > 0$ be a real number. Verify that A^ρ, obtained by multiplying the ith row of A by ρ and the ith column by $1/\rho$, has the same eigenvalues as A.
   c. Choose $\rho = 2\varepsilon/\delta$. Show that the ith Gershgorin disk of A^ρ contains exactly one eigenvalue. Deduce that the eigenvalues of A are simple and that

   $$d(\mathrm{Sp}(A), \mathrm{diag}(a_{11}, \dots, a_{nn})) \leq \frac{2n\varepsilon^2}{\delta}.$$

14. Let $A \in \mathbf{M}_n(\mathbb{C})$ be given. We define

   $$\mathscr{B}_{ij}(A) = \{z \in \mathbb{C} \,|\, |(z - a_{ii})(z - a_{jj})| \leq r_i(A)r_j(A)\}.$$

   These sets are *Cassini ovals*. Finally, set

   $$\mathscr{B}(A) := \bigcup_{1 \leq i < j \leq n} \mathscr{B}_{ij}(A).$$

    a. Show that $\mathrm{Sp}A \subset \mathscr{B}(A)$.
    b. Show that this result is sharper than Proposition 5.12.
    c. When $n = 2$, show that in fact $\mathrm{Sp}A$ is included in the boundary of $\mathscr{B}(A)$.

Note: It is tempting to make a generalization from the present exercise and Proposition 5.12, and conjecture that the spectrum is contained in the union of sets defined by inequalities

$$|(z - a_{ii})(z - a_{jj})(z - a_{kk})| \leq r_i(A)r_j(A)r_k(A)$$

and so on. However, the claim is already false with this third-order version.

15. Let I be an interval of $\mathbb{R}$ and $t \mapsto P(t)$ be a map of class $\mathscr{C}^1$ with values in $\mathbf{M}_n(\mathbb{R})$ such that for each t, $P(t)$ is a projector: $P(t)^2 = P(t)$.

    a. Show that the rank of $P(t)$ is constant.
    b. Show that $P(t)P'(t)P(t) = 0_n$.
    c. Let us define $Q(t) := [P'(t), P(t)]$. Show that $P'(t) = [Q(t), P(t)]$.
    d. Let $t_0 \in I$ be given. Show that the differential equation $U' = QU$ possesses a unique solution in I such that $U(t_0) = I_n$. Show that $P(t) = U(t)P(t_0)U(t)^{-1}$.

16. Show that the set of projectors of given rank p is a connected subset in $\mathbf{M}_n(\mathbb{C})$.
17. Let E be an invariant subspace of a matrix $M \in \mathbf{M}_n(\mathbb{R})$.

    a. Show that $E^\perp$ is invariant under M^T.
    b. Prove the following identity between characteristic polynomials:

$$P_M(X) = P_{M|E}(X)P_{M^T|E^\perp}(X). \tag{5.3}$$

18. Prove Proposition 5.9.
19. (Converse of Lemma 8.) Let A and B be 2×2 complex matrices, that have the same spectrum. We assume in addition that

$$\det[A^*, A] = \det[B^*, B].$$

Prove that A and B are unitarily similar. **Hint:** Prove that they both are unitarily similar to the same triangular matrix.
Deduce that two matrices in $\mathbf{M}_2(\mathbb{C})$ are unitarily similar if and only if they have the same numerical range.

20. Prove the following formula for complex matrices:

$$\log\det(I_n + zA) = \sum_{k=0}^{\infty} \frac{(-1)^{k+1}}{k} \mathrm{Tr}(A^k)z^k.$$

Hint: Use an analogous formula for $\log(1 + az)$.

Chapter 6
Hermitian Matrices

We recall that $\|\cdot\|_2$ denotes the usual Hermitian norm on $\mathbb{C}^n$:

$$\|x\|_2 := \left(\sum_{j=1}^{n} |x_j|^2 \right)^2.$$

6.1 The Square Root over HPD$_n$

Hermitian matrices do not form an algebra: the product of two Hermitian matrices is not in $\mathbf{H}_n$. One can even prove that $\mathbf{H}_n \cdot \mathbf{H}_n = \mathbf{M}_n(\mathbb{C})$. However, we notice the following important result.

Proposition 6.1 *Let $H \in \mathbf{HPD}_n$ and $K \in \mathbf{H}_n$ be given. Then the product HK (or KH as well) is diagonalizable with real eigenvalues. The number of positive (respectively, negative) eigenvalues of HK equals that for K.*

In terms of Hermitian forms, it means that given a positive-definite form Φ and another form ϕ, there exists a basis $\mathcal{B}$ which is orthogonal with respect to both forms.

Proof. Recall that H is unitary diagonalizable with real eigenvalues: there is a $U \in \mathbf{U}_n$ such that $H = U^* \operatorname{diag}(\mu_1, \ldots, \mu_n) U$. Because H is positive-definite, we have $\mu_j > 0$. Setting $h = U^* \operatorname{diag}(\sqrt{\mu_1}, \ldots, \sqrt{\mu_n}) U$, we have $h \in \mathbf{HPD}_n$ and $h^2 = H$.

Because $HK = h(hKh)h^{-1}$, HK is similar to $K' := hKh = h^*Kh$. Because K' is Hermitian, it is diagonalizable with real eigenvalues, and so is HK, with the same eigenvalues. The number of positive eigenvalues is the largest dimension of a subspace E on which the Hermitian form $x \mapsto x^*K'x$ is positive-definite. Equivalently, hE is a subspace on which the Hermitian form $x \mapsto y^*Ky$ is positive-definite. Inasmuch as h is nonsingular, this maximal dimension is the same for K' and K.

The same argument works for the negative eigenvalues. $\square$

We now have the following theorem.

Theorem 6.1 *Given $H \in \mathbf{HPD}_n$, there exists one and only one $h \in \mathbf{HPD}_n$ such that $h^2 = H$. The matrix h is denoted $\sqrt{H}$, and called the square root of H.*

Proof. Such a square root was constructed in the proof of Proposition 6.1. There remains to prove uniqueness. So let $h, h' \in HPD_n$ be such that $h'^2 = h^2$. Set $U := h'h^{-1}$. We have $U^*U = h^{-1}h'^2h^{-1} = h^{-1}h^2h^{-1} = I_n$, meaning that U is unitary. Thus its eigenvalues belong to the unit circle. However, Proposition 6.1 tells us that U is diagonalizable with real positive eigenvalues. Therefore $\lambda = 1$ is the only eigenvalue and U is similar, thus equal to, I_n. This gives $h' = h$. $\square$

The square root map can be extended by continuity to the closure of $\mathbf{HPD}_n$, the cone of positive-semidefinite matrices. This follows from the stronger global Hölderian property stated in Proposition 6.3.

6.2 Rayleigh Quotients

Let M be an $n \times n$ Hermitian matrix, and let $\lambda_1 \leq \cdots \leq \lambda_n$ be its eigenvalues arranged in increasing order and counted with multiplicity. We denote by $\mathcal{B} = \{v_1, \ldots, v_n\}$ an orthonormal eigenbasis $(Mv_j = \lambda_j v_j)$. If $x \in \mathbb{C}^n$, let $y_1, \ldots, y_n$ be its coordinates in the basis $\mathcal{B}$. Then

$$x^*Mx = \sum_j \lambda_j |y_j|^2 \leq \lambda_n \sum_j |y_j|^2 = \lambda_n \|x\|_2^2.$$

The above inequality is an equality for $x = v_n$, therefore we deduce a formula for the largest eigenvalue of M:

$$\lambda_n = \max_{x \neq 0} \frac{x^*Mx}{\|x\|_2^2} = \max\left\{x^*Mx \mid \|x\|_2^2 = 1\right\}. \tag{6.1}$$

Likewise, the smallest eigenvalue of an Hermitian matrix is given by

$$\lambda_1 = \min_{x \neq 0} \frac{x^*Mx}{\|x\|_2^2} = \min\{x^*Mx \mid \|x\|_2^2 = 1\}. \tag{6.2}$$

The expression

$$\frac{x^*Mx}{\|x\|_2^2}$$

is called a *Rayleigh Quotient*. For a symmetric matrix with real entries, the formulæ (6.1,6.2) remain valid when we replace x^* by x^T and take vectors with real coordinates.

We evaluate the other eigenvalues of $M \in \mathbf{H}_n$ in the following way. For every linear subspace F of $\mathbb{C}^n$ of dimension k, let us define

$$R(F) = \max_{x \in F \setminus \{0\}} \frac{x^*Mx}{\|x\|_2^2} = \max\left\{x^*Mx \mid x \in F, \|x\|_2^2 = 1\right\}.$$

Because of Corollary 1.1, the intersection of two subspaces of $\mathbb{C}^n$ whose dimensions sum up to $n+1$ is nontrivial. Therefore the intersection of F with the linear subspace spanned by $\{v_k, \ldots, v_n\}$ is of dimension greater than or equal to one: there exists a nonzero vector $x \in F$ such that $y_1 = \cdots = y_{k-1} = 0$. One then has

$$x^*Mx = \sum_{j=k}^{n} \lambda_j |y_j|^2 \geq \lambda_k \sum_j |y_j|^2 = \lambda_k \|x\|_2^2.$$

Hence, $R(F) \geq \lambda_k$. Furthermore, if G is the space spanned by $\{v_1, \ldots, v_k\}$, one has $R(G) = \lambda_k$. Thus, we have

$$\lambda_k = \min\{R(F) \mid \dim F = k\}.$$

Finally, we may state the following theorem.

Theorem 6.2 *Let M be an $n \times n$ Hermitian matrix and $\lambda_1, \ldots, \lambda_n$ its eigenvalues arranged in increasing order, counted with multiplicity. Then*

$$\lambda_k = \min_{\dim F = k} \max_{x \in F \setminus \{0\}} \frac{x^*Mx}{\|x\|_2^2}. \tag{6.3}$$

If M is real symmetric, one has similarly

$$\lambda_k = \min_{\dim F = k} \max_{x \in F \setminus \{0\}} \frac{x^T Mx}{\|x\|_2^2},$$

where F runs over subspaces of $\mathbb{R}^n$.

Equivalently, we have

$$\lambda_k = \max_{\dim F = n-k+1} \min_{x \in F \setminus \{0\}} \frac{x^*Mx}{\|x\|_2^2}. \tag{6.4}$$

These formulæ generalize (6.1) and (6.2).

Proof. There remains to prove (6.4). There are two possible strategies. Either we start from nothing and argue again about the intersection of subspaces whose dimensions sum up to $n+1$, or we remark that $\lambda_k(M) = -\lambda_{n-k+1}(-M)$, because M and $-M$ have opposite eigenvalues. Then using $\max\{-f\} = -\min\{f\}$ and $\min\{-f\} = -\max\{f\}$ for every quantity f, we see that (6.4) is a reformulation of (6.3). □

An easy application of Theorem 6.2 is the set of Weyl's inequalities. For this we denote by $\lambda_1(M) \leq \cdots \leq \lambda_n(M)$ the eigenvalues of M. We notice that $\lambda_k(\alpha M)$ equals $\alpha \lambda_k(M)$ if $\alpha \geq 0$, or $-\alpha \lambda_{n-k+1}(M)$ if $\alpha \leq 0$. In particular,

$$\lambda_n(\alpha M) - \lambda_1(\alpha M) = |\alpha|(\lambda_n(M) - \lambda_1(M)), \tag{6.5}$$

$$\max\{\lambda_n(\alpha M), -\lambda_1(\alpha M)\} = |\alpha| \max\{\lambda_n(M), -\lambda_1(M)\}. \tag{6.6}$$

Theorem 6.3 *Let A and B be $n \times n$ Hermitian matrices. Let $1 \leq i, j, k \leq n$ be indices.*

- *If $i + j = k + 1$, we have*

$$\lambda_k(A+B) \geq \lambda_i(A) + \lambda_j(B).$$

- *If $i + j = k + n$, we have*

$$\lambda_k(A+B) \leq \lambda_i(A) + \lambda_j(B).$$

Proof. Once again, any inequality can be deduced from the other ones by means of $(A,B) \leftrightarrow (-A,-B)$. Thus it is sufficient to treat the case where $i + j = k + n$.

From (6.4), we know that there exists an $(n - k + 1)$-dimensional subspace H such that

$$\lambda_k(A+B) = \min_{x \in H \setminus \{0\}} \frac{x^*(A+B)x}{\|x\|_2^2}.$$

From (6.3), there also exist i- and j-dimensional subspaces F, G such that

$$\lambda_i(A) = \max_{x \in F \setminus \{0\}} \frac{x^*Ax}{\|x\|_2^2}, \qquad \lambda_j(B) = \max_{x \in G \setminus \{0\}} \frac{x^*Bx}{\|x\|_2^2}.$$

We now use Corollary 1.1 twice: we have

$$\dim F \cap G \cap H \geq \dim F + \dim G \cap H - n \geq \dim F + \dim G + \dim H - 2n$$
$$= i + j + 1 - n - k = 1.$$

We deduce that there exists a unit vector z in $F \cap G \cap H$. From above, it satisfies

$$\lambda_k(A+B) \leq z^*(A+B)z, \qquad \lambda_i(A) \geq z^*Az, \qquad \lambda_j(B) \geq z^*Bz,$$

whence the conclusion. $\square$

The cases where $i = j = k$ are especially interesting. They give

$$\lambda_1(A+B) \geq \lambda_1(A) + \lambda_1(B), \qquad \lambda_n(A+B) \leq \lambda_n(A) + \lambda_n(B).$$

Because each function $M \mapsto \lambda_j(M)$ is positively homogeneous of degree one, we infer that λ_n is convex, while λ_1 is concave. Their maximum

$$[M] := \max\{\lambda_n(M), -\lambda_1(M)\}$$

is convex too. From (6.6), it is thus a *subnorm*. When $[M] \leq 0$, we have $\lambda_n(M) \leq \lambda_1(M)$; that is, $\mathrm{Sp}\, M = \{0\}$; because M, Hermitian, is diagonalizable, we deduce $M = 0$. Finally, $M \mapsto [M]$ is a *norm* over $\mathbf{H}_n$ (and over $\mathbf{Sym}_n(\mathbb{R})$ too). An interesting consequence of some of the Weyl's inequalities is the following.

Proposition 6.2 *The eigenvalues λ_k over $\mathbf{H}_n$ are Lipschitz functions, with Lipschitz ratio equal to one:*

$$|\lambda_k(M) - \lambda_k(N)| \le [M - N].$$

Proof. Taking $i = k$ and either $j = 1$ or $j = n$, then $A = N$ and $B = M - N$, we obtain

$$\lambda_1(M - N) \le \lambda_k(M) - \lambda_k(N) \le \lambda_n(M - N).$$

□

Comments

- The quantity $[M]$ is nothing but the largest modulus of an eigenvalue of M. It is a special case of what is called the *spectral radius* in Chapter 7. We warn the reader that the spectral radius is not a norm over $\mathbf{M}_n(\mathbb{C})$.
- The quantity $[M]$ can also be characterized as the supremum of $\|Mx\|_2/\|x\|_2$, which is described in Chapter 7 as a matrix norm, or operator norm. The standard notation is $\|M\|_2$: Proposition 6.2 reads

$$|\lambda_k(M) - \lambda_k(N)| \le \|M - N\|_2, \qquad \forall M, N \in \mathbf{H}_n. \tag{6.7}$$

- The Lipschitz quality of λ_k over $\mathbf{H}_n$ is intermediate between the continuity of the spectrum (Theorem 5.2) and the analyticity of simple eigenvalues (Theorem 5.3). It is, however, better than the standard behavior near a nonsemisimple eigenvalue of a (necessarily nonHermitian) matrix. In the latter case, the singularity always involves an algebraic branching, which is typically Hölderian of exponent $1/d$ where d is the size of the largest Jordan block associated with the eigenvalue under consideration.
- A remarkable fact happens when we restrict to a $\mathscr{C}^k$-curve $s \mapsto H(s)$ in $\mathbf{H}_n$. Rellich proved that the eigenvalues of $H(s)$ can be relabeled as $\mu_1(s), \ldots, \mu_n(s)$ in such a way that each function μ_j is of class $\mathscr{C}^k$. Mind the fact that this labeling does not coincide with the increasing order. For instance, if $H(s) = \text{diag}\{s, -s\}$, then $\lambda_1 = -|s|$ and $\lambda_2 = |s|$ are not more than Lipschitz, whereas $\mu_1 = -s$ and $\mu_2 = s$ are analytic. This regularity does not hold in general when the curve is replaced by a surface. For instance, there is no way to label the eigenvalues $\pm\sqrt{a^2 + b^2}$ of

$$H(a, b) := \begin{pmatrix} a & b \\ b & -a \end{pmatrix}, \qquad (a, b \in \mathbb{R}),$$

in such a way that they are smooth functions of (a, b). See Theorem 6.8 in [24].
- The description of the set of the $3n$-tuplets

$$(\lambda_1(A), \ldots, \lambda_n(A), \lambda_1(B), \ldots, \lambda_n(B), \lambda_1(A + B), \ldots, \lambda_n(A + B))$$

as A and B run over $\mathbf{H}_n$ is especially delicate. For a complete historical account of this question, one may read the first section of Fulton's and Bhatia's articles [16, 6]. For another partial result, see Exercise 13 of Chapter 8 (Lidskii's theorem).

6.3 Further Properties of the Square Root

Proposition 6.3 *The square root is Hölderian with exponent* $\frac{1}{2}$ *over* **HPD**$_n$*:*

$$\|\sqrt{H} - \sqrt{K}\|_2 \leq \sqrt{\|H - K\|_2}, \qquad \forall H, K \in \mathbf{HPD}_n.$$

We point out that the Hölderian property is global over **HPD**$_n$, and is uniform with respect to n because it does not involve a constant depending upon n.

Proof. Let $A, B \in \mathbf{HPD}_n$ be given. Let us develop

$$B^2 - A^2 = (B - A)^2 + (B - A)A + A(B - A). \tag{6.8}$$

Up to exchanging the roles of A and B, we may assume that $\lambda_n(B - A) \geq \lambda_n(A - B)$; that is, $\lambda_n(B - A) = [B - A] = \|B - A\|_2$. Let x be an eigenvector of $B - A$ associated with $\lambda_n(B - A)$. From the above, we infer

$$x^*(B^2 - A^2)x = \lambda_n^2 \|x\|_2^2 + 2\lambda_n x^* A x \geq \lambda_n^2 \|x\|_2^2.$$

By Cauchy–Schwarz and the fact that $[M] = \|M\|_2$ (see the comment above), we have

$$\|B - A\|_2^2 \|x\|_2^2 \leq \|x\|_2 \|(B^2 - A^2)x\|_2 \leq \|B^2 - A^2\|_2 \|x\|_2^2,$$

whence

$$\|B - A\|_2^2 \leq \|B^2 - A^2\|_2.$$

We apply this inequality to $B = \sqrt{K}$ and $A = \sqrt{H}$. □

The same expansion yields the monotonicity of the square root map.

Theorem 6.4 *The square root is* operator monotone *over the positive-semidefinite Hermitian matrices: if* $0_n \leq H \leq K$, *then* $\sqrt{H} \leq \sqrt{K}$.

More generally, if I is an interval in $\mathbb{R}$, a map $f : I \to \mathbb{R}$ is *operator monotone* if for every $H, K \in \mathbf{H}_n$ with spectra included in I, the inequality $H \leq K$ implies $f(H) \leq f(K)$. The study of operator monotone functions is the Loewner theory. It is intimately related to the theory of complex variable.

Proof. Let B and A be Hermitian positive-semidefinite. If $B^2 \leq A^2$, then (6.8) yields

$$(B - A)^2 + (B - A)A + A(B - A) \leq 0_n.$$

Let λ_1 be the smallest eigenvalue of $A - B$, and x an associate eigenvector. We obtain

$$\lambda_1^2 \|x\|_2^2 - 2\lambda_1 x^* A x = 0,$$

which shows that λ_1 lies between 0 and $2x^* A x / \|x\|_2^2$. Because A is nonnegative, this implies $\lambda_1 \geq 0$; that is, $A - B \geq 0_n$.

Apply this analysis to $A = \sqrt{K}$ and $B = \sqrt{H}$. □

6.4 Spectrum of Restrictions

In the following result called *Cauchy's interlacing theorem*, the block H is a *compression* of the larger matrix H'.

Theorem 6.5 *Let* $H \in \mathbf{H}_{n-1}$, $x \in \mathbb{C}^{n-1}$, *and* $a \in \mathbb{R}$ *be given. Let* $\lambda_1 \leq \cdots \leq \lambda_{n-1}$ *be the eigenvalues of* H *and* $\mu_1 \leq \cdots \leq \mu_n$ *those of the Hermitian matrix*

$$H' = \begin{pmatrix} H & x \\ x^* & a \end{pmatrix}.$$

One then has $\mu_1 \leq \lambda_1 \leq \cdots \leq \mu_j \leq \lambda_j \leq \mu_{j+1} \leq \cdots \leq \lambda_{n-1} \leq \mu_n$.

Proof. The inequality $\mu_j \leq \lambda_j$ follows from (6.3), because the infimum concerns the same quantity

$$\max_{x \in F, x \neq 0} \frac{x^* H' x}{\|x\|_2^2},$$

but is taken over a smaller set in the case of λ_j: that of subspaces of dimension j contained in $\mathbb{C}^{n-1} \times \{0\}$.

Conversely, let $\pi : x \mapsto (x_1, \ldots, x_{n-1})^T$ be the projection from $\mathbb{C}^n$ onto $\mathbb{C}^{n-1}$. If F is a linear subspace of $\mathbb{C}^n$ of dimension $j+1$, its image under π contains a linear subspace G of dimension j (it is often exactly of dimension j). One obviously has

$$\max_{x \in F, x \neq 0} \frac{x^* H' x}{\|x\|_2^2} \geq \max_{x \in G, x \neq 0} \frac{x^* H x}{\|x\|_2^2} \geq \lambda_j.$$

Taking the infimum, we obtain $\mu_{j+1} \geq \lambda_j$. □

Theorem 6.5 is optimal, in the following sense.

Theorem 6.6 *Let* $\lambda_1 \leq \cdots \leq \lambda_{n-1}$ *and* $\mu_1 \leq \cdots \leq \mu_n$ *be real numbers satisfying* $\mu_1 \leq \lambda_1 \leq \cdots \leq \mu_j \leq \lambda_j \leq \mu_{j+1} \leq \cdots$. *Then there exist a vector* $x \in \mathbb{R}^n$ *and* $a \in \mathbb{R}$ *such that the real symmetric matrix*

$$H = \begin{pmatrix} \Lambda & x \\ x^T & a \end{pmatrix},$$

where $\Lambda = \mathrm{diag}(\lambda_1, \ldots, \lambda_{n-1})$, *has the eigenvalues* μ_j.

Proof. Let us compute the characteristic polynomial of H from Schur's complement formula[1] (see Proposition 3.9):

$$p_n(X) = \left(X - a - x^T (XI_{n-1} - \Lambda)^{-1} x \right) \det(XI_{n-1} - \Lambda)$$

$$= \left(X - a - \sum_j \frac{x_j^2}{X - \lambda_j} \right) \prod_j (X - \lambda_j).$$

---

[1] One may equally (exercise) compute it by induction on n.

Let us assume for the moment that all the inequalities $\mu_j \leq \lambda_j \leq \mu_{j+1}$ hold strictly. In particular, the λ_js are distinct. Let us consider the partial fraction decomposition of the rational function

$$\frac{\prod_\ell (X - \mu_\ell)}{\prod_j (X - \lambda_j)} = X - a - \sum_j \frac{c_j}{X - \lambda_j}.$$

One thus obtains

$$a = \sum_\ell \mu_\ell - \sum_j \lambda_j,$$

a formula that could also have been found by comparing the traces of Λ and of H. The inequalities $\lambda_{j-1} < \mu_j < \lambda_j$ ensure that each c_j is positive, because

$$c_j = -\frac{\prod_\ell (\lambda_j - \mu_\ell)}{\prod_{k \neq j} (\lambda_j - \lambda_k)}.$$

Let us set $x_j = \sqrt{c_j}$. We obtain, as announced,

$$p_n(X) = \prod_\ell (X - \mu_\ell).$$

In the general case one may choose sequences $\mu_\ell^{(m)}$ and $\lambda_j^{(m)}$ that converge to the μ_ℓs and the λ_js as $m \to +\infty$ and that satisfy strictly the inequalities in the hypothesis. The first part of the proof (case with strict inequalities) provides matrices $H^{(m)}$. Because $K \mapsto [K]$, as defined above, is a norm over $\mathbf{Sym}_n(\mathbb{R})$, the sequence $(H^{(m)})_{m \in \mathbb{N}}$ is bounded. In other words, $(a^{(m)}, x^{(m)})$ remains bounded. Let us extract a subsequence that converges to a pair $(a, x) \in \mathbb{R} \times \mathbb{R}^{n-1}$. The matrix H associated with (a, x) solves our problem, because the eigenvalues depend continuously on the entries of the matrix. $\square$

Corollary 6.1 *Let $H \in \mathbf{H}_{n-1}(\mathbb{R})$ be given, with eigenvalues $\lambda_1 \leq \cdots \leq \lambda_{n-1}$. Let $\mu_1, \ldots, \mu_n$ be real numbers satisfying $\mu_1 \leq \lambda_1 \leq \cdots \leq \mu_j \leq \lambda_j \leq \mu_{j+1} \leq \cdots$. Then there exist a vector $x \in \mathbb{C}^n$ and $a \in \mathbb{R}$ such that the Hermitian matrix*

$$H' = \begin{pmatrix} H & x \\ x^T & a \end{pmatrix}$$

has the eigenvalues μ_j.

The proof consists in diagonalizing H through a unitary matrix $U \in \mathbf{U}_{n-1}$, then applying Theorem 6.6, and conjugating the resulting matrix by $\mathrm{diag}(U^*, 1)$.

6.5 Spectrum versus Diagonal

Let us begin with an order relation between finite sequences of real numbers. If $a = (a_1, \ldots, a_n)$ is a sequence of n real numbers, and if $1 \leq l \leq n$, we denote by $s_k(a)$ the number

$$\min \left\{ \sum_{j \in J} a_j \mid \operatorname{card} J = k \right\}.$$

This is nothing but the sum of the k smallest elements of a. In particular

$$s_n(a) = a_1 + \cdots + a_n.$$

One may always restrict attention to the case of nondecreasing sequences $a_1 \leq \cdots \leq a_n$, in which case we have $s_k(a) = a_1 + \cdots + a_k$.

Definition 6.1 *Let $a = (a_1, \ldots, a_n)$ and $b = (b_1, \ldots, b_n)$ be two sequences of n real numbers. One says that b majorizes a, and one writes $a \prec b$, if*

$$s_k(a) \leq s_k(b), \quad \forall 1 \leq k \leq n, \quad s_n(a) = s_n(b).$$

The relation $a \prec b$ for nondecreasing sequences can now be written as

$$a_1 + \cdots + a_k \leq b_1 + \cdots + b_k, \quad \forall k = 1, \ldots, n-1,$$
$$a_1 + \cdots + a_n = b_1 + \cdots + b_n.$$

The latter equality plays a crucial role in the analysis below. The relation $\prec$ is a partial ordering.

Proposition 6.4 *Let $x, y \in \mathbb{R}^n$ be given. Then $x \prec y$ if and only if for every real number t,*

$$\sum_{j=1}^n |x_j - t| \geq \sum_{j=1}^n |y_j - t|. \tag{6.9}$$

Proof. We may assume that x and y are nondecreasing. If the inequality (6.9) holds, we write it first for t outside the interval I containing the x_js and the y_js. This gives $s_n(x) = s_n(y)$. Then we write it for $t = x_k$. Using $s_n(x) = s_n(y)$, one has

$$\sum_j |x_j - x_k| = \sum_1^k (x_k - y_j) + \sum_{k+1}^n (y_j - x_k) + 2(s_k(y) - s_k(x))$$
$$\leq \sum_j |y_j - x_k| + 2(s_k(y) - s_k(x)),$$

which with (6.9) gives $s_k(x) \leq s_k(y)$.

Conversely, let us assume that $x \prec y$. Let us define $\phi(t) := \sum_j |x_j - t| - \sum_j |y_j - t|$. This is a piecewise linear function, zero outside I. Its derivative is integer-valued, piecewise constant; it increases at the points x_js and decreases at the points y_js only. If $\min\{\phi(t); t \in \mathbb{R}\} < 0$, this minimum is thus reached at some x_k, with $\phi'(x_k - 0) \leq$

$0 \leq \phi'(x_k + 0)$, from which one obtains $y_{k-1} \leq x_k \leq y_{k+1}$. There are two cases, depending on the position of y_k with respect to x_k. For example, if $y_k \leq x_k$, we compute

$$\sum_j |x_j - x_k| = \sum_{k+1}^n (x_j - x_k) + \sum_1^k (x_k - x_j).$$

From the assumption, it follows that

$$\sum_j |x_j - x_k| \geq \sum_{k+1}^n (y_j - x_k) + \sum_1^k (x_k - y_j) = \sum_{j \neq k} |y_j - x_k|,$$

which means that $\phi(x_k) \geq 0$, which contradicts the hypothesis. Hence, ϕ is a non-negative function. $\square$

Our first statement expresses an order between the diagonal and the spectrum of an Hermitian matrix.

Theorem 6.7 (Schur) *Let H be an Hermitian matrix with diagonal a and spectrum λ. Then $a \succ \lambda$.*

Proof. Let n be the size of H. We argue by induction on n. We may assume that a_n is the largest component of a. Because $s_n(\lambda) = \text{Tr}A$, one has $s_n(\lambda) = s_n(a)$. In particular, the theorem holds true for $n = 1$. Let us assume that it holds for order $n - 1$. Let A be the matrix obtained from H by deleting the nth row and the nth column. Let $\mu = (\mu_1, \ldots, \mu_{n-1})$ be the spectrum of A. Let us arrange λ and μ in increasing order. From Theorem 6.5, one has $\lambda_1 \leq \mu_1 \leq \lambda_2 \leq \cdots \leq \mu_{n-1} \leq \lambda_n$. It follows that $s_k(\mu) \geq s_k(\lambda)$ for every $k < n$. The induction hypothesis tells us that $s_k(\mu) \leq s_k(a')$, where $a' = (a_1, \ldots, a_{n-1})$. Finally, we have $s_k(a') = s_k(a)$, and $s_k(\lambda) \leq s_k(a)$ for every $k < n$, which ends the induction. $\square$

Here is the converse.

Theorem 6.8 *Let a and λ be two sequences of n real numbers such that $a \succ \lambda$. Then there exists a real symmetric matrix of size $n \times n$ whose diagonal is a and spectrum is λ.*

Proof. We proceed by induction on n. The statement is trivial if $n = 1$. If $n \geq 2$, we use the following lemma, which is proved afterwards.

Lemma 9. *Let $n \geq 2$ and α, β be two nondecreasing sequences of n real numbers, satisfying $\alpha \prec \beta$. Then there exists a sequence γ of $n - 1$ real numbers such that*

$$\alpha_1 \leq \gamma_1 \leq \alpha_2 \leq \cdots \leq \gamma_{n-1} \leq \alpha_n$$

and $\gamma \prec \beta' = (\beta_1, \ldots, \beta_{n-1})$.

We apply the lemma to the sequences $\alpha = \lambda$, $\beta = a$. Because $\gamma \prec a'$, the induction hypothesis tells us that there exists a real symmetric matrix S of size $(n-1) \times (n-1)$

with diagonal a' and spectrum γ. From Corollary 6.1, there exist a vector $y \in \mathbb{R}^n$ and a real number b such that the matrix

$$\Sigma = \begin{pmatrix} S & y^T \\ y & b \end{pmatrix}$$

has spectrum λ. Inasmuch as $s_n(a) = s_n(\lambda) = \mathrm{Tr}\,\Sigma = \mathrm{Tr}\,S + b = s_{n-1}(a') + b$, we have $b = a_n$. Hence, a is the diagonal of Σ. □

We now prove Lemma 9. Let Δ be the set of sequences δ of $n-1$ real numbers satisfying

$$\alpha_1 \le \delta_1 \le \alpha_2 \le \cdots \le \delta_{n-1} \le \alpha_n \qquad (6.10)$$

together with

$$\sum_{j=1}^{k} \delta_j \le \sum_{j=1}^{k} \beta_j, \quad \forall k \le n-2. \qquad (6.11)$$

We must show that there exists $\delta \in \Delta$ such that $s_{n-1}(\delta) = s_{n-1}(\beta')$. Because Δ is convex and compact (it is closed and bounded in $\mathbb{R}^n$), it is enough to show that

$$\inf_{\delta \in \Delta} s_{n-1}(\delta) \le s_{n-1}(\beta') \le \sup_{\delta \in \Delta} s_{n-1}(\delta). \qquad (6.12)$$

On the one hand, $\alpha' = (\alpha_1, \ldots, \alpha_{n-1})$ belongs to Δ and $s_{n-1}(\alpha') \le s_{n-1}(\beta')$ from the hypothesis, which proves the first inequality in (6.12).

Let us now choose a δ that achieves the supremum of s_{n-1} over Δ. Let r be the largest index less than or equal to $n-2$ such that $s_r(\delta) = s_r(\beta')$, with $r = 0$ if all the inequalities are strict. From $s_j(\delta) < s_j(\beta')$ for $r < j < n-1$, one has $\delta_j = \alpha_{j+1}$, because otherwise, there would exist $\varepsilon > 0$ such that $\hat{\delta} := \delta + \varepsilon e^j$ belong to Δ, and one would have $s_{n-1}(\hat{\delta}) = s_{n-1}(\delta) + \varepsilon$, contrary to the maximality of δ. Now let us compute

$$\begin{aligned} s_{n-1}(\delta) - s_{n-1}(\beta') &= s_r(\beta) - s_{n-1}(\beta) + \alpha_{r+2} + \cdots + \alpha_n \\ &= s_r(\beta) - s_{n-1}(\beta) + s_n(\alpha) - s_{r+1}(\alpha) \\ &\ge s_r(\beta) - s_{n-1}(\beta) + s_n(\beta) - s_{r+1}(\beta) \\ &= \beta_n - \beta_{r+1} \ge 0. \end{aligned}$$

This proves (6.12) and completes the proof of the lemma.

6.6 The Determinant of Nonnegative Hermitian Matrices

The determinant of nonnegative Hermitian matrices enjoys several nice properties. We present two of them below.

6.6.1 Hadamard's Inequality

Proposition 6.5 *Let* $H \in \mathbf{H}_n$ *be a positive-semidefinite Hermitian matrix. Then*

$$\det H \leq \prod_{j=1}^{n} h_{jj}.$$

If $H \in \mathbf{HPD}_n$, *the equality holds only if* H *is diagonal.*

Proof. If $\det H = 0$, there is nothing to prove, because the h_{jj} are nonnegative (these are numbers $(\mathbf{e}^j)^* H \mathbf{e}^j$). Otherwise, H is positive-definite and one has $h_{jj} > 0$. We restrict our attention to the case with a constant diagonal by letting $D :=$ $\mathrm{diag}(h_{11}^{-1/2}, \ldots, h_{nn}^{-1/2})$ and writing $(\det H)/(\prod_j h_{jj}) = \det DHD = \det H'$, where the diagonal entries of H' equal one. There remains to prove that $\det H' \leq 1$.

The eigenvalues $\mu_1, \ldots, \mu_n$ of H' are strictly positive, of sum $\mathrm{Tr}\, H' = n$. Inasmuch as the logarithm is concave, one has

$$\frac{1}{n}\log \det H' = \frac{1}{n}\sum_j \log \mu_j \leq \log \frac{1}{n}\sum_j \mu_j = \log 1 = 0,$$

which proves the inequality. The concavity being strict, the equality holds only if $\mu_1 = \cdots = \mu_n = 1$, but then H' is similar, thus equal to I_n. In that case, H is diagonal. $\square$

Applying Proposition 6.5 to matrices of the form M^*M or MM^*, one obtains the following result.

Theorem 6.9 *For* $M \in \mathbf{M}_n(\mathbb{C})$, *one has*

$$|\det M| \leq \prod_{i=1}^{n}\left(\sum_{j=1}^{n}|m_{ij}|^2\right)^{1/2}, \quad |\det M| \leq \prod_{j=1}^{n}\left(\sum_{i=1}^{n}|m_{ij}|^2\right)^{1/2}.$$

When $M \in \mathbf{GL}_n(\mathbb{C})$, *the first (respectively, the second) inequality is an equality if and only if the rows (respectively, the columns) of* M *are pairwise orthogonal.*

6.6.2 A Concavity Result

We now turn towards an ubiquitous result due to Gårding, who first proved it in the context of hyperbolic differential operators. It is meaningful in convex geometry, combinatorics, and optimization too.

Theorem 6.10 *The map*

$$H \mapsto (\det H)^{1/n}$$

is concave over the cone of positive semidefinite $n \times n$ *Hermitian matrices.*

Because the logarithm is a concave increasing function, we deduce the following.

Corollary 6.2 *The map*

$$H \mapsto \log \det H$$

is concave over **HPD**$_n$.

The corollary has the advantage of being independent of the size of the matrices. However, at fixed size n, it is obviously weaker than the theorem.

Proof. Let $H, K \in$ **HPD**$_n$ be given. From Proposition 6.1, the eigenvalues $\mu_1, \ldots, \mu_n$ of HK are real and positive, even though HK is not Hermitian. We thus have

$$(\det H)^{1/n}(\det K)^{1/n} = (\det HK)^{1/n} = \left(\prod_{j=1}^{n} \mu_j \right)^{1/n} \leq \frac{1}{n} \sum_{j=1}^{n} \mu_j = \frac{1}{n} \operatorname{Tr}(HK),$$

$$(6.13)$$

where we have applied the arithmetico-geometric inequality to $(\mu_1, \ldots, \mu_n)$.

When choosing $K = H^{-1}$, (6.13) becomes an equality. We therefore have

$$(\det H)^{1/n} = \min \left\{ \frac{\operatorname{Tr} HK}{n(\det K)^{1/n}} \,\middle|\, K \in \mathbf{HPD}_n \right\}.$$

Because of the homogeneity, this is equivalent to

$$(\det H)^{1/n} = \min \left\{ \frac{1}{n} \operatorname{Tr} HK \,\middle|\, K \in \mathbf{HPD}_n \text{ and } \det K = 1 \right\}.$$

Therefore the function $H \mapsto (\det H)^{1/n}$ appears as the infimum of linear functions

$$H \mapsto \frac{1}{n} \operatorname{Tr} HK.$$

This ensures concavity. $\square$

We point out that because $H \mapsto (\det H)^{1/n}$ is homogeneous of degree one, it is linear along rays originating from 0_n. However, it is strictly concave along all other segments.

Exercises

1. For $A \in \mathbf{M}_n(\mathbb{R})$, symmetric positive-definite, show that

$$\max_{i,j \leq n} |a_{ij}| = \max_{i \leq n} a_{ii}.$$

2. Let $(a_1, \ldots, a_n)$ and $(b_1, \ldots, b_n)$ be two sequences of real numbers. Find the supremum and the infimum of $\operatorname{Tr}(AB)$ as A (respectively, B) runs over the Hermitian matrices with spectrum equal to $(a_1, \ldots, a_n)$ (respectively, $(b_1, \ldots, b_n)$).

3. (Kantorovich inequality)

   a. Let $a_1 \leq \cdots \leq a_n$ be a list of real numbers, with $a_n^{-1} = a_1 > 0$. Define

$$l(u) := \sum_{j=1}^{n} a_j u_j, \quad L(u) := \sum_{j=1}^{n} \frac{u_j}{a_j}.$$

Let K_n be the simplex of $\mathbb{R}^n$ defined by the constraints $u_j \geq 0$ for every $j = 1, \ldots, n$, and $\sum_j u_j = 1$. Show that there exists an element $v \in K_n$ that maximizes $l + L$ and minimizes $|L - l|$ on K_n simultaneously.

   b. Deduce that

$$\max_{u \in K_n} l(u)L(u) = \left(\frac{a_1 + a_n}{2} \right)^2.$$

   c. Let $A \in \mathbf{HPD}_n$ and let a_1, a_n be the smallest and largest eigenvalues of A. Show that for every $x \in C^n$,

$$(x^*Ax)(x^*A^{-1}x) \leq \frac{(a_1 + a_n)^2}{4a_1 a_n} \|x\|^4.$$

4. Let A be an Hermitian matrix of size $n \times n$ whose eigenvalues are $\alpha_1 \leq \cdots \leq \alpha_n$. Let B be an Hermitian positive-semidefinite matrix. Let $\gamma_1 \leq \cdots \leq \gamma_n$ be the eigenvalues of $A + B$. Show that $\gamma_k \geq \alpha_k$.

5. Let M, N be two Hermitian matrices such that N and $M - N$ are positive-semidefinite. Show that $\det N \leq \det M$.

6. Let $A \in \mathbf{M}_p(\mathbb{C})$, $C \in \mathbf{M}_q(\mathbb{C})$ be given with $p, q \geq 1$. Assume that

$$M := \begin{pmatrix} A & B \\ B^* & C \end{pmatrix}$$

is Hermitian positive-definite. Show that $\det M \leq (\det A)(\det C)$. Use the previous exercise and Proposition 3.9.

7. For $M \in \mathbf{HPD}_n$, we denote by $P_k(M)$ the product of all the principal minors of order k of M. There are

$$\binom{n}{k}$$

such minors.

Applying Proposition 6.5 to the matrix M^{-1}, show that $P_n(M)^{n-1} \leq P_{n-1}(M)$, and then in general that $P_{k+1}(M)^k \leq P_k(M)^{n-k}$.

8. Describe every positive-semidefinite matrix $M \in \mathbf{Sym}_n(\mathbb{R})$ such that $m_{jj} = 1$ for every j and possessing the eigenvalue $\lambda = n$ (first show that M has rank one).

9. If $A, B \in M_{n \times m}(\mathbb{C})$, define the *Hadamard product* of A and B by

$$A \circ B := (a_{ij}b_{ij})_{1 \leq i \leq n, 1 \leq j \leq m}.$$

   a. Let A, B be two Hermitian matrices. Verify that $A \circ B$ is Hermitian.

b. Assume that A and B are positive-semidefinite, of respective ranks p and q. Using Proposition 5.6, show that there exist pq vectors $z_{\alpha\beta}$ such that

$$A \circ B = \sum_{\alpha,\beta} z_{\alpha\beta} z_{\alpha\beta}^* .$$

Deduce that $A \circ B$ is positive-semidefinite.

c. If A and B are positive-definite, show that $A \circ B$ also is positive-definite.

d. Construct an example for which $p, q < n$, but $A \circ B$ is positive-definite.

10. In the previous exercise, show that the matrix $A \circ B$ is extracted from $A \otimes B$, where we choose the same row and column indices. Deduce another proof that if A and B are positive-semidefinite, then $A \circ B$ is also.

11. Recall that the Hadamard product of two matrices $A, B \in \mathbf{M}_{p \times q}(k)$ is the matrix $A \circ B \in \mathbf{M}_{p \times q}(k)$ of entries $a_{ij} b_{ij}$ with $1 \le i \le p$ and $1 \le j \le q$. If $A \in \mathbf{M}_n(k)$ is given blockwise

$$A = \begin{pmatrix} a_{11} & A_{12} \\ A_{21} & A_{22} \end{pmatrix},$$

and if a_{11} is invertible, then the Schur complement $A_{22} - A_{21} a_{11}^{-1} A_{12}$ is denoted $A|a_{11}$ and we have the formula $\det A = a_{11} \det(A|a_{11})$.

a. Let $A, B \in \mathbf{M}_n(k)$ be given blockwise as above, with $a_{11}, b_{11} \in k^*$ (and therefore $A_{22}, B_{22} \in \mathbf{M}_{n-1}(k)$.) Prove that

$$(A \circ B)|a_{11} b_{11} = A_{22} \circ (B|b_{11}) + (A|a_{11}) \circ E, \qquad E := \frac{1}{b_{11}} B_{21} B_{12}.$$

b. From now on, A and B are positive-definite Hermitian matrices. Show that

$$\det(A \circ B) \ge a_{11} b_{11} \det(A_{22} \circ (B|b_{11})).$$

Deduce Oppenheim's inequality:

$$\det(A \circ B) \ge \left(\prod_{i=1}^{n} a_{ii} \right) \det B.$$

Hint: Argue by induction over n.

c. In case of equality, prove that B is diagonal.

d. Verify that Oppenheim's inequality is valid when A and B are only positive-semidefinite.

e. Deduce that

$$\det(A \circ B) \ge \det A \det B.$$

12. (Pusz and Woronowicz). Let $A, B \in \mathbf{H}_n^+$ two given positive-semidefinite matrices. We show here that among the positive-semidefinite matrices $X \in \mathbf{H}_n^+$ such that

$$H(X) := \begin{pmatrix} A & X \\ X & B \end{pmatrix} \geq 0_{2n},$$

there exists a maximal one. The latter is called the *geometric mean* of A and B, and is denoted by $A\#B$. Then we extend properties that were well known for scalars.

a. We begin with the case where A is positive-definite.
  i. Prove that $H(X) \geq 0_{2n}$ is equivalent to $XA^{-1}X \leq B$.
  ii. Deduce that $A^{-1/2}XA^{-1/2} \leq (A^{-1/2}BA^{-1/2})^{1/2}$. **Hint:** Use Theorem 6.4.
  iii. Deduce that among the matrices $X \in \mathbf{H}_n^+$ such that $H(X) \geq 0_{2n}$, there exists a maximal one, denoted by $A\#B$. Write the explicit formula for $A\#B$.
  iv. If both A, B are positive-definite, prove that $(A\#B)^{-1} = A^{-1}\#B^{-1}$.
b. We now consider arbitrary elements A, in $\mathbf{H}_n^+$.
  i. Let $\varepsilon > 0$ be given. Show that $H(X) \geq 0_{2n}$ implies $X \leq (A + \varepsilon I_n)\#B$.
  ii. Prove that $\varepsilon \mapsto (A + \varepsilon I_n)\#B$ is nondecreasing.
  iii. Deduce that $A\#B := \lim_{\varepsilon \to 0^+} (A + \varepsilon I_n)\#B$ exists, and that it is the largest matrix in $\mathbf{H}_n^+$ among those satisfying $H(X) \geq 0_{2n}$. In particular,

$$\lim_{\varepsilon \to 0^+} (A + \varepsilon I_n)\#B = \lim_{\varepsilon \to 0^+} A\#(B + \varepsilon I_n).$$

  The matrix $A\#B$ is called the *geometric mean* of A and B.
c. Prove the following identities. **Hint:** Don't use the explicit formula. Use instead the definition of $A\#B$ by means of $H(X)$.
  • $A\#B = B\#A$.
  • If $M \in \mathbf{GL}_n(\mathbb{C})$, then $M(A\#B)M^* = (MAM^*)\#(MBM^*)$.
d. Prove the following inequality among harmonic, geometric, and arithmetic means.

$$2(A^{-1} + B^{-1}) \leq A\#B \leq \frac{1}{2}(A + B).$$

Hint: Just check that

$$H\left(2(A^{-1} + B^{-1})\right) \leq 0_{2n} \quad \text{and} \quad H\left(\frac{1}{2}(A + B)\right) \geq 0_{2n}.$$

In the latter case, use again the fact that $s \mapsto \sqrt{s}$ is operator monotone.
e. Prove that the geometric mean is "operator monotone":

$$(A_1 \leq A_2 \text{ and } B_1 \leq B_2) \implies (A_1\#B_1 \leq A_2\#B_2),$$

and that it is "operator concave", in the sense that for every $\theta \in (0,1)$, there holds

$$(\theta A_1 + (1-\theta)A_2)\#(\theta B_1 + (1-\theta)B_2) \geq \theta(A_1\#B_1) + (1-\theta)(A_2\#B_2).$$

Note that the latter property is accurate, because the geometric mean is positively homogeneous of order one. Note also that the concavity gives another proof of the arithmetico-geometric inequality, by taking $A_1 = B_2 = A$, $A_2 = B_1 = B$ and $\theta = 1/2$.

f. Prove the identity among arithmetic, harmonic, and geometric means.

$$\left(2(A^{-1} + B^{-1})^{-1}\right) \# \frac{A + B}{2} = A \# B.$$

Hint: Use the fact that $M \# N$ is the unique solution in $\mathbf{H}_n^+$ of the Ricatti equation $XM^{-1}X = N$. Use it thrice.

13. Let A, B, C be three Hermitian matrices such that $ABC \in \mathbf{H}_n$. Show that if three of the matrices A, B, C, ABC are positive-definite, then the fourth is positive-definite too.

14. Let $H \in \mathbf{H}_n$ be written blockwise

$$H = \begin{pmatrix} A & B \\ B^* & C \end{pmatrix}, \qquad A \in \mathbf{H}_p,$$

with $0 < p < n$. A matrix $Z \in \mathbf{M}_{n-p}(\mathbb{C})$ is given, such that $\Re Z > 0_n$, and we form

$$M := \begin{pmatrix} 0_p & 0 \\ 0 & Z \end{pmatrix}.$$

a. We assume that H is nonsingular. Show that $H + i\xi Z$ is nonsingular for every $\xi \in \mathbb{R}$.

b. If in addition A is nonsingular, prove that $\xi \mapsto (H + i\xi M)^{-1}$ is bounded over $\mathbb{R}$. Actually, prove that for every $x \in \mathbb{C}^n$, $\xi \mapsto (H + i\xi M)^{-1}x$ is bounded. **Hint:** for ξ large enough, $C + i\xi Z$ is nonsingular and we may eliminate the last block in x.

15. Let $J \in \mathbf{M}_{2n}(\mathbb{R})$ be the standard skew-symmetric matrix:

$$J = \begin{pmatrix} 0_n & -I_n \\ I_n & 0_n \end{pmatrix}.$$

Let $S \in \mathbf{Sym}_{2n}(\mathbb{R})$ be given. We assume that $\dim \ker S = 1$.

a. Show that 0 is an eigenvalue of JS, geometrically simple, but not algebraically.

b. We assume, moreover, that there exists a vector $x \neq 0$ in $\mathbb{R}^{2n}$ such that the quadratic form $y \mapsto y^T S y$, restricted to $\{x, Jx\}^\perp$ is positive-definite. Prove that the eigenvalues of JS are purely imaginary. **Hint:** Use Proposition 6.1.

c. In the previous question, S has a zero eigenvalue and may have a negative one. On the contrary, assume that S has one negative eigenvalue and is invertible. Show that JS has a pair of real opposite eigenvalues. **Hint:** What is the sign of $\det(JS)$?

Chapter 7
Norms

In this chapter, the field K is always $\mathbb{R}$ or $\mathbb{C}$ and E denotes K^n. The scalar (if $K = \mathbb{R}$) or Hermitian (if $K = \mathbb{C}$) product on E is denoted by $\langle x, y \rangle := \sum_j \bar{x}_j y_j$.

Definition 7.1 *If $A \in \mathbf{M}_n(K)$, the* spectral radius *of A, denoted by $\rho(A)$, is the largest modulus of the eigenvalues of A:*

$$\rho(A) = \max\{|\lambda|; \lambda \in \mathrm{Sp}(A)\}.$$

When $K = \mathbb{R}$, this takes into account the complex eigenvalues when computing $\rho(A)$.

7.1 A Brief Review

7.1.1 The ℓ^p Norms

The vector space E is endowed with various norms, pairwise equivalent because E has finite dimension (Proposition 7.3 below). Among these, the most used norms are the l^p norms:

$$\|x\|_p = \left(\sum_j |x_j|^p \right)^{1/p}, \quad \|x\|_\infty = \max_j |x_j|.$$

Proposition 7.1 *For $1 \leq p \leq \infty$, the map $x \mapsto \|x\|_p$ is a norm on E. In particular, one has* Minkowski's inequality

$$\|x + y\|_p \leq \|x\|_p + \|y\|_p. \tag{7.1}$$

Furthermore, one has Hölder's inequality

$$|\langle x,y\rangle| \le \|x\|_p \|y\|_{p'}, \quad \frac{1}{p} + \frac{1}{p'} = 1. \tag{7.2}$$

The numbers p, p' are called *conjugate exponents*.

Proof. Everything is obvious except perhaps the Hölder and Minkowski inequalities. When $p = 1$ or $p = \infty$, these inequalities are trivial. We thus assume that $1 < p < \infty$.

Let us begin with (7.2). If x or y is null, it is obvious. Indeed, one can even assume, by decreasing the value of n, that none of the x_j, y_js is null. Likewise, because $|\langle x,y\rangle| \le \sum_j |x_j||y_j|$, one can also assume that the x_j, y_j are real and positive. Dividing by $\|x\|_p$ and by $\|y\|_{p'}$, one may restrict attention to the case where $\|x\|_p = \|y\|_{p'} = 1$. Hence, $x_j, y_j \in (0, 1]$ for every j. Let us define

$$a_j = p \log x_j, \quad b_j = p' \log y_j.$$

Because the exponential function is convex,

$$e^{a_j/p + b_j/p'} \le \frac{1}{p} e^{a_j} + \frac{1}{p'} e^{b_j};$$

that is,

$$x_j y_j \le \frac{1}{p} x_j^p + \frac{1}{p'} y_j^{p'}.$$

Summing over j, we obtain

$$\langle x,y\rangle \le \frac{1}{p}\|x\|_p^p + \frac{1}{p'}\|y\|_{p'}^{p'} = \frac{1}{p} + \frac{1}{p'} = 1,$$

which proves (7.2).

We now turn to (7.1). First, we have

$$\|x+y\|_p^p = \sum_k |x_k + y_k|^p \le \sum_k |x_k||x_k + y_k|^{p-1} + \sum_k |y_k||x_k + y_k|^{p-1}.$$

Let us apply Hölder's inequality to each of the two terms of the right-hand side. For example,

$$\sum_k |x_k||x_k + y_k|^{p-1} \le \|x\|_p \left(\sum_k |x_k + y_k|^{(p-1)p'} \right)^{1/p'},$$

which amounts to

$$\sum_k |x_k||x_k + y_k|^{p-1} \le \|x\|_p \|x+y\|_p^{p-1}.$$

Finally,

$$\|x+y\|_p^p \le (\|x\|_p + \|y\|_p)\|x+y\|_p^{p-1},$$

which gives (7.1). $\square$

For $p = 2$, the norm $\|\cdot\|_2$ is given by an Hermitian form and thus satisfies the Cauchy–Schwarz inequality:

$$|\langle x, y \rangle| \leq \|x\|_2 \|y\|_2.$$

This is a particular case of Hölder's inequality.

Proposition 7.2 *For conjugate exponents p, p', one has*

$$\|x\|_p = \sup_{y \neq 0} \frac{\Re \langle x, y \rangle}{\|y\|_{p'}} = \sup_{y \neq 0} \frac{|\langle x, y \rangle|}{\|y\|_{p'}}.$$

Proof. The inequality $\geq$ is a consequence of Hölder's. The reverse inequality is obtained by taking $y_j = \bar{x}_j |x_j|^{p-2}$ if $p < \infty$. If $p = \infty$, choose $y_j = \bar{x}_j$ for an index j such that $|x_j| = \|x\|_\infty$. For $k \neq j$, take $y_k = 0$. $\square$

7.1.2 Equivalent Norms

Definition 7.2 *Two norms N and N' on a (real or complex) vector space are said to be equivalent if there exist two numbers $c, c' \in \mathbb{R}$ such that*

$$N \leq cN', \quad N' \leq c'N.$$

The equivalence between norms is obviously an equivalence relation, as its name implies. As announced above, we have the following result.

Proposition 7.3 *All norms on $E = K^n$ are equivalent. For example,*

$$\|x\|_\infty \leq \|x\|_p \leq n^{1/p} \|x\|_\infty.$$

Proof. It is sufficient to show that every norm is equivalent to $\|\cdot\|_1$.

Let N be a norm on E. If $x \in E$, the triangle inequality gives

$$N(x) \leq \sum_i |x_i| N(\mathbf{e}^i),$$

where $(\mathbf{e}^1, \ldots, \mathbf{e}^n)$ is the canonical basis. One thus has $N \leq c \|\cdot\|_1$ for $c := \max_i N(\mathbf{e}^i)$. Observe that this first inequality expresses the fact that N is Lipschitz (hence continuous) on the metric space $X = (E, \|\cdot\|_1)$.

For the reverse inequality, we reduce *ad absurdum*. Let us assume that the supremum of $\|x\|_1 / N(x)$ is infinite for $x \neq 0$. By homogeneity, there would then exist a sequence of vectors $(x^m)_{m \in \mathbb{N}}$ such that $\|x^m\|_1 = 1$ and $N(x^m) \to 0$ when $m \to +\infty$. The unit sphere of X is compact, thus one may assume (up to the extraction of a subsequence) that x^m converges to a vector x such that $\|x\|_1 = 1$. In particular, $x \neq 0$. Because N is continuous on X, one has also $N(x) = \lim_{m \to +\infty} N(x^m) = 0$ and because N is a norm, we deduce $x = 0$, a contradiction. $\square$

7.1.3 Duality

Definition 7.3 *Given a norm* $\| \cdot \|$ *on* K^n $(K = \mathbb{R}$ *or* $\mathbb{C})$, *its dual norm on* K^n *is defined by*

$$\|x\|' := \sup_{y \neq 0} \frac{\Re \langle x, y \rangle}{\|y\|},$$

or equivalently

$$\|x\|' := \sup_{y \neq 0} \frac{|\langle x, y \rangle|}{\|y\|}.$$

The fact that $\| \cdot \|'$ is a norm is obvious. For every $x, y \in K^n$, one has

$$|\langle x, y \rangle| \leq \|x\| \cdot \|y\|'. \tag{7.3}$$

Proposition 7.2 shows that the dual norm of $\| \cdot \|_p$ is $\| \cdot \|_q$ for $1/p + 1/q = 1$. Because $p \mapsto q$ is an involution, this suggests the following property:

Proposition 7.4 *The bi-dual (dual of the dual norm) of a norm is this norm itself:*

$$\left(\| \cdot \|' \right)' = \| \cdot \|.$$

Proof. From (7.3), one has $\left(\| \cdot \|' \right)' \leq \| \cdot \|$. The converse is a consequence of the Hahn–Banach theorem: the unit ball B of $\| \cdot \|$ is convex and compact. If x is a point of its boundary (i.e., $\|x\| = 1$), there exists an $\mathbb{R}$-affine (i.e., of the form constant plus $\mathbb{R}$-linear) function that vanishes at x and is nonpositive on B. Such a function can be written in the form $z \mapsto \Re \langle z, y \rangle + c$, where c is a constant, necessarily equal to $-\Re \langle x, y \rangle$. Without loss of generality, one may assume that $\langle y, x \rangle$ is real and non-negative. Hence

$$\|y\|' = \sup_{\|z\| = 1} \Re \langle y, z \rangle = \langle y, x \rangle.$$

One deduces

$$\left(\|x\|' \right)' \geq \frac{\langle y, x \rangle}{\|y\|'} = 1 = \|x\|.$$

By homogeneity, this is true for every $x \in \mathbb{C}^n$. $\square$

7.1.4 Matrix Norms

Let us recall that $\mathbf{M}_n(K)$ can be identified with the set of endomorphisms of $E = K^n$ by

$$A \mapsto (x \mapsto Ax).$$

Definition 7.4 *If* $\| \cdot \|$ *is a norm on* E *and if* $A \in \mathbf{M}_n(K)$, *we define*

$$\|A\| := \sup_{x \neq 0} \frac{\|Ax\|}{\|x\|}.$$

Equivalently,

$$\|A\| = \sup_{\|x\| \leq 1} \|Ax\| = \max_{\|x\| \leq 1} \|Ax\|.$$

One verifies easily that $A \mapsto \|A\|$ is a norm on $\mathbf{M}_n(K)$. It is called the *norm induced* by that of E, or the norm *subordinated* to that of E. Although we adopted the same notation $\| \cdot \|$ for the two norms, that on E and that on $\mathbf{M}_n(K)$, these are, of course, distinct objects. In many places, one finds the notation $\|| \cdot \||$ for the induced norm. When one does not wish to mention by which norm on E a given norm on $\mathbf{M}_n(K)$ is induced, one says that $A \mapsto \|A\|$ is a *matrix norm*. The main properties of matrix norms are

$$\|AB\| \leq \|A\| \|B\|, \quad \|I_n\| = 1.$$

These properties are those of any *algebra norm*. In particular, one has $\|A^k\| \leq \|A\|^k$ for every $k \in \mathbb{N}$.

Examples

Three l^p-matrix norms can be computed in closed form:

$$\|A\|_1 = \max_{1 \leq j \leq n} \sum_{i=1}^{i=n} |a_{ij}|,$$

$$\|A\|_\infty = \max_{1 \leq i \leq n} \sum_{j=1}^{j=n} |a_{ij}|,$$

$$\|A\|_2 = \|A^*\|_2 = \rho(A^*A)^{1/2}.$$

To prove these formulæ, we begin by proving the inequalities $\geq$, selecting a suitable vector x, and writing $\|A\|_p \geq \|Ax\|_p / \|x\|_p$. For $p = 1$ we choose an index j such that the maximum in the above formula is achieved. Then we let $x_j = 1$, and $x_k = 0$ otherwise. For $p = \infty$, we let $x_j = \bar{a}_{i_0 j}/|a_{i_0 j}|$, where i_0 achieves the maximum in the above formula. For $p = 2$ we choose an eigenvector of A^*A associated with an eigenvalue of maximal modulus. We thus obtain three inequalities. The reverse inequalities are direct consequences of the definitions. The similarity of the formulæ for $\|A\|_1$ and $\|A\|_\infty$, as well as the equality $\|A\|_2 = \|A^*\|_2$ illustrate the general formula

$$\|A^*\|_{p'} = \|A\|_p = \sup_{x \neq 0} \sup_{y \neq 0} \frac{\Re(y^*Ax)}{\|x\|_p \cdot \|y\|_{p'}} = \sup_{x \neq 0} \sup_{y \neq 0} \frac{|(y^*Ax)|}{\|x\|_p \cdot \|y\|_{p'}},$$

where again p and p' are conjugate exponents.

We point out that if H is Hermitian, then $\|H\|_2 = \rho(H^2)^{1/2} = \rho(H)$. Therefore the spectral radius is a norm over $\mathbf{H}_n$, although it is not over $\mathbf{M}_n(\mathbb{C})$. We already mentioned this fact, as a consequence of the Weyl inequalities.

Proposition 7.5 *For an induced norm, the condition $\|B\| < 1$ implies that $I_n - B$ is invertible, with the inverse given by the sum of the series*

$$\sum_{k=0}^{\infty} B^k.$$

Proof. The series $\sum_k B^k$ is normally convergent, because $\sum_k \|B^k\| \leq \sum_k \|B\|^k$, where the latter series converges because $\|B\| < 1$. Because $\mathbf{M}_n(K)$ is complete, the series $\sum_k B^k$ converges. Furthermore, $(I_n - B)\sum_{k \leq N} B^k = I_n - B^{N+1}$, which tends to I_n. The sum of the series is thus the inverse of $I_n - B$. One has, moreover,

$$\|(I_n - B)^{-1}\| \leq \sum_k \|B\|^k = \frac{1}{1 - \|B\|}.$$

$\square$

One can also deduce Proposition 7.5 from the following statement.

Proposition 7.6 *For every induced norm, one has*

$$\rho(A) \leq \|A\|.$$

Proof. The case $K = \mathbb{C}$ is easy, because there exists an eigenvector $X \in E$ associated with an eigenvalue of modulus $\rho(A)$:

$$\rho(A)\|X\| = \|\lambda X\| = \|AX\| \leq \|A\|\,\|X\|.$$

If $K = \mathbb{R}$, one needs a more involved trick.

Let us choose a norm on $\mathbb{C}^n$ and let us denote by N the induced norm on $\mathbf{M}_n(\mathbb{C})$. We still denote by N its restriction to $\mathbf{M}_n(\mathbb{R})$; it is a norm. This space has finite dimension, thus any two norms are equivalent: there exists $C > 0$ such that $N(B) \leq C\|B\|$ for every B in $\mathbf{M}_n(\mathbb{R})$. Using the result already proved in the complex case, one has for every $m \in \mathbb{N}$ that

$$\rho(A)^m = \rho(A^m) \leq N(A^m) \leq C\|A^m\| \leq C\|A\|^m.$$

Taking the mth root and letting m tend to infinity, and noticing that $C^{1/m}$ tends to 1, one obtains the announced inequality. $\square$

In general, the equality does not hold. For example, if A is nilpotent although nonzero, one has $\rho(A) = 0 < \|A\|$ for every matrix norm.

Proposition 7.7 *Let $\|\cdot\|$ be a norm on K^n and $P \in \mathbf{GL}_n(K)$. Hence, $N(x) := \|Px\|$ defines a norm on K^n. Denoting still by $\|\cdot\|$ and N the induced norms on K^n, one has $N(A) = \|PAP^{-1}\|$.*

Proof. Using the change of dummy variable $y = Px$, we have

$$N(A) = \sup_{x \neq 0} \frac{\|PAx\|}{\|Px\|} = \sup_{y \neq 0} \frac{\|PAP^{-1}y\|}{\|y\|} = \|PAP^{-1}\|.$$

$\square$

7.2 Householder's Theorem

Householder's theorem is a kind of converse of the inequality $\rho(B) \leq \|B\|$.

Theorem 7.1 *For every $B \in \mathbf{M}_n(\mathbb{C})$ and all $\varepsilon > 0$, there exists a norm on $\mathbb{C}^n$ such that for the induced norm,*

$$\|B\| \leq \rho(B) + \varepsilon.$$

In other words, $\rho(B)$ is the infimum of $\|B\|$, as $\|\cdot\|$ ranges over the set of matrix norms.

Proof. From Theorem 3.5 there exists $P \in \mathbf{GL}_n(\mathbb{C})$ such that $T := PBP^{-1}$ is upper-triangular. From Proposition 7.7, one has

$$\inf\|B\| = \inf\|PBP^{-1}\| = \inf\|T\|,$$

where the infimum is taken over the set of induced norms. Because B and T have the same spectra, hence the same spectral radius, it is enough to prove the theorem for upper-triangular matrices.

For such a matrix T, Proposition 7.7 still gives

$$\inf\|T\| \leq \inf\{\|QTQ^{-1}\|_2; Q \in \mathbf{GL}_n(\mathbb{C})\}.$$

Let us now take $Q(\mu) = \mathrm{diag}(1, \mu, \mu^2, \ldots, \mu^{n-1})$. The matrix $Q(\mu)TQ(\mu)^{-1}$ is upper-triangular, with the same diagonal as that of T. Indeed, the entry with indices (i, j) becomes $\mu^{i-j}t_{ij}$. Hence,

$$\lim_{\mu \to \infty} Q(\mu)TQ(\mu)^{-1}$$

is simply the matrix $D = \mathrm{diag}(t_{11}, \ldots, t_{nn})$. Because $\|\cdot\|_2$ is continuous (as is every norm), one deduces

$$\inf\|T\| \leq \lim_{\mu \to \infty} \|Q(\mu)TQ(\mu)^{-1}\|_2 = \|D\|_2 = \sqrt{\rho(D^*D)} = \max|t_{jj}| = \rho(T).$$

$\square$

Remark

The theorem tells us that $\rho(A) = \Lambda(A)$, where

$$\Lambda(A) := \inf \|A\|,$$

the infimum being taken over the set of matrix norms. The first part of the proof tells us that ρ and Λ coincide on the set of diagonalizable matrices, which is a dense subset of $\mathbf{M}_n(\mathbb{C})$. But this is insufficient to conclude, since Λ is a priori only upper semicontinuous, as the infimum of continuous functions. The continuity of Λ is actually a consequence of the theorem.

An interesting consequence of Householder's theorem is the following link between matrix norms and the spectral radius.

Proposition 7.8 *If $A \in \mathbf{M}_n(k)$ (with $k = \mathbb{R}$ or $\mathbb{C}$), then*

$$\rho(A) = \lim_{m \to \infty} \|A^m\|^{1/m}$$

for every matrix norm.

Proof. Let $\|\cdot\|$ be a matrix norm over $\mathbf{M}_n(k)$. From Proposition 7.6 and the fact that

$$\mathrm{Sp}(A^m) = \{\lambda^m \mid \lambda \in \mathrm{Sp}A\},$$

we have

$$\rho(A) = \rho(A^m)^{1/m} \leq \|A^m\|^{1/m}.$$

Passing to the limit, we have

$$\rho(A) \leq \liminf_{m \to +\infty} \|A^m\|^{1/m}. \tag{7.4}$$

Conversely, let $\varepsilon > 0$ be given. From Theorem 7.1, there exists a matrix norm N over $\mathbf{M}_n(\mathbb{C})$ such that $N(A) < \rho(A) + \varepsilon$. By finite-dimensionality, the restriction of N over $\mathbf{M}_n(k)$ (just in case that $k = \mathbb{R}$) is equivalent to $\|\cdot\|$: there exists a constant c such that $\|\cdot\| \leq cN$. If $m \geq 1$, we thus have

$$\|A^m\| \leq cN(A^m) \leq cN(A)^m < c(\rho(A) + \varepsilon)^m,$$

whence

$$\|A^m\|^{1/m} \leq c^{1/m}(\rho(A) + \varepsilon).$$

Passing to the limit, we have

$$\limsup_{m \to +\infty} \|A^m\|^{1/m} \leq \rho(A) + \varepsilon.$$

Letting ε tend to zero, there remains

$$\limsup_{m \to +\infty} \|A^m\|^{1/m} \leq \rho(A).$$

Comparing with (7.4), we see that the sequence $\|A^m\|^{1/m}$ is convergent and its limit equals $\rho(A)$.

□

Comments

- In the first edition, we proved Proposition 7.8 by applying the Hadamard formula for the convergence radius of the series $\sum_{j\geq 0} z^n A^n$.
- A classical lemma in calculus tells us that the limit of $\|A^m\|^{1/m}$ is also its infimum over m, because the sequence $\|A^m\|$ is submultiplicative.

7.3 An Interpolation Inequality

Theorem 7.2 (case $K = \mathbb{C}$) *Let $\|\cdot\|_p$ be the norm on $\mathbf{M}_n(\mathbb{C})$ induced by the norm l^p on $\mathbb{C}^n$. The function*

$$1/p \mapsto \log\|A\|_p,$$
$$[0,1] \to \mathbb{R},$$

is convex. In other words, if $1/r = \theta/p + (1-\theta)/q$ with $\theta \in (0,1)$, then

$$\|A\|_r \leq \|A\|_p^\theta \|A\|_q^{1-\theta}.$$

Remarks

1. The proof uses the fact that $K = \mathbb{C}$. However, the norms induced by the $\|\cdot\|_p$s on $\mathbf{M}_n(\mathbb{R})$ and $\mathbf{M}_n(\mathbb{C})$ take the same values on real matrices, even although their definitions are different (see Exercise 6). The statement is thus still true in $\mathbf{M}_n(\mathbb{R})$.
2. The case $(p,q,r) = (1,\infty,2)$ admits a direct proof. See the exercises.
3. The result still holds true in infinite dimension, at the expense of some functional analysis. One can even take different L^p norms at the source and target spaces. Here is an example.

Theorem 7.3 (Riesz–Thorin) *Let Ω be an open set in $\mathbb{R}^D$ and ω an open set in $\mathbb{R}^d$. Let p_0, p_1, q_0, q_1 be four numbers in $[1,+\infty]$. Let $\theta \in [0,1]$ and p,q be defined by*

$$\frac{1}{p} = \frac{1-\theta}{p_0} + \frac{\theta}{p_1}, \quad \frac{1}{q} = \frac{1-\theta}{q_0} + \frac{\theta}{q_1}.$$

Consider a linear operator T defined on $L^{p_0} \cap L^{p_1}(\Omega)$, taking values in $L^{q_0} \cap L^{q_1}(\omega)$. Assume that T can be extended as a continuous operator from $L^{p_j}(\Omega)$ to $L^{q_j}(\omega)$, with norm M_j, $j = 1,2$:

$$M_j := \sup_{f \neq 0} \frac{\|Tf\|_{q_j}}{\|f\|_{p_j}}.$$

Then T can be extended as a continuous operator from $L^p(\Omega)$ to $L^q(\omega)$, and its norm is bounded above by

$$M_0^{1-\theta} M_1^{\theta}.$$

4. A fundamental application is the continuity of the Fourier transform from $L^p(\mathbb{R}^d)$ into its dual $L^{p'}(\mathbb{R}^d)$ when $1 \leq p \leq 2$. We have only to observe that

$$(p_0, p_1, q_0, q_1) = (1, 2, +\infty, 2)$$

is suitable. It can be proved by inspection that every pair (p, q) such that the Fourier transform is continuous from $L^p(\mathbb{R}^d)$ into $L^q(\mathbb{R}^d)$ has the form (p, p') with $1 \leq p \leq 2$.
5. One has analogous results for the Fourier series. Therein lies the origin of the Riesz–Thorin theorem.

Proof. (Due to Riesz)
Let us fix x and y in K^n. We have to bound

$$|\langle y, Ax \rangle| = \left| \sum_{j,k} a_{jk} x_j \bar{y}_k \right|.$$

Let B be the strip in the complex plane defined by $\Re z \in [0, 1]$. Given $z \in B$, define "conjugate" exponents $r(z)$ and $r'(z)$ by

$$\frac{1}{r(z)} = \frac{z}{p} + \frac{1-z}{q}, \quad \frac{1}{r'(z)} = \frac{z}{p'} + \frac{1-z}{q'}.$$

Set

$$X_j(z) := |x_j|^{-1+r/r(z)} x_j = x_j \exp\left(\left(\frac{r}{r(z)} - 1 \right) \log |x_j| \right),$$

$$Y_j(z) := |y_j|^{-1+r'/r'(\bar{z})} y_j.$$

We then have

$$\|X(z)\|_{r(\Re z)} = \|x\|_r^{r/r(\Re z)}, \quad \|Y(z)\|_{r'(\Re z)} = \|y\|_{r'}^{r'/r'(\Re z)}.$$

Next, define a holomorphic map in the strip B by $f(z) := \langle Y(z), AX(z) \rangle$. It is bounded, because the numbers $X_j(z)$ and $Y_k(z)$ are. For example,

$$|X_j(z)| = |x_j|^{r/r(\Re z)}$$

lies between $|x_j|^{r/p}$ and $|x_j|^{r/q}$.

Let us set $M(\theta) = \sup\{|f(z)|; \Re z = \theta\}$. Hadamard's *three-line lemma* (see [33], Chapter 12, Exercise 8) tells us that, f being bounded and holomorphic in the strip,

$$\theta \mapsto \log M(\theta)$$

is convex on $(0,1)$. However, $r(0) = q$, $r(1) = p$, $r'(0) = q'$, $r'(1) = p'$, $r(\theta) = r$, $r'(\theta) = r'$, $X(\theta) = x$, and $Y(\theta) = y$. Hence

$$|\langle y, Ax \rangle| = |f(\theta)| \leq M(\theta) \leq M(1)^{\theta} M(0)^{1-\theta}.$$

Now we have

$$
\begin{aligned}
M(1) &= \sup\{|f(z)|; \Re z = 1\} \\
&\leq \sup\{\|AX(z)\|_{r(1)}\|Y(z)\|_{r(1)'}; \Re z = 1\} \\
&= \sup\{\|AX(z)\|_{p}\|Y(z)\|_{p'}; \Re z = 1\} \\
&\leq \|A\|_{p} \sup\{\|X(z)\|_{p}\|Y(z)\|_{p'}; \Re z = 1\} \\
&= \|A\|_{p}\|x\|_{r}^{r/p}\|y\|_{r'}^{r'/p'}.
\end{aligned}
$$

Likewise, $M(0) \leq \|A\|_{q}\|x\|_{r}^{r/q}\|y\|_{r'}^{r'/q'}$. Hence

$$
\begin{aligned}
|\langle y, Ax \rangle| &\leq \|A\|_{p}^{\theta}\|A\|_{q}^{1-\theta}\|x\|_{r}^{r(\theta/p+(1-\theta)/q)}\|y\|_{r'}^{r'(\theta/p'+(1-\theta)/q')} \\
&= \|A\|_{p}^{\theta}\|A\|_{q}^{1-\theta}\|x\|_{r}\|y\|_{r'}.
\end{aligned}
$$

Finally,

$$\|Ax\|_{r} = \sup_{y \neq 0} \frac{|\langle y, Ax \rangle|}{\|y\|_{r'}} \leq \|A\|_{p}^{\theta}\|A\|_{q}^{1-\theta}\|x\|_{r},$$

which proves the theorem. $\square$

7.4 Von Neumann's Inequality

We say that a matrix $M \in \mathbf{M}_n(\mathbb{C})$ is a *contraction* if $\|Mx\|_2 \leq \|x\|_2$ for every vector x, or equivalently $\|M\|_2 \leq 1$. Developing in the form $x^* M^* M x \leq x^* x$, this translates as $M^* M \leq I_n$ in the sense of Hermitian matrices. Inasmuch as $\|M^*\|_2 = \|M\|_2$, the Hermitian conjugate of a contraction is a contraction. When U and V are unitary matrices, $U^* M V$ is a contraction if and only if M is a contraction.

The following statement is due to von Neumann.

Theorem 7.4 *Let $M \in \mathbf{M}_n(\mathbb{C})$ be a contraction and $P \in \mathbb{C}[X]$ be a polynomial. Then*

$$\|P(M)\|_2 \leq \sup\{|P(z)| \,|\, z \in \mathbb{C}, |z| = 1\}.$$

Proof. We begin with the easy case, where M is normal. Then $M = U^*DU$ with $U \in \mathbf{U}_n$ and D is diagonal (Theorem 5.4). Because U is unitary, D is a contraction, meaning that its diagonal entries d_j belong to the unit disk. We have $P(M) = U^*P(D)U$, whence

$$\|P(M)\|_2 = \|P(D)\|_2 = \max_j |P(d_j)| \leq \sup\{|P(z)| \,|\, z \in \mathbb{C}, |z| = 1\},$$

using the maximum principle.

We now turn to the general case of a nonnormal contraction. Using the square root of nonnegative Hermitian matrices, and thanks to $M^*M \leq I_n$ and $MM^* \leq I_n$, we denote

$$S = \sqrt{I_n - M^*M}, \qquad T := \sqrt{I_n - MM^*}.$$

Let us choose a real polynomial $Q \in \mathbb{R}[X]$ with the interpolation property that $Q(t) = \sqrt{1-t}$ over the spectrum of M^*M (which is a finite set and equals the spectrum of MM^*). Then $S = Q(M^*M)$ and $T = Q(MM^*)$. Because $M(M^*M)^r = (MM^*)^rM$ for every $r \in \mathbb{N}$, we infer $MS = TM$. Likewise, we have $SM^* = M^*T$.

Given an integer $k \geq 1$ and $\ell = 2k+1$, we define a matrix $V_k \in \mathbf{M}_{\ell n}(\mathbb{C})$ blockwise:

$$V_k = \begin{pmatrix} \ddots & & & & & & \\ & I_n & & & & & \\ & & I_n & & & & \\ & & & S & -M^* & & \\ & & & M & T & & \\ & & & & & I_n & \\ & & & & & & I_n \\ & & & & & & & \ddots \\ I_n & & & & & & \end{pmatrix},$$

where the dots represent blocks I_n. The column and row indices range from $-k$ to k. The central block indexed by $(0,0)$ is M. All the missing (apart from dots) entries are blocks 0_n. In particular, the diagonal blocks are equal to 0_n, except for the central one.

Computing the product $V_k^*V_k$, and using $SM^* = M^*T$ and $MS = TM$, we find $I_{\ell n}$. In other words, V_k is a unitary matrix. This is a special case of normal contraction, therefore the first step above thus applies:

$$\|P(V_k)\|_2 \leq \sup\{|P(z)| \,|\, z \in \mathbb{C}, |z| = 1\}. \tag{7.5}$$

Let us now observe that in the qth power of V_k, the central block is M^k provided that $q \leq 2k$. This because V_k is block-triangular[1] up to the lower-left block I_n. Therefore the central block of $p(V_k)$ for a polynomial p with $d^o p \leq 2k$ is precisely $p(M)$.

---

[1] Compare with Exercise 12 of Chapter 5, which states that normal matrices are far from triangular.

We now choose $k \geq (d^o P)/2$, so that $P(M)$ is a block of $P(V_k)$. We have $P(M) = \Pi P(V_k)\Pi*$ where

$$\Pi = (\ldots, 0_n, I_n, 0_n, \ldots).$$

If $x \in \mathbb{C}^n$, let $X \in \mathbb{C}^{\ell n}$ denote the vector $\Pi^* x$. We have $\|X\|_2 = \|x\|_2$. Because Π is an orthogonal projection, we find

$$\|P(M)x\|_2 = \|\Pi P(V_k)X\|_2 \leq \|P(V_k)X\|_2 \leq \sup\{|P(z)|\,|\,z \in \mathbb{C}, |z| = 1\}\|X\|_2$$
$$= \sup\{|P(z)|\,|\,z \in \mathbb{C}, |z| = 1\}\|x\|_2,$$

where we have used (7.5). This is exactly

$$\|P(M)\|_2 \leq \sup\{|P(z)|\,|\,z \in \mathbb{C}, |z| = 1\}.$$

□

Exercises

1. Under what conditions on the vectors $a, b \in \mathbb{C}^n$ does the matrix M defined by $m_{ij} = a_i b_j$ satisfy $\|M\|_p = 1$ for every $p \in [1, \infty]$?
2. Under what conditions on x, y, and p does the equality in (7.2) or (7.1) hold?
3. Show that
$$\lim_{p \to +\infty} \|x\|_p = \|x\|_\infty, \quad \forall x \in E.$$

4. A norm on K^n is a *strictly convex* norm if $\|x\| = \|y\| = 1$, $x \neq y$, and $0 < \theta < 1$ imply $\|\theta x + (1-\theta)y\| < 1$.

   a. Show that $\|\cdot\|_p$ is strictly convex for $1 < p < \infty$, but is not so for $p = 1, \infty$.
   b. Deduce from Corollary 8.1 that the induced norm $\|\cdot\|_p$ is not strictly convex on $\mathbf{M}_n(\mathbb{R})$.

5. Let N be a norm on $\mathbb{R}^n$.

   a. For $x \in \mathbb{C}^n$, define

   $$N_1(x) := \inf\left\{\sum_\ell |\alpha_\ell| N(x^\ell)\right\},$$

   where the infimum is taken over the set of decompositions $x = \sum_\ell \alpha_\ell x^\ell$ with $\alpha_\ell \in \mathbb{C}$ and $x^\ell \in \mathbb{R}^n$. Show that N_1 is a norm on $\mathbb{C}^n$ (as a $\mathbb{C}$-vector space) whose restriction to $\mathbb{R}^n$ is N. **Note:** N_1 is called the *complexification* of N.
   b. Same question as above for N_2, defined by

   $$N_2(x) := \frac{1}{2\pi}\int_0^{2\pi} [e^{i\theta}x]d\theta,$$

where

$$[x] := \sqrt{N(\Re x)^2 + N(\Im x)^2}.$$

c. Show that $N_2 \leq N_1$.
d. If $N(x) = \|x\|_1$, show that $N_1(x) = \|x\|_1$. Considering then the vector

$$x = \begin{pmatrix} 1 \\ i \end{pmatrix},$$

show that $N_2 \neq N_1$.

6. (Continuation of Exercise 5)
   The norms N (on $\mathbb{R}^n$) and N_1 (on $\mathbb{C}^n$) lead to induced norms on $\mathbf{M}_n(\mathbb{R})$ and $\mathbf{M}_n(\mathbb{C})$, respectively. Show that if $M \in \mathbf{M}_n(\mathbb{R})$, then $N(M) = N_1(M)$. Deduce that Theorem 7.2 holds true in $\mathbf{M}_n(\mathbb{R})$.

7. Let $\| \cdot \|$ be an algebra norm on $\mathbf{M}_n(K)$ ($K = \mathbb{R}$ or $\mathbb{C}$), that is, a norm satisfying $\|AB\| \leq \|A\| \cdot \|B\|$. Show that $\rho(A) \leq \|A\|$ for every $A \in \mathbf{M}_n(K)$.

8. Let $B \in \mathbf{M}_n(\mathbb{C})$ be given. Assume that there exists an induced norm such that $\|B\| = \rho(B)$. Let λ be an eigenvalue of maximal modulus and X a corresponding eigenvector. Show that X does not belong to the range of $B - \lambda I_n$. Deduce that the Jordan block associated with λ is diagonal (Jordan reduction is presented in Chapter 9).

9. (Continuation of Exercise 8)
   Conversely, show that if the Jordan blocks of B associated with the eigenvalues of maximal modulus of B are diagonal, then there exists a norm on $\mathbb{C}^n$ such that, using the induced norm, $\rho(B) = \|B\|$.

10. Here is another proof of Theorem 7.1. Let $K = \mathbb{R}$ or $\mathbb{C}$, $A \in \mathbf{M}_n(K)$, and let N be a norm on K^n. If $\varepsilon > 0$, we define for all $x \in K^n$

$$\|x\| := \sum_{k \in \mathbb{N}} (\rho(A) + \varepsilon)^{-k} N(A^k x).$$

a. Show that this series is convergent (use Proposition 7.8).
b. Show that $\| \cdot \|$ is a norm on K^n.
c. Show that for the induced norm, $\|A\| \leq \rho(A) + \varepsilon$.

11. A norm $\| \cdot \|$ on $\mathbf{M}_n(\mathbb{C})$ is said to be *unitarily invariant* if $\|UAV\| = \|A\|$ for every $A \in \mathbf{M}_n(\mathbb{C})$ and all unitary matrices U, V.

a. Find, among the most classical norms, two examples of unitarily invariant norms.
b. Given a unitarily invariant norm, show that there exists a norm N on $\mathbb{R}^n$ such that

$$\|A\| = N(s_1(A), \dots, s_n(A)),$$

where the $s_j(A)$s, the eigenvalues of H in the polar decomposition $A = QH$ (see Section 11.4 for this notion), are called the *singular values* of A.

12. (Bhatia [5]) Suppose we are given a norm $\| \cdot \|$ on $\mathbf{M}_n(\mathbb{C})$ that is unitarily invariant (see the previous exercise). If $A \in \mathbf{M}_n(\mathbb{C})$, we denote by $D(A)$ the diagonal matrix obtained by keeping only the a_{jj} and setting all the other entries to zero. If σ is a permutation, we denote by A^σ the matrix whose entry of index (j,k) equals a_{jk} if $k = \sigma(j)$, and zero otherwise. For example, $A^{id} = D(A)$, where id is the identity permutation. If r is an integer between $1-n$ and $n-1$, we denote by $D_r(A)$ the matrix whose entry of index (j,k) equals a_{jk} if $k-j=r$, and zero otherwise. For example, $D_0(A) = D(A)$.

a. Let $\omega = \exp(2i\pi/n)$ and let U be the diagonal matrix whose diagonal entries are the roots of unity $1, \omega, \ldots, \omega^{n-1}$. Show that

$$D(A) = \frac{1}{n} \sum_{j=0}^{n-1} U^{*j} A U^j.$$

Deduce that $\|D(A)\| \leq \|A\|$.

b. Show that $\|A^\sigma\| \leq \|A\|$ for every $\sigma \in S_n$. Observe that $\|P\| = \|I_n\|$ for every permutation matrix P. Show that $\|M\| \leq \|I_n\|$ for every bistochastic matrix M (see Section 8.5 for this notion).

c. If $\theta \in \mathbb{R}$, let us denote by U_θ the diagonal matrix, whose kth diagonal term equals $\exp(ik\theta)$. Show that

$$D_r(A) = \frac{1}{2\pi} \int_0^{2\pi} e^{ir\theta} U_\theta A U_\theta^* d\theta.$$

d. Deduce that $\|D_r(A)\| \leq \|A\|$.

e. Let p be an integer between zero and $n-1$, and set $r = 2p+1$. Let us denote by $T_r(A)$ the matrix whose entry of index (j,k) equals a_{jk} if $|k-j| \leq p$, and zero otherwise. For example, $T_3(A)$ is a tridiagonal matrix. Show that

$$T_r(A) = \frac{1}{2\pi} \int_0^{2\pi} d_p(\theta) U_\theta A U_\theta^* d\theta,$$

where

$$d_p(\theta) = \sum_{-p}^{p} e^{ik\theta}$$

is the *Dirichlet kernel*.

f. Deduce that $\|T_r(A)\| \leq L_p \|A\|$, where

$$L_p = \frac{1}{2\pi} \int_0^{2\pi} |d_p(\theta)| d\theta$$

is the *Lebesgue constant* (**Note:** $L_p = 4\pi^{-2} \log p + O(1)$).

g. Let $\Delta(A)$ be the upper-triangular matrix whose entries above the diagonal coincide with those of A. Using the matrix

$$B = \begin{pmatrix} 0_n & \Delta(A)^* \\ \Delta(A) & 0_n \end{pmatrix},$$

show that $\|\Delta(A)\|_2 \le L_n \|A\|_2$ (observe that $\|B\|_2 = \|\Delta(A)\|_2$).
  h. What inequality do we obtain for $\Delta_0(A)$, the strictly upper-triangular matrix whose entries lying strictly above the diagonal coincide with those of A?

13. We endow $\mathbb{C}^n$ with the usual Hermitian structure, so that $\mathbf{M}_n(\mathbb{C})$ is equipped with the norm $\|A\|_2 = \rho(A^*A)^{1/2}$.
  Suppose we are given a sequence of matrices $(A_j)_{j \in \mathbb{Z}}$ in $\mathbf{M}_n(\mathbb{C})$ and a summable sequence $\gamma \in l^1(\mathbb{Z})$ of positive real numbers. Assume, finally, that for every pair $(j,k) \in \mathbb{Z} \times \mathbb{Z}$,

$$\|A_j^* A_k\|_2 \le \gamma(j-k)^2, \quad \|A_j A_k^*\|_2 \le \gamma(j-k)^2.$$

  a. Let F be a finite subset of $\mathbb{Z}$. Let B_F denote the sum of the A_js as j runs over F. Show that

$$\|(B_F^* B_F)^{2m}\|_2 \le \text{card } F \, \|\gamma\|_1^{2m}, \quad \forall m \in \mathbb{N}.$$

  b. Deduce that $\|B_F\|_2 \le \|\gamma\|_1$.
  c. Show (*Cotlar's lemma*) that for every $x,y \in \mathbb{C}^n$, the series

$$y^T \sum_{j \in \mathbb{Z}} A_j x$$

  is convergent, and that its sum $y^T A x$ defines a matrix $A \in \mathbf{M}_n(\mathbb{C})$ that satisfies

$$\|A\| \le \sum_{j \in \mathbb{Z}} \gamma(j).$$

  Hint: For a sequence $(u_j)_{j \in \mathbb{Z}}$ of real numbers, the series $\sum_j u_j$ is absolutely convergent if and only if there exists $M < +\infty$ such that $\sum_{j \in F} |u_j| \le M$ for every finite subset F.
  d. Deduce that the series $\sum_j A_j$ converges in $\mathbf{M}_n(\mathbb{C})$. May one conclude that it converges normally?

14. Let $\|\cdot\|$ be an induced norm on $\mathbf{M}_n(\mathbb{R})$. We wish to characterize the matrices $B \in \mathbf{M}_n(\mathbb{R})$ such that there exist $\varepsilon_0 > 0$ and $\omega > 0$ with

$$(0 < \varepsilon < \varepsilon_0) \Longrightarrow (\|I_n - \varepsilon B\| \le 1 - \omega \varepsilon).$$

  a. For the norm $\|\cdot\|_\infty$, it is equivalent that B be strictly diagonally dominant.
  b. What is the characterization for the norm $\|\cdot\|_1$?
  c. For the norm $\|\cdot\|_2$, it is equivalent that $B^T + B$ be positive-definite.

15. Let $B \in \mathbf{M}_n(\mathbb{C})$ be given.

   a. Returning to the proof of Theorem 7.1, show that for every $\varepsilon > 0$ there exists on $\mathbb{C}^n$ an Hermitian norm $\|\cdot\|$ such that for the induced norm $\|B\| \leq \rho(B) + \varepsilon$.

   b. Deduce that $\rho(B) < 1$ holds if and only if there exists a matrix $A \in \mathbf{HPD}_n$ such that $A - B^*AB \in \mathbf{HPD}_n$.

16. Let $A \in \mathbf{M}_n(\mathbb{C})$ be a diagonalizable matrix: $A = S\,\mathrm{diag}(d_1,\ldots,d_n)S^{-1}$. Let $\|\cdot\|$ be an induced norm for which $\|D\| = \max_j |d_j|$ holds, where

$$D := \mathrm{diag}(d_1,\ldots,d_n).$$

Show that for every $E \in \mathbf{M}_n(\mathbb{C})$ and for every eigenvalue λ of $A + E$, there exists an index j such that

$$|\lambda - d_j| \leq \|S\| \cdot \|S^{-1}\| \cdot \|E\|.$$

17. Let $A \in \mathbf{M}_n(K)$, with $K = \mathbb{R}$ or $\mathbb{C}$. Give another proof, using the Cauchy–Schwarz inequality, of the following particular case of Theorem 7.2:

$$\|A\|_2 \leq \|A\|_1^{1/2}\|A\|_\infty^{1/2}.$$

18. Show that if $A \in \mathbf{M}_n(\mathbb{C})$ is normal, then $\rho(A) = \|A\|_2$. Deduce that if A and B are normal, $\rho(AB) \leq \rho(A)\rho(B)$.

19. Let N_1 and N_2 be two norms on $\mathbb{C}^n$. Denote by $\mathscr{N}_1$ and $\mathscr{N}_2$ the induced norms on $\mathbf{M}_n(\mathbb{C})$. Let us define

$$R := \max_{x\neq 0} \frac{N_1(x)}{N_2(x)}, \quad S := \max_{x\neq 0} \frac{N_2(x)}{N_1(x)}.$$

   a. Show that

$$\max_{A\neq 0} \frac{\mathscr{N}_1(A)}{\mathscr{N}_2(A)} = RS = \max_{A\neq 0} \frac{\mathscr{N}_2(A)}{\mathscr{N}_1(A)}.$$

   b. Deduce that if $\mathscr{N}_1 = \mathscr{N}_2$, then N_2/N_1 is constant.

   c. Show that if $\mathscr{N}_1 \leq \mathscr{N}_2$, then N_2/N_1 is constant and therefore $\mathscr{N}_2 = \mathscr{N}_1$.

20. (Continuation of Exercise 19)
Let $\|\cdot\|$ be an algebra norm on $\mathbf{M}_n(\mathbb{C})$. If $y \in \mathbb{C}^n$ is nonzero, we define $\|x\|_y := \|xy^*\|$.

   a. Show that $\|\cdot\|_y$ is a norm on $\mathbb{C}^n$ for every $y \neq 0$.

   b. Let $\mathscr{N}_y$ be the norm induced by $\|\cdot\|_y$. Show that $\mathscr{N}_y \leq \|\cdot\|$.

   c. We say that $\|\cdot\|$ is *minimal* if there exists no other algebra norm less than or equal to $\|\cdot\|$. Show that the following assertions are equivalent.

     i. $\|\cdot\|$ is an induced norm on $\mathbf{M}_n(\mathbb{C})$.

     ii. $\|\cdot\|$ is a minimal norm on $\mathbf{M}_n(\mathbb{C})$.

     iii. For all $y \neq 0$, one has $\|\cdot\| = \mathscr{N}_y$.

21. (Continuation of Exercise 20)

Let $\|\cdot\|$ be an induced norm on $\mathbf{M}_n(\mathbb{C})$.

    a. Let $y, z \neq 0$ be two vectors in $\mathbb{C}^n$. Show that (with the notation of the previous exercise) $\|\cdot\|_y / \|\cdot\|_z$ is constant.

    b. Prove the equality

$$\|xy^*\| \cdot \|zt^*\| = \|xt^*\| \cdot \|zy^*\|.$$

22. Let $M \in \mathbf{M}_n(\mathbb{C})$ and $H \in \mathbf{HPD}_n$ be given. Show that

$$\|HMH\|_2 \leq \frac{1}{2}\|H^2 M + M H^2\|_2.$$

23. We endow $\mathbb{R}^2$ with the Euclidean norm $\|\cdot\|_2$, and $\mathbf{M}_2(\mathbb{R})$ with the induced norm, also denoted by $\|\cdot\|_2$. We denote by Σ the unit sphere of $\mathbf{M}_2(\mathbb{R})$: $M \in \Sigma$ is equivalent to $\|M\|_2 = 1$, that is, to $\rho(M^T M) = 1$. Likewise, B denotes the unit ball of $\mathbf{M}_2(\mathbb{R})$.

Recall that if C is a convex set and if $P \in C$, then P is called an *extremal* point if $P \in [Q, R]$ and $Q, R \in C$ imply either $Q = P$ or $R = P$.

    a. Show that the set of extremal points of B is equal to $\mathbf{O}_2(\mathbb{R})$.

    b. Show that $M \in \Sigma$ if and only if there exist two matrices $P, Q \in \mathbf{O}_2(\mathbb{R})$ and a number $a \in [0, 1]$ such that

$$M = P \begin{pmatrix} a & 0 \\ 0 & 1 \end{pmatrix} Q.$$

    c. We denote by $\mathscr{R} = \mathbf{SO}_2(\mathbb{R})$ the set of rotation matrices, and by $\mathscr{S}$ that of matrices of planar symmetry. Recall that $\mathbf{O}_2(\mathbb{R})$ is the disjoint union of $\mathscr{R}$ and $\mathscr{S}$. Show that Σ is the union of the segments $[r, s]$ as r runs over $\mathscr{R}$ and s runs over $\mathscr{S}$.

    d. Show that two such "open" segments (r, s) and (r', s') are either disjoint or equal.

    e. Let $M, N \in \Sigma$. Show that $\|M - N\|_2 = 2$ (i.e., (M, N) is a diameter of B) if and only if there exists a segment $[r, s]$ ($r \in \mathscr{R}$ and $s \in \mathscr{S}$) such that $M \in [r, s]$ and $N \in [-r, -s]$.

24. (The Banach–Mazur distance)

    a. Let N and N' be two norms on k^n ($k = \mathbb{R}$ or $\mathbb{C}$). If $A \in \mathbf{GL}_n(k)$, we may define norms

$$\|A\|_{\rightarrow} := \sup_{x \neq 0} \frac{N'(Ax)}{N(x)}, \qquad \|A^{-1}\|_{\leftarrow} := \sup_{x \neq 0} \frac{N(A^{-1}x)}{N'(x)}.$$

Show that $A \mapsto \|A\|_{\rightarrow}\|A^{-1}\|_{\leftarrow}$ achieves its upper bound. We denote by $\delta(N, N')$ the minimum value. Verify

$$0 \leq \log \delta(N,N'') \leq \log \delta(N,N') + \log \delta(N',N'').$$

When $N = \|\cdot\|_p$, we write ℓ^p instead. If in addition $N' = \|\cdot\|_q$, we write $\|\cdot\|_{p,q}$ for $\|\cdot\|_\rightarrow$.

b. In the set $\mathcal{N}$ of norms on k^n, let us consider the following equivalence relation: $N \sim N'$ if and only if there exists an $A \in \mathbf{GL}_n(k)$ such that $N' = N \circ A$. Show that $\log \delta$ induces a metric d on the quotient set $\mathbf{Norm} := \mathcal{N}/\sim$. This metric is called the *Banach–Mazur distance*. How many classes of Hermitian norms are there ?

c. Compute $\|I_n\|_{p,q}$ for $1 \leq p,q \leq n$ (there are two cases, depending on the sign of $q - p$). Deduce that

$$\delta(\ell^p, \ell^q) \leq n^\kappa, \qquad \kappa := \left| \frac{1}{p} - \frac{1}{q} \right|.$$

d. Show that $\delta(\ell^p, \ell^q) = \delta(\ell^{p'}, \ell^{q'})$, where p', q' are the conjugate exponents.

e. i. When $H \in \mathbf{H}_n$ is positive-semidefinite, find that the average of $x^* H x$, as x runs over the set defined by $|x_j| = 1$ for all js, is $\operatorname{Tr} H$ (the measure is the product of n copies of the normalized Lebesgue measure on the unit disk). Deduce that

$$\sqrt{\operatorname{Tr} M^* M} \leq \|M\|_{\infty,2} := \sup_{x \neq 0} \frac{\|Mx\|_2}{\|x\|_\infty}$$

for every $M \in \mathbf{M}_n(k)$.

  ii. Prove also that

$$\|A\|_{p,\infty} = \max_{1 \leq i \leq n} \|A^{(i)}\|_{p'},$$

where $A^{(i)}$ denotes the ith row vector of A.

  iii. Deduce that $\delta(\ell^2, \ell^\infty) = \sqrt{n}$.

  iv. Using the triangle inequality for $\log \delta$, deduce that

$$\delta(\ell^p, \ell^q) = n^\kappa$$

whenever $p,q \geq 2$, and then for every p,q such that $(p-2)(q-2) \geq 0$. **Note:** The exact value of $\delta(\ell^p, \ell^q)$ is not known when $(p-2)(q-2) < 0$.

  v. Remark that the "curves" $\{\ell^p \mid 2 \leq p \leq \infty\}$ and $\{\ell^p \mid 1 \leq p \leq 2\}$ are geodesics, in the sense that the restrictions of the Banach–Mazur distance to these curves satisfy the triangular *equality*.

f. When $n = 2$, prove that $\delta(\ell^1, \ell^\infty) = 1$. On the contrary, if $n \geq 3$, then prove $\delta(\ell^1, \ell^\infty) > 1$.

g. A theorem proven by John states that the diameter of $(\mathbf{Norm}, d)$ is precisely $\frac{1}{2} \log n$. Show that this metric space is compact. **Note:** One may consider the norm whose unit ball is an m-agon in $\mathbb{R}^2$, with m even. Denote its class by N_m. It seems that $d(\ell^1, N_m) = \frac{1}{2} \log 2$ when $8 | m$.

25. Given three matrices $A \in \mathbf{M}_{p \times q}(k)$, $B \in \mathbf{M}_{p \times s}(k)$, and $C \in \mathbf{M}_{r \times q}(k)$, we consider the affine set $\mathscr{W}$ of matrices $W \in \mathbf{M}_{n \times m}(k)$ of the form

$$W = \begin{pmatrix} A & B \\ C & D \end{pmatrix},$$

where D runs over $\mathbf{M}_{r \times s}(k)$. Thus $n = p + r$ and $m = q + s$. Denoting

$$P = \begin{pmatrix} I \\ 0 \end{pmatrix}, \quad Q = (I\ 0)$$

the projection matrices, we are going to prove (Parrott's lemma) that

$$\min\{\|W\|_2 \,|\, W \in \mathscr{W}\} = \max\{\|QW\|_2, \|WP\|_2\}, \qquad (7.6)$$

where the right-hand side does not depend on D:

$$WP = \begin{pmatrix} A \\ C \end{pmatrix}, \quad QW = (A\ B)$$

a. Check the inequality

$$\inf\{\|W\|_2 \,|\, W \in \mathscr{W}\} \geq \max\{\|QW\|_2, \|WP\|_2\}.$$

b. Denote $\mu(D) := \|W\|_2$. Show that the infimum of μ on $\mathscr{W}$ is attained.
c. Show that it is sufficient to prove (7.6) when $s = 1$.
d. From now on, we assume that $s = 1$, and we consider a matrix $D_0 \in \mathbf{M}_{r \times 1}(k)$ such that μ is minimal at D_0. We denote by W_0 the associated matrix. Let us introduce a function $D \mapsto \eta(D) = \mu(D)^2$. Recall that η is the largest eigenvalue of $W^* W$. We denote f_0 its multiplicity when $D = D_0$.
   i. If $f_0 \geq 2$, show that $W_0^* W_0$ has an eigenvector v with $v_m = 0$. Deduce that $\mu(D_0) \leq \|WP\|_2$. Conclude in this case.
   ii. From now on, we suppose $f_0 = 1$. Show that $\eta(D)$ is a simple eigenvalue for every D in a small neighbourhood of D_0. Show that $D \mapsto \eta(D)$ is differentiable at D_0, and that its differential is given by

$$\Delta \mapsto \frac{2}{\|y\|_2^2} \Re\left[(QW_0 y)^* \Delta Q y\right],$$

   where y is an associated eigenvector:

$$W_0^* W_0 y = \eta(D_0) y.$$

   iii. Deduce that either $Qy = 0$ or $QW_0 y = 0$.
   iv. In the case where $Qy = 0$, show that $\mu(D_0) \leq \|WP\|_2$ and conclude.
   v. In the case where $QW_0 y = 0$, prove that $\mu(D_0) \leq \|QW\|_2$ and conclude.

26. Let k be $\mathbb{R}$ or $\mathbb{C}$. Given a bounded subset F of $\mathbf{M}_n(k)$, let us denote by F_k the set of all possible products of k elements in F. Given a matrix norm $\|\cdot\|$, we denote $\|F_k\|$ the supremum of the norms of elements of F_k.

    a. Show that $\|F_{k+l}\| \leq \|F_k\| \cdot \|F_l\|$.

    b. Deduce that the sequence $\|F_k\|^{1/k}$ converges, and that its limit is the infimum of the sequence.

    c. Prove that this limit does not depend on the choice of the matrix norm. This limit is called the *joint spectral radius* of the family F, and denoted $\rho(F)$. This notion is due to Rota and Strang.

    d. Let $\hat{\rho}(F)$ denote the infimum of $\|F\|$ when $\|\cdot\|$ runs over all matrix norms. Show that $\rho(F) \leq \hat{\rho}(F)$.

    e. Given a norm N on k^n and a number $\varepsilon > 0$, we define for every $x \in k^n$

$$\|x\| := \sum_{l=0}^{\infty} (\rho(F) + \varepsilon)^{-l} \max\{N(Bx) \mid B \in F_\ell\}.$$

      i. Show that the series converges, and that it defines a norm on k^n.

      ii. For the matrix norm associated with $\|\cdot\|$, show that $\|A\| \leq \rho(F) + \varepsilon$ for every $A \in F$.

      iii. Deduce that actually $\rho(F) = \hat{\rho}(F)$. Compare with Householder's theorem.

27. (Rota & Strang.) Let k be $\mathbb{R}$ or $\mathbb{C}$. Given a subset F of $\mathbf{M}_n(k)$, we consider the semi-group $\mathscr{F}$ generated by F. It is the union of sets F_k defined in the previous exercise, as k runs over $\mathbb{N}$. We have $F_0 = \{I_n\}$, $F_1 = F$, $F_2 = F \cdot F,\ldots$
If $\mathscr{F}$ is bounded, prove that there exists a matrix norm $\|\cdot\|$ such that $\|A\| \leq 1$ for every $A \in F$. **Hint:** In the previous exercise, take a sup instead of a series.

28. Let $A \in \mathbf{M}_n(\mathbb{C})$ be given. Let $\sigma(A)$ be the spectrum of A and $\rho(A)$ its complement (the *resolvent set*). For $\varepsilon > 0$, we define the ε-pseudospectrum of A as

$$\sigma_\varepsilon(A) := \sigma(A) \cup \left\{ z \in \rho(A);\, \|(z-A)^{-1}\|_2 \geq \frac{1}{\varepsilon} \right\}.$$

    a. Prove that

$$\sigma_\varepsilon(A) = \bigcup_{\|B\|_2 \leq \varepsilon} \sigma(A+B).$$

    b. Prove also that

$$\sigma_\varepsilon(A) \subset \{z \in \mathbb{C};\, \operatorname{dist}(z; \mathscr{H}(A)) \leq \varepsilon\},$$

where $\mathscr{H}(A)$ is the numerical range of A.

Note: the notion of pseudo-spectrum is fundamental in several scientific domains, including dynamical systems, numerical analysis, and quantum mechanics (in semiclassical analysis, one speaks of *quasi-modes*). The reader interested in this subject should consult the book by Trefethen and Embree [38].

29. We recall that the numerical radius of $A \in \mathbf{M}_n(\mathbb{C})$ is defined by

$$w(A) := \sup\{|z| \, ; z \in \mathscr{H}(A)\} = \sup\{|x^*Ax| \, ; x \in \mathbb{C}^n, \|x\|_2 = 1\}.$$

Prove that

$$w(A) \leq \|A\|_2 \leq 2w(A).$$

Hint: Use the polarization principle to prove the second inequality.

30. Let $A \in \mathbf{M}_n(\mathbb{C})$ be a nilpotent matrix of order two: $A^2 = 0_n$.

   a. Using standard properties of the norm $\|\cdot\|_2$, verify that $\|M\|_2^2 \leq \|MM^* + M^*M\|_2$ for every $M \in \mathbf{M}_n(\mathbb{C})$.

   b. When k is a positive integer, compute $(AA^* + A^*A)^k$ in closed form. Deduce that

$$\|AA^* + A^*A\|_2 \leq 2^{1/k}\|A\|_2^2.$$

   c. Passing to the limit as $k \to +\infty$, prove that

$$\|A\|_2 = \|AA^* + A^*A\|_2^{1/2}. \tag{7.7}$$

Chapter 8
Nonnegative Matrices

In this chapter matrices have real entries in general. In a few specified cases, entries might be complex.

8.1 Nonnegative Vectors and Matrices

Definition 8.1 *A vector $x \in \mathbb{R}^n$ is* nonnegative, *and we write $x \geq 0$, if its coordinates are nonnegative. It is* positive, *and we write $x > 0$, if its coordinates are (strictly) positive. Furthermore, a matrix $A \in M_{n \times m}(\mathbb{R})$ (not necessarily square) is* nonnegative *(respectively,* positive*) if its entries are nonnegative (respectively, positive); we again write $A \geq 0$ (respectively, $A > 0$). More generally, we define an order relation $x \leq y$ whose meaning is $y - x \geq 0$.*

Definition 8.2 *Given $x \in \mathbb{C}^n$, we let $|x|$ denote the nonnegative vector whose coordinates are the numbers $|x_j|$. Likewise, if $A \in \mathbf{M}_n(\mathbb{C})$, the matrix $|A|$ has entries $|a_{ij}|$.*

Observe that given a matrix and a vector (or two matrices), the triangle inequality implies

$$|Ax| \leq |A| \cdot |x|.$$

Proposition 8.1 *A matrix is nonnegative if and only if $x \geq 0$ implies $Ax \geq 0$. It is positive if and only if $x \geq 0$ and $x \neq 0$ imply $Ax > 0$.*

Proof. Let us assume that $Ax \geq 0$ (respectively, > 0) for every $x \geq 0$ (respectively, ≥ 0 and $\neq 0$). Then the ith column $A^{(i)}$ is nonnegative (respectively, positive), since it is the image of the ith vector of the canonical basis. Hence $A \geq 0$ (respectively, > 0).

Conversely, $A \geq 0$ and $x \geq 0$ imply trivially $Ax \geq 0$. If $A > 0$, $x \geq 0$, and $x \neq 0$, there exists an index ℓ such that $x_\ell > 0$. Then

$$(Ax)_i = \sum_j a_{ij} x_j \geq a_{i\ell} x_\ell > 0,$$

and hence $Ax > 0$. $\square$

An important point is the following:

Proposition 8.2 *If $A \in \mathbf{M}_n(\mathbb{R})$ is nonnegative and irreducible, then $(I+A)^{n-1} > 0$.*

Proof. Let $x \neq 0$ be nonnegative, and define $x^m = (I+A)^m x$, which is nonnegative too. Let us denote by P_m the set of indices of the nonzero components of x^m: P_0 is nonempty. Because $x_i^{m+1} \geq x_i^m$, one has $P_m \subset P_{m+1}$. Let us assume that the cardinality $|P_m|$ of P_m is strictly less than n. There are thus one or more zero components, whose indices form a nonempty subset I, complement of P_m. Because A is irreducible, there exists some nonzero entry a_{ij}, with $i \in I$ and $j \in P_m$. Then $x_i^{m+1} \geq a_{ij} x_j^m > 0$, which shows that P_{m+1} is not equal to P_m, and thus $|P_{m+1}| > |P_m|$. By induction, we deduce that $|P_m| \geq \min\{m+1, n\}$. Hence $|P_{n-1}| = n$, meaning that $x^{n-1} > 0$. We conclude with Proposition 8.1. $\square$

8.2 The Perron–Frobenius Theorem: Weak Form

The following result is not very impressive. We prove much more in the next section, with elementary calculus. It has, however, its own interest, as an elegant consequence of Brouwer's fixed point theorem.

Theorem 8.1 *Let $A \in \mathbf{M}_n(\mathbb{R})$ be a nonnegative matrix. Then $\rho(A)$ is an eigenvalue of A associated with a nonnegative eigenvector.*

Proof. Let λ be an eigenvalue of maximal modulus and v an eigenvector, normalized by $\|v\|_1 = 1$. Then

$$\rho(A)|v| = |\lambda v| = |Av| \leq A|v|.$$

Let us denote by C the subset of $\mathbb{R}^n$ (actually a subset of the unit simplex K_n) defined by the (in)equalities $\sum_i x_i = 1$, $x \geq 0$, and $Ax \geq \rho(A)x$. This is a closed convex set, nonempty, inasmuch as it contains $|v|$. Finally, it is bounded, because $x \in C$ implies $0 \leq x_j \leq 1$ for every j; thus it is compact. Let us distinguish two cases.

1. There exists $x \in C$ such that $Ax = 0$. Then $\rho(A)x \leq 0$ furnishes $\rho(A) = 0$. The theorem is thus proved in this case.
2. For every x in C, $Ax \neq 0$. Then let us define on C a continuous map f by

$$f(x) = \frac{1}{\|Ax\|_1} Ax.$$

It is clear that $f(x) \geq 0$ and that $\|f(x)\|_1 = 1$. Finally,

$$Af(x) = \frac{1}{\|Ax\|_1} AAx \geq \frac{1}{\|Ax\|_1} A\rho(A)x = \rho(A)f(x),$$

so that $f(C) \subset C$. Then Brouwer's theorem (see [3], p. 217) asserts that a continuous function from a compact convex subset of $\mathbb{R}^N$ into itself has a fixed point. Thus let y be a fixed point of f. It is a nonnegative eigenvector, associated with the eigenvalue $r = \|Ay\|_1$. Because $y \in C$, we have $ry = Ay \geq \rho(A)y$ and thus $r \geq \rho(A)$, which implies $r = \rho(A)$.

$\Box$

That proof can be adapted to the case where a real number r and a nonzero vector y are given satisfying $y \geq 0$ and $Ay \geq ry$. Just take for C the set of vectors x such that $\sum_i x_i = 1$, $x \geq 0$, and $Ax \geq rx$. We then conclude that $\rho(A) \geq r$.

8.3 The Perron–Frobenius Theorem: Strong Form

Theorem 8.2 *Let $A \in \mathbf{M}_n(\mathbb{R})$ be a nonnegative irreducible matrix. Then $\rho(A)$ is a simple eigenvalue of A, associated with a positive eigenvector. Moreover, $\rho(A) > 0$.*

8.3.1 Remarks

1. Alhough the Perron–Frobenius theorem says that $\rho(A)$ is a simple eigenvalue, it does not tell us anything about the other eigenvalues of maximal modulus. The following example shows that such other eigenvalues may exist:

$$\begin{pmatrix} 0 & 1 \\ 1 & 0 \end{pmatrix}.$$

   The existence of several eigenvalues of maximal modulus is studied in Section 8.4.

2. One obtains another proof of the weak form of the Perron–Frobenius theorem by applying the strong form to $A + \alpha J$, where $J > 0$ and $\alpha > 0$, letting α tend to zero and using Theorem 5.2.

3. Without the irreducibility assumption, $\rho(A)$ may be a multiple eigenvalue, and a nonnegative eigenvector may not be positive. This holds for a matrix of size $n = 2m$ that reads blockwise

$$A = \begin{pmatrix} B & 0_m \\ I_m & B \end{pmatrix}.$$

Here, $\rho(A) = \rho(B)$, and every eigenvalue has an even algebraic multiplicity, because $P_A = (P_B)^2$.

Let us assume that B is nonnegative and irreducible. Then $\rho(B)$ is a simple eigenvalue of B, associated with the eigenvector $r > 0$. In addition, B^T is irreducible (Proposition 3.25) and thus has a positive eigenvector ℓ associated with $\rho(B)$.

Let

$$X = \begin{pmatrix} y \\ z \end{pmatrix}$$

belong to the kernel of $A - \rho(A)I_n$. We have

$$By = \rho(B)y, \qquad y + Bz = \rho(B)z.$$

The first equality tells us that $y = \alpha r$ for some $\alpha \in \mathbb{R}$. Multiplying the second equality by ℓ^T, we obtain $\ell^T y = 0$; that is, $\alpha \ell^T r = 0$. Because $\ell > 0$ and $r > 0$, this gives $\alpha = 0$; that is, $y = 0$. Then $Bz = \rho(B)z$, meaning that $z \parallel r$. Finally, the eigenspace is spanned by

$$X = \begin{pmatrix} 0_m \\ r \end{pmatrix},$$

which is nonnegative, but not positive.
4. As a matter of fact, not only the eigenvector associated with $\rho(A)$ is positive, but it is the only one to be positive. For let ℓ be the positive eigenvector of A^T associated with $\rho(A)$. Then every eigenvector x of A associated with an eigenvalue $\lambda \neq \rho(A)$ satisfies $\ell^T x = 0$, which prevents x from being positive.

Proof. For $r \geq 0$, we denote by C_r the set of vectors of $\mathbb{R}^n$ defined by the conditions

$$x \geq 0, \quad \|x\|_1 = 1, \quad Ax \geq rx.$$

Each C_r is a convex compact set. We saw in the previous section that if λ is an eigenvalue associated with an eigenvector x of unit norm $\|x\|_1 = 1$, then $|x| \in C_{|\lambda|}$. In particular, $C_{\rho(A)}$ is nonempty. Conversely, if C_r is nonempty, then for $x \in C_r$,

$$r = r\|x\|_1 \leq \|Ax\|_1 \leq \|A\|_1 \|x\|_1 = \|A\|_1,$$

and therefore $r \leq \|A\|_1$. Furthermore, the map $r \mapsto C_r$ is nonincreasing with respect to inclusion, and is "left continuous" in the following sense. If $r > 0$, one has

$$C_r = \bigcap_{s < r} C_s.$$

Let us then define

$$R = \sup\{r \,|\, C_r \neq \emptyset\},$$

so that $R \in [\rho(A), \|A\|_1]$. The monotonicity with respect to inclusion shows that $r < R$ implies $C_r \neq \emptyset$.

If $x > 0$ and $\|x\|_1 = 1$, then $Ax > 0$ because A is nonnegative and irreducible. Setting $r := \min_j (Ax)_j / x_j > 0$, we have $C_r \neq \emptyset$, whence $R \geq r > 0$. The set C_R, being the intersection of a totally ordered family of nonempty compacts sets, is nonempty.

Let $x \in C_R$ be given. Lemma 10 below shows that x is an eigenvector of A associated with the eigenvalue R. We observe that this eigenvalue is not less than $\rho(A)$ and infer that $\rho(A) = R$. Hence $\rho(A)$ is an eigenvalue associated with the eigenvector x. Lemma 11 below ensures that $x > 0$ and $\rho(A) > 0$.

The proof of the simplicity of the eigenvalue $\rho(A)$ is given in Section 8.3.3.

8.3.2 A Few Lemmas

Lemma 10. *Let $r \geq 0$ and $x \geq 0$ such that $Ax \geq rx$ and $Ax \neq rx$. Then there exists $r' > r$ such that $C_{r'}$ is nonempty.*

Proof. Set $y := (I_n + A)^{n-1}x$. Because A is irreducible and $x \geq 0$ is nonzero, one has $y > 0$. Likewise, $Ay - ry = (I_n + A)^{n-1}(Ax - rx) > 0$. Let us define $r' := \min_j(Ay)_j/y_j$, which is strictly larger than r. We then have $Ay \geq r'y$, so that $C_{r'}$ contains the vector $y/\|y\|_1$. $\square$

Lemma 11. *The nonnegative eigenvectors of A are positive. The corresponding eigenvalue is positive too.*

Proof. Given such a vector x with $Ax = \lambda x$, we observe that $\lambda \in \mathbb{R}^+$. Then

$$x = \frac{1}{(1+\lambda)^{n-1}}(I_n + A)^{n-1}x,$$

and the right-hand side is strictly positive, from Proposition 8.2.

Inasmuch as A is irreducible and nonnegative, we infer $Ax \neq 0$. Thus $\lambda \neq 0$; that is, $\lambda > 0$. $\square$

Finally, we can state the following result.

Lemma 12. *Let $M, B \in \mathbf{M}_n(\mathbb{C})$ be matrices, with M irreducible and $|B| \leq M$. Then $\rho(B) \leq \rho(M)$.*
In the case of equality ($\rho(B) = \rho(M)$), the following hold.

- $|B| = M$.
- *For every eigenvector x of B associated with an eigenvalue of modulus $\rho(M)$, $|x|$ is an eigenvector of M associated with $\rho(M)$.*

Proof. In order to establish the inequality, we proceed as above. If λ is an eigenvalue of B, of modulus $\rho(B)$, and if x is a normalized eigenvector, then $\rho(B)|x| \leq |B| \cdot |x| \leq M|x|$, so that $C_{\rho(B)}$ is nonempty. Hence $\rho(B) \leq R = \rho(M)$.

Let us investigate the case of equality. If $\rho(B) = \rho(M)$, then $|x| \in C_{\rho(M)}$, and therefore $|x|$ is an eigenvector: $M|x| = \rho(M)|x| = \rho(B)|x| \leq |B| \cdot |x|$. Hence, $(M - |B|)|x| \leq 0$. Because $|x| > 0$ (from Lemma 11) and $M - |B| \geq 0$, this gives $|B| = M$. $\square$

8.3.3 The Eigenvalue $\rho(\mathbf{A})$ Is Simple

Let $P_A(X)$ be the characteristic polynomial of A. It is given as the composition of an n-linear form (the determinant) with polynomial vector-valued functions (the columns of $XI_n - A$). If ϕ is p-linear and if $V_1(X), \ldots, V_p(X)$ are polynomial vector-valued functions, then the derivative of the polynomial $P(X) := \phi(V_1(X), \ldots, V_p(X))$ is given by

$$P'(X) = \phi(V_1', V_2, \ldots, V_p) + \phi(V_1, V_2', \ldots, V_p) + \cdots + \phi(V_1, \ldots, V_{p-1}, V_p').$$

One therefore has

$$P_A'(X) = \det(\mathbf{e}^1, a_2, \ldots, a_n) + \det(a_1, \mathbf{e}^2, \ldots, a_n) + \cdots + \det(a_1, \ldots, a_{n-1}, \mathbf{e}^n),$$

where a_j is the jth column of $XI_n - A$ and $\{\mathbf{e}^1, \ldots, \mathbf{e}^n\}$ is the canonical basis of $\mathbb{R}^n$. Developing the jth determinant with respect to the jth column, one obtains

$$P_A'(X) = \sum_{j=1}^{n} P_{A_j}(X), \qquad (8.1)$$

where $A_j \in \mathbf{M}_{n-1}(\mathbb{R})$ is obtained from A by deleting the jth row and the jth column. Let us now denote by $B_j \in \mathbf{M}_n(\mathbb{R})$ the matrix obtained from A by replacing the entries of the jth row and column by zeroes. This matrix is block-diagonal, the two diagonal blocks being $A_j \in \mathbf{M}_{n-1}(\mathbb{R})$ and $0 \in \mathbf{M}_1(\mathbb{R})$. Hence, the eigenvalues of B_j are those of A_j, together with zero, and therefore $\rho(B_j) = \rho(A_j)$. Furthermore, $|B_j| \leq A$, but $|B_j| \neq A$ because A is irreducible and B_j is block-diagonal, hence reducible. It follows (Lemma 12) that $\rho(B_j) < \rho(A)$. Hence P_{A_j} does not vanish over $[\rho(A), +\infty)$. Because $P_{A_j}(t) \approx t^{n-1}$ at infinity, we deduce that $P_{A_j}(\rho(A)) > 0$. Finally, $P_A'(\rho(A))$ is positive and $\rho(A)$ is a simple root. $\square$

8.4 Cyclic Matrices

The following statement completes Theorem 8.2.

Theorem 8.3 *Under the assumptions of Theorem 8.2, let p be the cardinality of the set $\mathrm{Sp}_{\max}(A)$ of eigenvalues of A of maximal modulus $\rho(A)$.*

Then we have $\mathrm{Sp}_{\max}(A) = \rho(A)\mathscr{U}_p$, where $\mathscr{U}_p$ is the group of pth roots of unity. Every such eigenvalue is simple. The spectrum of A is invariant under multiplication by $\mathscr{U}_p$. Finally, A is conjugated via a permutation matrix to the following cyclic form. In this cyclic matrix each element is a block, and the diagonal blocks (which all vanish) are square with nonzero sizes:

$$\begin{pmatrix} 0 & M_1 & 0 & \cdots & 0 \\ \vdots & \ddots & \ddots & \ddots & \vdots \\ \vdots & & \ddots & \ddots & 0 \\ 0 & & & \ddots & M_{p-1} \\ M_p & 0 & \cdots & \cdots & 0 \end{pmatrix}.$$

Remark

The converse is true. The characteristic polynomial of a cyclic matrix is

$$X \mapsto \det(X^p I_m - M_1 M_2 \cdots M_p),$$

up to a factor X^ν (with ν possibly negative). Its spectrum is thus stable under multiplication by $\exp(2i\pi/p)$.

Proof. Let us denote by X the unique nonnegative eigenvector of A normalized by $\|X\|_1 = 1$. If Y is a unitary eigenvector, associated with an eigenvalue μ of maximal modulus $\rho(A)$, the inequality $\rho(A)|Y| = |AY| \le A|Y|$ implies (Lemma 12) $|Y| = X$. Hence there is a diagonal matrix $D = \mathrm{diag}(e^{i\alpha_1}, \ldots, e^{i\alpha_n})$ such that $Y = DX$. Let us define a unimodular complex number $e^{i\gamma} = \mu/\rho(A)$ and set $B := e^{-i\gamma} D^{-1} A D$. One has $|B| = A$ and $BX = X$. For every j, one therefore has

$$\left| \sum_{k=1}^n b_{jk} x_k \right| = \sum_{k=1}^n |b_{jk}| x_k.$$

Because $X > 0$, one deduces that B is real-valued and nonnegative. Therefore $B = A$; that is, $D^{-1} A D = e^{i\gamma} A$. The spectrum of A is thus invariant under multiplication by $e^{i\gamma}$.

Let $\mathcal{U} = \rho(A)^{-1} \mathrm{Sp}_{max}(A)$, which is included in S^1, the unit circle. The previous discussion shows that $\mathcal{U}$ is stable under multiplication. Because $\mathcal{U}$ is finite, it follows that its elements are roots of unity. The inverse of a dth root of unity is its own $(d-1)$th power, therefore $\mathcal{U}$ is stable under inversion. Hence it is a finite subgroup of S^1. With p its cardinal, we have $\mathcal{U} = \mathcal{U}_p$.

Let P_A be the characteristic polynomial and let $\omega = \exp(2i\pi/p)$. One may apply the first part of the proof to $\mu = \omega \rho(A)$. One has thus $D^{-1} A D = \omega A$, and it follows that $P_A(X) = \omega^n P_A(X/\omega)$. Therefore, multiplication by ω sends eigenvalues to eigenvalues of the same multiplicities. In particular, the eigenvalues of maximal modulus are simple.

Iterating the conjugation, one obtains $D^{-p} A D^p = A$. Let us set

$$D^p = \mathrm{diag}(d_1, \ldots, d_n).$$

One has thus $d_j = d_k$, provided that $a_{jk} \ne 0$. Because A is irreducible, one can link any two indices j and k by a chain $j_0 = j, \ldots, j_r = k$ such that $a_{j_{s-1}, j_s} \ne 0$ for every

s. It follows that $d_j = d_k$ for every j,k. But because one may choose $Y_1 = X_1$, that is $\alpha_1 = 0$, one also has $d_1 = 1$ and hence $D^p = I_n$. The α_j are thus pth roots of unity. Applying a conjugation by a permutation matrix we may limit ourselves to the case where D has the block-diagonal form $\mathrm{diag}(J_0, \omega J_1, \ldots, \omega^{p-1} J_{p-1})$, where the J_ℓ are identity matrices of respective sizes $n_0, \ldots, n_{p-1}$. Decomposing A into blocks A_{lm} of sizes $n_\ell \times n_m$, one obtains $\omega^k A_{jk} = \omega^{j+1} A_{jk}$ directly from the conjugation identity. Hence $A_{jk} = 0$, except for the pairs (j,k) of the form $(0,1),(1,2),\ldots,(p-2,p-1),(p-1,0)$. This is the announced cyclic form. $\square$

8.5 Stochastic Matrices

Definition 8.3 *A matrix $M \in \mathbf{M}_n(\mathbb{R})$ is said to be* stochastic *if $M \geq 0$ and if for every $i = 1, \ldots, n$, one has*

$$\sum_{j=1}^n m_{ij} = 1.$$

One says that M is bistochastic *(or* doubly stochastic*) if both M and M^T are stochastic.*

Denoting by $\mathbf{e} \in \mathbb{R}^n$ the vector all of whose coordinates equal one, one sees that M is stochastic if and only if $M \geq 0$ and $M\mathbf{e} = \mathbf{e}$. Likewise, M is bistochastic if $M \geq 0$, $M\mathbf{e} = \mathbf{e}$, and $\mathbf{e}^T M = \mathbf{e}^T$. If M is stochastic, one has $\|Mx\|_\infty \leq \|x\|_\infty$ for every $x \in \mathbb{C}^n$, and therefore $\rho(M) \leq 1$. But because $M\mathbf{e} = \mathbf{e}$, one has in fact $\rho(M) = 1$.

The stochastic matrices play an important role in the study of Markov chains. A special instance of a bistochastic matrix is a permutation matrix $P(\sigma)$ ($\sigma \in S_n$), whose entries are

$$p_{ij} = \delta^j_{\sigma(i)}.$$

The following theorem enlightens the role of permutation matrices.

Theorem 8.4 (Birkhoff) *A matrix $M \in \mathbf{M}_n(\mathbb{R})$ is bistochastic if and only if it is a center of mass (i.e., a barycenter with nonnegative weights) of permutation matrices.*

The fact that a center of mass of permutation matrices is a doubly stochastic matrix is obvious, because the set $\mathbf{DS}_n$ of doubly stochastic matrices is convex. The interest of the theorem lies in the statement that if $M \in \mathbf{DS}_n$, there exist permutation matrices $P_1, \ldots, P_r$ and positive real numbers $\alpha_1, \ldots, \alpha_r$ with $\alpha_1 + \cdots + \alpha_r = 1$ such that $M = \alpha_1 P_1 + \cdots + \alpha_r P_r$.

Let us recall that a point x of a convex set C is an *extremal point* if $x \in [y,z] \subset C$ implies $x = y = z$. The permutation matrices are extremal points of $\mathbf{M}_n([0,1]) \sim [0,1]^{n^2}$, thus they are extremal points of the smaller convex set $\mathbf{DS}_n$.

The Krein–Milman theorem (see [34], Theorem 3.23) says that a convex compact subset of $\mathbb{R}^n$ is the convex hull, that is, the set of centers of mass of its extremal points. Because $\mathbf{DS}_n$ is closed and bounded, hence compact, we may apply this statement. Theorem 8.4 thus amounts to saying that the extremal points of Δ_n are precisely the permutation matrices.

Proof. Let $M \in \mathbf{DS}_n$ be given. If M is not a permutation matrix, there exists an entry $m_{i_1 j_1} \in (0, 1)$. Inasmuch as M is stochastic, there also exists $j_2 \neq j_1$ such that $m_{i_1 j_2} \in (0, 1)$. Because M^T is stochastic, there exists $i_2 \neq i_1$ such that $m_{i_2 j_2} \in (0, 1)$. By this procedure one constructs a sequence $(j_1, i_1, j_2, i_2, \ldots)$ such that $m_{i_\ell j_\ell} \in (0, 1)$ and $m_{i_{\ell-1} j_\ell} \in (0, 1)$. The set of indices is finite, therefore it eventually happens that one of the indices (a row index or a column index) is repeated.

Therefore, one can assume that the sequence $(j_s, i_s, \ldots, j_r, i_r, j_{r+1} = j_s)$ has the above property, and that $j_s, \ldots, j_r$ are pairwise distinct, as well as $i_s \ldots, i_r$. Let us define a matrix $B \in \mathbf{M}_n(\mathbb{R})$ by $b_{i_\ell j_\ell} = 1$, $b_{i_\ell j_{\ell+1}} = -1$, $b_{ij} = 0$ otherwise. By construction, $Be = 0$ and $\mathbf{e}^T B = 0$. If $\alpha \in \mathbb{R}$, one therefore has $(M \pm \alpha B)\mathbf{e} = \mathbf{e}$ and $\mathbf{e}^T (M \pm \alpha B) = \mathbf{e}^T$. If $\alpha > 0$ is small enough, $M \pm \alpha B$ turns out to be nonnegative. Finally, $M + \alpha B$ and $M - \alpha B$ are bistochastic, and

$$M = \frac{1}{2}(M - \alpha B) + \frac{1}{2}(M + \alpha B).$$

Hence M is not an extremal point of $\mathbf{DS}_n$. □

Here is a nontrivial consequence (Stoer and Witzgall [36]):

Corollary 8.1 *Let $\| \cdot \|$ be a norm on $\mathbb{R}^n$, invariant under permutation of the coordinates. Then $\|M\| = 1$ for every bistochastic matrix (where as usual we have denoted $\| \cdot \|$ the induced norm on $\mathbf{M}_n(\mathbb{R})$).*

Proof. To begin with, $\|P\| = 1$ for every permutation matrix, by assumption. Because the induced norm is convex (true for every norm), one deduces from Birkhoff's theorem that $\|M\| \leq 1$ for every bistochastic matrix. Furthermore, $M\mathbf{e} = \mathbf{e}$ implies $\|M\| \geq \|M\mathbf{e}\|/\|\mathbf{e}\| = 1$. □

This result applies, for instance, to the norm $\| \cdot \|_p$, providing a nontrivial convex set on which the map $1/p \mapsto \log \|M\|_p$ is constant (compare with Theorem 7.2).

The bistochastic matrices are intimately related to the relation $\prec$ (see Section 6.5). In fact, we have the following theorem.

Theorem 8.5 *A matrix A is bistochastic if and only if $Ax \succ x$ for every $x \in \mathbb{R}^n$.*

Proof. If A is bistochastic, then $\|Ax\|_1 \leq \|A\|_1 \|x\|_1 = \|x\|_1$, because A^T is stochastic. Because A is stochastic, $A\mathbf{e} = \mathbf{e}$. Applying the inequality to $x - t\mathbf{e}$, one therefore has $\|Ax - t\mathbf{e}\|_1 \leq \|x - t\mathbf{e}\|_1$. Proposition 6.4 then shows that $x \prec Ax$.

Conversely, let us assume that $x \prec Ax$ for every $x \in \mathbb{R}^n$. Choosing x as the jth vector of the canonical basis, $\mathbf{e}^j$, the inequality $s_1(\mathbf{e}^j) \leq s_1(A\mathbf{e}^j)$ expresses that A is a nonnegative matrix, and $s_n(\mathbf{e}^j) = s_n(A\mathbf{e}^j)$ yields

$$\sum_{i=1}^n a_{ij} = 1. \tag{8.2}$$

One then chooses $x = \mathbf{e}$. The inequality $s_1(\mathbf{e}) \leq s_1(A\mathbf{e})$ expresses[1] that $A\mathbf{e} \geq \mathbf{e}$. Finally, $s_n(\mathbf{e}) = s_n(A\mathbf{e})$ and $A\mathbf{e} \geq \mathbf{e}$ give $A\mathbf{e} = \mathbf{e}$. Hence, A is bistochastic. □

This statement is completed by the following.

Theorem 8.6 *Let* $x, y \in \mathbb{R}^n$. *Then* $x \prec y$ *if and only if there exists a bistochastic matrix* A *such that* $y = Ax$.

Proof. From the previous theorem, it is enough to show that if $x \prec y$, there exists A, a bistochastic matrix, such that $y = Ax$. To do so, one applies Theorem 6.8: there exists an Hermitian matrix H whose diagonal and spectrum are y and x, respectively. Let us diagonalize H by a unitary conjugation: $H = U^*DU$, with $D = \mathrm{diag}(x_1,\ldots,x_n)$. Then $y = Ax$, where $a_{ij} = |u_{ij}|^2$. Because U is unitary, A is bistochastic.[2] □

Exercises

1. We consider the following three properties for a matrix $M \in \mathbf{M}_n(\mathbb{R})$.

   P1 M is nonnegative.
   P2 $M^T e = e$, where $e = (1,\ldots,1)^T$.
   P3 $\|M\|_1 \leq 1$.

   a. Show that **P2** and **P3** imply **P1**.
   b. Show that **P2** and **P1** imply **P3**.
   c. Do **P1** and **P3** imply **P2** ?

2. Here is another proof of the simplicity of $\rho(A)$ in the Perron–Frobenius theorem, which does not require Lemma 12. We assume that A is irreducible and nonnegative, and we denote by x a positive eigenvector associated with the eigenvalue $\rho(A)$.

   a. Let K be the set of nonnegative eigenvectors y associated with $\rho(A)$ such that $\|y\|_1 = 1$. Show that K is compact and convex.
   b. Show that the geometric multiplicity of $\rho(A)$ equals 1. (**Hint:** Otherwise, K would contain a vector with at least one zero component.)
   c. Show that the algebraic multiplicity of $\rho(A)$ equals 1. (**Hint:** Otherwise, there would be a nonnegative vector y such that $Ay - \rho(A)y = x > 0$.)

3. Let $M \in \mathbf{M}_n(\mathbb{R})$ be either strictly diagonally dominant or irreducible and strongly diagonally dominant. Assume that $m_{jj} > 0$ for every $j = 1,\ldots,n$ and $m_{ij} \leq 0$ otherwise. Show that M is invertible and that the solution of $Mx = b$, when $b \geq 0$, satisfies $x \geq 0$. Deduce that $M^{-1} \geq 0$.

---

[1] For another vector y, $s_1(y) \leq s_1(Ay)$ does not imply $Ay \geq y$.
[2] This kind of bistochastic matrix is called *orthostochastic*.

4. Here is another proof of Theorem 8.2, due to Perron himself. We proceed by induction on the size n of the matrix. The statement is obvious if $n = 1$. We therefore assume that it holds for matrices of size n. We give ourselves an irreducible nonnegative matrix $A \in M_{n+1}(\mathbb{R})$, which we decompose blockwise as

$$A = \begin{pmatrix} a & \xi^T \\ \eta & B \end{pmatrix}, \quad a \in \mathbb{R}, \quad \xi, \eta \in \mathbb{R}^n, \quad B \in M_n(\mathbb{R}).$$

   a. Applying the induction hypothesis to the matrix $B + \varepsilon J$, where $\varepsilon > 0$ and $J > 0$ is a matrix, then letting ε go to zero, shows that $\rho(B)$ is an eigenvalue of B, associated with a nonnegative eigenvector (this avoids the use of Theorem 8.1).
   b. Using the formula

$$(\lambda I_n - B)^{-1} = \sum_{k=1}^{\infty} \lambda^{-k} B^{k-1},$$

   valid for $\lambda \in (\rho(B), +\infty)$, deduce that the function $h(\lambda) := \lambda - a - \xi^T(\lambda I_n - B)^{-1}\eta$ is strictly increasing on this interval, and that on the same interval the vector $x(\lambda) := (\lambda I_n - B)^{-1}\eta$ is positive.
   c. Using the Schur complement formula, prove $P_A(\lambda) = P_B(\lambda)h(\lambda)$.
   d. Deduce that the matrix A has one and only one eigenvalue in $(\rho(B), +\infty)$, and that it is a simple one, associated with a positive eigenvector. One denotes this eigenvalue by λ_0.
   e. Applying the previous results to A^T, show that there exists $\ell \in \mathbb{R}^n$ such that $\ell > 0$ and $\ell^T(A - \lambda_0 I_n) = 0$.
   f. Let μ be an eigenvalue of A, associated with an eigenvector X. Show that $(\lambda_0 - |\mu|)\ell^T|X| \geq 0$. Conclusion?

5. Let $A \in M_n(\mathbb{R})$ be a matrix satisfying $a_{ij} \geq 0$ for every pair (i, j) of distinct indices.

   a. Let us define
$$\sigma := \sup\{\Re\lambda; \lambda \in \text{Sp}\, A\}.$$

   Among the eigenvalues of A whose real parts equal σ, let us denote by μ the one with the largest imaginary part. Show that for every positive large enough real number τ, $\rho(A + \tau I_n) = |\mu + \tau|$.
   b. Deduce that $\mu = \sigma = \rho(A)$ (apply Theorem 8.1).

6. Let $B \in M_n(\mathbb{R})$ be a matrix whose off-diagonal entries are positive and such that the eigenvalues have strictly negative real parts. Show that there exists a nonnegative diagonal matrix D such that $B' := D^{-1}BD$ is strictly diagonally dominant, namely,
$$b'_{ii} < - \sum_{j \neq i} b'_{ij}.$$

7.   a. Let $B \in \mathbf{M}_n(\mathbb{R})$ be given, with $\rho(B) = 1$. Assume that the eigenvalues of B of modulus one are (algebraically) simple. Show that the sequence $(B^m)_{m \geq 1}$ is bounded.

   b. Let $M \in \mathbf{M}_n(\mathbb{R})$ be a nonnegative irreducible matrix, with $\rho(M) = 1$. We denote by x and y^T the left and right eigenvectors for the eigenvalue 1 ($Mx = x$ and $y^T M = y^T$), normalized by $y^T x = 1$. We define $L := xy^T$ and $B = M - L$.

      i. Verify that $B - I_n$ is invertible. Determine the spectrum and the invariant subspaces of B by means of those of M.

      ii. Show that the sequence $(B^m)_{m \geq 1}$ is bounded. Express M^m in terms of B^m.

      iii. Deduce that

$$\lim_{N \to +\infty} \frac{1}{N} \sum_{m=0}^{N-1} M^m = L.$$

      iv. Under what additional assumption do we have the stronger convergence

$$\lim_{N \to +\infty} M^N = L?$$

8. Let $B \in \mathbf{M}_n(\mathbb{R})$ be a nonnegative irreducible matrix and let $C \in \mathbf{M}_n(\mathbb{R})$ be a nonzero nonnegative matrix. For $t > 0$, we define $r_t := \rho(B + tC)$ and we let X_t denote the nonnegative unitary eigenvector associated with the eigenvalue r_t.

   a. Show that $t \mapsto r_t$ is strictly increasing.
   Define $r := \lim_{t \to +\infty} r_t$. We wish to show that $r = +\infty$. Let X be a cluster point of the sequence X_t. We may assume, up to a permutation of the indices, that

$$X = \begin{pmatrix} Y \\ 0 \end{pmatrix}, \quad Y > 0.$$

   b. Suppose that in fact, $r < +\infty$. Show that $BX \leq rX$. Deduce that $B'Y = 0$, where B' is a matrix extracted from B.
   c. Deduce that $X = Y$; that is, $X > 0$.
   d. Show, finally, that $CX = 0$. Conclude that $r = +\infty$.
   e. Assume, moreover, that $\rho(B) < 1$. Show that there exists one and only one $t \in \mathbb{R}$ such that $\rho(B + tC) = 1$.

9. Verify that $\mathbf{DS}_n$ is stable under multiplication. In particular, if M is bistochastic, the sequence $(M^m)_{m \geq 1}$ is bounded.

10. Let $M \in \mathbf{M}_n(\mathbb{R})$ be a bistochastic irreducible matrix. Show that

$$\lim_{N \to +\infty} \frac{1}{N} \sum_{m=0}^{N-1} M^m = \frac{1}{n} \begin{pmatrix} 1 & \cdots & 1 \\ \vdots & & \vdots \\ 1 & \cdots & 1 \end{pmatrix} =: J_n$$

(use Exercise 7). Show by an example that the sequence $(M^m)_{m \geq 1}$ may or may not converge.

11. Show directly that for every $p \in [1, \infty]$, $\|J_n\|_p = 1$, where J_n was defined in the previous exercise.

12. Let $P \in \mathbf{GL}_n(\mathbb{R})$ be given such that $P, P^{-1} \in \mathbf{DS}_n$. Show that P is a permutation matrix.

13. Let $A, B \in \mathbf{H}_n$ be given and $C := A + B$.

   a. If $t \in [0, 1]$ consider the matrix $S(t) := A + tB$, so that $S(0) = A$ and $S(1) = C$. Arrange the eigenvalues of $S(t)$ in increasing order $\lambda_1(t) \leq \cdots \leq \lambda_n(t)$. For each value of t there exists an orthonormal eigenbasis $\{X_1(t), \ldots, X_n(t)\}$. We admit the fact that it can be chosen continuously with respect to t, so that $t \mapsto X_j(t)$ is continuous with a piecewise continuous derivative (see [24], Chapter 2, Section 6.) Show that $\lambda'_j(t) = (BX_j(t), X_j(t))$.

   b. Let $\alpha_j, \beta_j, \gamma_j$ ($j = 1, \ldots, n$) be the eigenvalues of A, B, C, respectively. Deduce from part (a) that

$$\gamma_j - \alpha_j = \int_0^1 (BX_j(t), X_j(t)) \, dt.$$

   c. Let $\{Y_1, \ldots, Y_n\}$ be an orthonormal eigenbasis for B. Define

$$\sigma_{jk} := \int_0^1 |(X_j(t), Y_k)|^2 dt.$$

   Show that the matrix $\Sigma := (\sigma_{jk})_{1 \leq j, k \leq n}$ is bistochastic.

   d. Show that $\gamma_j - \alpha_j = \sum_k \sigma_{jk} \beta_k$. Deduce (Lidskiĭ's theorem) that the vector $(\gamma_1 - \alpha_1, \ldots, \gamma_n - \alpha_n)$ belongs to the convex hull of the vectors obtained from the vector $(\beta_1, \ldots, \beta_n)$ by all possible permutations of the coordinates.

14. Let $a \in \mathbb{R}^n$ be given, $a = (a_1, \ldots, a_n)$.

   a. Show that $C(a) := \{b \in \mathbb{R}^n \mid b \succ a\}$ is a convex compact set. Characterize its extremal points.

   b. Show that $Y(a) := \{M \in \mathbf{Sym}_n(\mathbb{R}) \mid \mathrm{Sp}\, M \succ a\}$ is a convex compact set. Characterize its extremal points.

   c. Deduce that $Y(a)$ is the closed convex hull (actually the convex hull) of the set $X(a) := \{M \in \mathbf{Sym}_n(\mathbb{R}) \mid \mathrm{Sp}\, M = a\}$.

   d. Set $\alpha = s_n(a)/n$ and $a' := (\alpha, \ldots, \alpha)$. Show that $a' \in C(a)$, and that $b \in C(a) \Longrightarrow b \prec a'$.

   e. Characterize the set $\{M \in \mathbf{Sym}_n(\mathbb{R}) \mid \mathrm{Sp}\, M \prec a'\}$.

15. Use Exercise 14 to prove the theorem of Horn and Schur. The set of diagonals $(h_{11}, \ldots, h_{nn})$ of Hermitian matrices with given spectrum $(\lambda_1, \ldots, \lambda_n)$ is the convex hull of the points $(\lambda_{\sigma(1)}, \ldots, \lambda_{\sigma(n)})$ as σ runs over the permutations of $\{1, \ldots, n\}$.

16. (Boyd, Diaconis, Sun and Xiao.) Let P be a symmetric stochastic $n \times n$ matrix:

$$p_{ij} = p_{ji} \geq 0, \qquad \sum_j p_{ij} = 1 \quad (i = 1, \ldots, n).$$

We recall that $\lambda_1 = 1$ is an eigenvalue of P, which is the largest in modulus (Perron–Frobenius). We are interested in the second largest modulus $\mu(P) = \max\{\lambda_2, -\lambda_n\}$ where $\lambda_1 \geq \cdots \geq \lambda_n$ is the spectrum of P; $\mu(P)$ is the second singular value of P.

  a. Let $y \in \mathbb{R}^n$ be such that $\|y\|_2 = 1$ and $\sum_j y_j = 1$. Let $z \in \mathbb{R}^n$ be such that

$$(p_{ij} \neq 0) \Longrightarrow \left(\frac{1}{2}(z_i + z_j) \leq y_i y_j \right).$$

  Show that $\lambda_2 \geq \sum_j z_j$. **Hint:** Use Rayleigh ratio.
  b. Likewise, if y is as above and w such that

$$(p_{ij} \neq 0) \Longrightarrow \left(\frac{1}{2}(w_i + w_j) \geq y_i y_j \right),$$

  show that $\lambda_n \leq \sum_j w_j$.
  c. Taking

$$y_j = \sqrt{\frac{2}{n}} \cos \frac{(2j-1)\pi}{2n}, \quad z_j = \frac{1}{n} \left(\cos \frac{\pi}{n} + \frac{\cos \frac{(2j-1)\pi}{n}}{\cos \frac{\pi}{n}} \right),$$

  deduce that $\mu(P) \geq \cos \dfrac{\pi}{n}$ for every symmetric stochastic $n \times n$ matrix.
  d. Find a symmetric stochastic $n \times n$ matrix Q such that $\mu(Q) = \cos \dfrac{\pi}{n}$. **Hint:** Exploit the equality case in the analysis, with the y and z given above.
  e. Prove that $P \mapsto \mu(P)$ is a convex function over symmetric stochastic $n \times n$ matrices.

17. Prove the equivalence of the following properties for real $n \times n$ matrices A:

  Strong Perron–Frobenius. The spectral radius is a simple eigenvalue of A, the only one of this modulus; it is associated with positive left and right eigenvectors.
  Eventually positive matrix. There exists an integer $k \geq 1$ such that $A^k > 0_n$.

18. Let A be a cyclic matrix, as in Theorem 8.3. If $1 \leq j \leq p$, prove that

$$P_A(X) = X^{n-pm_j} \det(X^p I_{m_j} - M_j M_{j+1} \cdots M_p M_1 \cdots M_{j-1}),$$

where m_j is the number of rows of M_j (or columns in M_{j-1}) (**Hint:** argue by induction over p, using the Schur complement formula). Deduce a lower bound of the multiplicity of the eigenvalue $\lambda = 0$.

Chapter 9
Matrices with Entries in a Principal Ideal Domain; Jordan Reduction

9.1 Rings, Principal Ideal Domains

In this chapter we consider only *commutative integral domains* A (see Chapter 3). Such a ring A can be embedded in its field of fractions, which is the quotient of $A \times (A \setminus \{0\})$ by the equivalence relation $(a,b)\mathscr{R}(c,d) \Leftrightarrow ad = bc$. The embedding is the map $a \mapsto (a,1)$.

In a ring A, the set of invertible elements, denoted by A^*, is a multiplicative group. If $a,b \in A$ are such that $b = ua$ with $u \in A^*$, we say that a and b are *associated*, and we write $a \sim b$, which amounts to saying that $aA = bA$. If there exists $c \in A$ such that $ac = b$, we say that a divides b and write $a|b$. Then the quotient c is unique and is denoted by b/a. Divisibility is an order relation. We say that b is a prime, or irreducible, element if the equality $b = ac$ implies that one of the factors is invertible.

An *ideal* I in a ring A is an additive subgroup of A such that $A \cdot I \subset I : a \in A$ and $x \in I$ imply $ax \in I$. For example, if $b_1, \ldots, b_r \in A$, the subset

$$b_1 A + \cdots + b_r A$$

is an ideal, denoted by $(b_1, \ldots, b_r)$. We say that this ideal is generated by $b_1, \ldots, b_r$. Ideals of the form (b), thus generated by a single element, are called *principal ideals*.

9.1.1 Facts About Principal Ideal Domains

Definition 9.1 *A commutative integral domain A is a* principal ideal domain *if every ideal in A is principal: For every ideal $\mathscr{I}$ there exists $a \in A$ such that $\mathscr{I} = (a)$.*

A field is a principal ideal domain that has only two ideals, $(0) = \{0\}$ and $(1) = A$. The set $\mathbb{Z}$ of rational integers and the polynomial algebra over a field k, denoted by $k[X]$, are also principal ideal domains.

In a commutative integral domain one says that d is a *greatest common divisor* (*gcd*) of a and b if d divides a and b, and if every common divisor of a and b divides d. In other words, the set of common divisors of a and b, ordered by divisibility, admits d as a greatest element. The gcd of a and b, whenever it exists, is unique up to multiplication by an invertible element. We say that a and b are coprime if all their common divisors are invertible; in that case, $\gcd(a,b) = 1$.

Proposition 9.1 *In a principal ideal domain, every pair of elements has a greatest common divisor. The gcd satisfies the* Bézout *identity: for every* $a, b \in A$, *there exist* $u, v \in A$ *such that*

$$\gcd(a,b) = ua + vb.$$

Such u *and* v *are coprime.*

Proof. Let A be a principal ideal domain. If $a, b \in A$, the ideal $\mathscr{I} =: (a,b)$ is principal: $\mathscr{I} = (d)$. Because $a, b \in \mathscr{I}$, d divides a and b. Furthermore, $d = ua + vb$ because $d \in \mathscr{I}$. If c divides a and b, then c divides $ua + vb$; hence divides d, which happens to be a gcd of a and b.

If m divides u and v, then $md | ua + vb$; hence $d = smd$. If $d \neq 0$, one has $sm = 1$, which means that $m \in A^*$. Thus u and v are coprime. If $d = 0$, then $a = b = 0$, and one may take $u = v = 1$, which are coprime. $\square$

As a generator of the ideal (a,b), a gcd of a and b is nonunique. Every element associated with it is another gcd. In certain rings one can choose the gcd in a canonical way, such as a positive element in $\mathbb{Z}$, or a monic polynomial in $k[X]$.

The gcd is associative: $\gcd(a, \gcd(b,c)) = \gcd(\gcd(a,b), c)$. It is therefore possible to speak of the gcd of an arbitrary finite subset of A. In the above example we denote it by $\gcd(a,b,c)$. We have a generalized Bézout formula: there exist elements $u_1, \ldots, u_r \in A$ such that

$$\gcd(a_1, \ldots, a_r) = a_1 u_1 + \cdots + a_r u_r.$$

Definition 9.2 *A ring A is* Noetherian *if every nondecreasing (for inclusion) sequence of ideals is constant beyond some index:* $I_0 \subset I_1 \subset \cdots \subset I_m \subset \cdots$ *implies that there is an l such that* $I_l = I_{l+1} = \cdots$.

Proposition 9.2 *The principal ideal domains are Noetherian.*

Observe that in the case of principal ideal domains the Noetherian property means exactly that if a sequence $a_1, \ldots$ of elements of A is such that every element is divisible by the next one, then there exists an index J such that the a_js are pairwise associated for every $j \geq J$.

This property seems natural because it is shared by all the rings encountered in number theory. But the ring of entire holomorphic functions is not Noetherian: just take for a_n the function

$$z \mapsto \left(\prod_{k=1}^{n} (z-k)^{-1} \right) \sin 2\pi z.$$

Proof. Let A be a principal ideal domain and let $(I_j)_{j \geq 0}$ be a nondecreasing sequence of ideals in A. Let $\mathscr{I}$ be their union, which happens to be an ideal. Let a be a generator: $\mathscr{I} = (a)$. Then a belongs to one of the ideals, say $a \in I_k$. Hence $\mathscr{I} \subset I_k$, which implies $I_j = \mathscr{I}$ for $j \geq k$. $\square$

The proof above works with slight changes if we know that every ideal in A is spanned by a finite set. For example, the ring of polynomials over a Noetherian ring is itself Noetherian: $\mathbb{Z}[X]$ and $k[X, Y]$ are Noetherian rings, although not principal.

The principal ideal domains are also *factorial* (a shortcut for *unique factorization domain*): every element of A admits a factorization consisting of prime factors. This factorization is unique up to ambiguities, which may be of three types: the order of factors, the presence of invertible elements, and the replacement of factors by associated ones. This property is fundamental to the arithmetic in A.

9.1.2 Euclidean Domains

Definition 9.3 *A* Euclidean domain *is a ring A endowed with a map $N : A \setminus \{0\} \mapsto \mathbb{N}$ such that for every $a, b \in A$ with $b \neq 0$, there exists a unique pair $(q, r) \in A \times A$ such that $a = qb + r$ with either $N(r) < N(b)$ or $r = 0$. The map $(a, b) \mapsto (q, r)$ is the* Euclidean division. *We call q the* quotient *and r the* remainder.

When b divides a, we have $q = b/a$ and $r = 0$.

Classical examples of Euclidean domains are the ring of the rational integers $\mathbb{Z}$, with $N(a) = |a|$, the ring $k[X]$ of polynomials over a field k, with $N(P) = \deg P$, and the ring of Gaussian integers $\mathbb{Z}[\sqrt{-1}]$, with $N(z) = |z|^2$.

Proposition 9.3 *Euclidean domains are principal ideal domains.*

Proof. Let $\mathscr{I}$ be an ideal of a Euclidean domain A. If $\mathscr{I} = (0)$, there is nothing to show. Otherwise, let us select an element a for which $N(a)$ is minimal in $\mathscr{I} \setminus \{0\}$. If $b \in \mathscr{I}$, the remainder r of the Euclidean division of b by a is an element of $\mathscr{I}$ and satisfies either $r = 0$ or $N(r) < N(a)$. The minimality of $N(a)$ implies $r = 0$; that is, $a | b$. Finally, $\mathscr{I} = (a)$. $\square$

The converse of Proposition 9.3 is not true. For example, the quadratic ring $\mathbb{Z}[\sqrt{14}]$ is a principal ideal domain, although not Euclidean. More information about rings of quadratic integers can be found in Cohn's monograph [10].

9.1.2.1 Calculation of the GCD in a Euclidean Domain

Given a, b in a Euclidean domain A, with say $b \neq 0$, let us define a sequence $(a_j)_{j \geq 0}$ by $a_0 = a$, $a_1 = b$ and then a_{j+1} the remainder of the division of a_{j-1} by a_j. The sequence is well defined until some a_{k+1} equals 0, that is, $a_k | a_{k-1}$. This must necessarily happen, because otherwise the sequence $(N(a_j))_{j \geq 0}$ would decrease endlessly, which contradicts the fact that the range of N is in $\mathbb{N}$. Because $a_{k-2} = q_{k-2}a_{k-1} + a_k$, a_k divides a_{k-2} too. By backward induction over j, a_k divides all the a_js, and in particular $a_k | \gcd(a, b)$. Conversely, by forward induction, $\gcd(a, b)$ divides all the a_js, and in particular a_k. Therefore

$$a_k = \gcd(a, b).$$

The moral of this analysis is that the Euclidean division is a practical tool for computing the gcd. We say that the calculation of the gcd is *effective* in a Euclidean domain. We can design a software to implement it on a computer.

9.1.3 Elementary Matrices

An *elementary matrix* of order n is a matrix of one of the following forms:

- The permutation matrices: for $\sigma \in S_n$, the matrix P_σ has entries $p_{ij} = \delta^j_{\sigma(i)}$.
- The matrices $I_n + aJ_{ik}$, for $a \in A$ and $1 \leq i \neq k \leq n$, with

$$(J_{ik})_{lm} = \delta^\ell_i \delta^m_k.$$

- The diagonal invertible matrices, that is, those whose diagonal entries are invertible in A.

We observe that the inverse of an elementary matrix is again elementary. For example, $(I_n + aJ_{ik})(I_n - aJ_{ik}) = I_n$.

Theorem 9.1 *A square invertible matrix of size n with entries in a Euclidean domain A is a product of elementary matrices with entries in A.*

Proof. We prove the theorem for $n = 2$. The general case is deduced from that particular one and from the proof of Theorem 9.2 below, which uses multiplications by block-diagonal matrices with 1×1 and 2×2 diagonal blocks.

Let

$$M = \begin{pmatrix} a & a_1 \\ c & d \end{pmatrix}$$

be given in $\mathbf{GL}_2(A)$: we have $ad - a_1c \in A^*$. If $N(a) < N(a_1)$ or $a = 0$, we multiply M on the right by

$$\begin{pmatrix} 0 & 1 \\ 1 & 0 \end{pmatrix}.$$

We are now in the case where $a \neq 0$ and either $N(a_1) \leq N(a)$ or $a_1 = 0$. Assuming the former, we let $a = a_1q + a_2$ be the Euclidean division of a by a_1. Then

$$M \begin{pmatrix} 1 & 0 \\ -q & 1 \end{pmatrix} =: M' = \begin{pmatrix} a_2 & a_1 \\ \cdot & d \end{pmatrix}.$$

Next, we have

$$M' \begin{pmatrix} 0 & 1 \\ 1 & 0 \end{pmatrix} =: M_1 = \begin{pmatrix} a_1 & a_2 \\ \cdot & \cdot \end{pmatrix},$$

with $N(a_2) < N(a_1)$ or $a_2 = 0$. We thus construct a sequence of matrices M_k of the form

$$\begin{pmatrix} a_{k-1} & a_k \\ \cdot & \cdot \end{pmatrix},$$

with $a_{k-1} \neq 0$, each one the product of the previous one by elementary matrices. Furthermore, either $N(a_k) < N(a_{k-1})$ or $a_k = 0$. This sequence is thus finite, and there is a step for which $a_k = 0$. The matrix M_k, being triangular and invertible, has an invertible diagonal D. Then $M_k D^{-1}$ has the form

$$\begin{pmatrix} 1 & 0 \\ \cdot & 1 \end{pmatrix},$$

which is an elementary matrix. $\quad\square$

Theorem 9.1 is false in general for principal ideal domains. Whether $\mathbf{GL}_n(A)$ equals the group spanned by elementary matrices is a difficult question of K-theory.

9.2 Invariant Factors of a Matrix

Theorem 9.2 *Let $M \in \mathbf{M}_{n \times m}(A)$ be a matrix with entries in a principal ideal domain. Then there exist two invertible matrices $P \in \mathbf{GL}_n(A)$, $Q \in \mathbf{GL}_m(A)$ and a quasi-diagonal matrix $D \in \mathbf{M}_{n \times m}(A)$ (i.e., $d_{ij} = 0$ for $i \neq j$) such that:*

- *$M = PDQ$.*
- *On the other hand, $d_1 | d_2, \ldots, d_i | d_{i+1}, \ldots$, where the d_j are the diagonal entries of D.*

Furthermore, if $M = P'D'Q'$ is another decomposition with these two properties, the scalars d_j and d'_j are associated. Up to invertible elements, they are thus unique.

In other words, every matrix in $\mathbf{M}_{n \times m}(A)$ is equivalent to a quasi-diagonal matrix with $d_1 | d_2 | \ldots$

Definition 9.4 *The scalars $d_1, \ldots, d_r$ ($r = \min(n, m)$) are called the* invariant factors *of M.*

Proof. Uniqueness: For $k \leq r$, let us denote by $D_k(N)$ the gcd of minors of order k of the matrix $N \in \mathbf{M}_{n \times m}(A)$. From Corollary 3.2, we have $D_k(M) = D_k(D) = D_k(D')$. It is immediate that $D_k(D) = d_1 \cdots d_k$ (because the minors of order k are either null, or products of k terms d_j with distinct subscripts), so that

$$d_1 \cdots d_k = u_k d'_1 \cdots d'_k, \quad 1 \leq k \leq r,$$

for some $u_k \in A^*$. Hence, d_1 and d'_1 are associated. Because A is an integral domain, we also have $d'_k = u_k^{-1} u_{k-1} d_k$. In other words, d_k and d'_k are associated.

Existence: We have seen above that the d_js are determined by the equalities $d_1 \cdots d_j = D_j(M)$. In particular, d_1 is the gcd of the entries of M. Hence the first step consists in finding a matrix M', equivalent to M, such that m'_{11} is equal to this gcd. Our construction is based upon an auxiliary result.

Lemma 13. *There exists a map* $T : \mathbf{M}_{n \times m}(A) \to \mathbf{M}_{n \times m}(A)$ *with the following properties.*

- $N' := T(N)$ *is equivalent to* N.
- $n'_{11} | n_{11}$,
- *This divisibility is either strict (i.e., n'_{11} is not associated with n_{11}), or n_{11} divides all the entries n_{ij}. Notice that in this latter case, we have $n_{11} = D_1(N) = d_1(N)$.*

If A is a Euclidean domain, the map T is given by an effective algorithm; $T(N)$ is obtained by multiplications of N at left and/or right by elementary matrices.

We argue by induction over r. Using the lemma, we construct a sequence of matrices by $M^{(0)} = M$ and then $M^{(k+1)} = T(M^{(k)})$. By induction, each $M^{(k)}$ is equivalent to M. In particular, $D_\ell(M^{(k)}) = D_\ell(M)$ and $d_\ell(M^{(k)}) = d_\ell(M)$ for every $1 \le \ell \le r$.

The sequence of upper-left entries is ordered by divisibility: $m_{11}^{(k+1)} | m_{11}^{(k)}$. From Proposition 9.2, the elements of the sequence $(m_{11}^{(p)})_{p \ge 0}$ are pairwise associated, once p is large enough. From Lemma 13, we see that for some q, $m_{11}^{(q)}$ divides all the $m_{ij}^{(q)}$s.

We have $m_{i1}^{(q)} = a_i m_{11}^{(q)}$ and $m_{1j}^{(q)} = b_j m_{11}^{(q)}$, therefore we set $P \in \mathbf{GL}_n(A)$ and $Q \in \mathbf{GL}_m(A)$ as follows.

- $p_{ii} = 1$, $p_{i1} = -a_i$ if $i \ge 2$, $p_{ij} = 0$ otherwise.
- $q_{jj} = 1$, $q_{1j} = -b_j$ if $j \ge 2$, $q_{ij} = 0$ otherwise.

The matrix $M' := PM^{(q)}Q$ is equivalent to $M^{(q)}$, and hence to M. It has the form

$$M' = \begin{pmatrix} m & 0 & \cdots & 0 \\ 0 & & & \\ \vdots & & M'' & \\ 0 & & & \end{pmatrix},$$

where m divides all the entries of M''. Obviously, $m = D_1(M') = D_1(M)$.

From the induction assumption, there exist nonsingular matrices P'' (of size $n - 1$) and Q'' (of size $m - 1$) such that $P''M''Q'' = D''$ is quasi-diagonal with $d''_1 | d''_2 | \cdots$. Because m divides the entries of M'', that is, $d_1(M'')$, we have $m | d''_1$. Forming the nonsingular matrices $\bar{P} := \mathrm{diag}(1, P'')P$ and $\bar{Q} := Q\,\mathrm{diag}(1, Q'')$, we infer

$$\bar{P}M^{(q)}\bar{Q} = \operatorname{diag}(m, D'').$$

Therefore M is equivalent to $\operatorname{diag}(1, D'')$, which is a quasi-diagonal matrix with the required divisibility property of its diagonal elements.

□

Proof. (of the lemma).

We intentionally write this proof in the algorithmic style. Given the matrix $N \in \mathbf{M}_{n \times m}(A)$, we distinguish four cases.

1. IF n_{11} does not divide some n_{1j}, THEN take the smallest such index j. Set $d := \gcd(n_{11}, n_{1j})$, which reads $d = un_{11} + vn_{1j}$. Define $w := -n_{1j}/d$ and $z := n_{11}/d$ and a matrix $Q \in \mathbf{GL}_m(A)$ by:

   - $q_{11} = u, q_{j1} = v, q_{1j} = w, q_{jj} = z$.
   - $q_{kl} = \delta_k^\ell$, otherwise.

   Then $M^{(p)} := M^{(p-1)}Q$ is an equivalence, and $m_{11}^{(p)} = d | n_{11} = m_{11}^{(p-1)}$, where the division is strict.

2. ELSE n_{11} divides each n_{1j}. IF it does not divide n_{i1}, THEN take the smallest such index i. This case is symmetric to the previous one. Multiplication on the right by a suitable $P \in \mathbf{GL}_n(A)$ furnishes $M^{(p)}$, with $m_{11}^{(p)} = \gcd(n_{11}, n_{i1}) | m_{11}^{(p-1)}$ being a strict division.

3. ELSE n_{11} divides each n_{1j} and each n_{i1}. IF it does not divide some n_{ij} with $i, j \geq 2$, THEN take the smallest such pair (i, j) in the lexicographic ordering. We have $n_{i1} = an_{11}$. Define a matrix $P \in \mathbf{GL}_n(A)$ by

   - $p_{11} = a + 1, p_{i1} = 1, p_{1i} = -1, p_{ii} = 0$.
   - $p_{kl} = \delta_k^\ell$, otherwise.

   Set $\bar{N} = PN$. We have $\bar{n}_{11} = n_{11}$ and $\bar{n}_{1j} = (a+1)n_{1j} - n_{ij}$. Observe that $\bar{n}_{11}$ does not divide n_{1j}. GOTO Step 1.

4. ELSE n_{11} divides all the entries of the matrix N. In that case, $M^{(p)} := M^{(p-1)}$.

   □

9.2.1 Comments

- In the list of invariant factors of a matrix some d_js may equal zero. In that case, $d_j = 0$ implies $d_{j+1} = \cdots = d_r = 0$. Moreover, some invariant factor may occur several times in the list $d_1, \ldots, d_r$, up to association. The number of times that a factor d or its associates occurs is its *multiplicity*.
- If $m = n$ and if the invariant factors of a matrix M are $(1, \ldots, 1)$, then $D = I_n$, and $M = PQ$ is invertible. Conversely, if M is invertible, then the decomposition $M = MI_nI_n$ shows that $d_1 = \cdots = d_n = 1$.
- If A is a field, then there are only two ideals: $A = (1)$ itself and (0). The list of invariant factors of a matrix is thus of the form $(1, \ldots, 1, 0, \ldots, 0)$. Of course, there

may be no 1s (for the matrix $0_{m\times n}$), or no 0s. There are thus exactly $\min(n,m)+1$ classes of equivalent matrices in $\mathbf{M}_n(A)$, two matrices being equivalent if and only if they have the same rank q. The rank is then the number of 1s among the invariant factors. The decomposition $M = PDQ$ is then called the *rank decomposition*.

Theorem 9.3 *Let k be a field and $M \in \mathbf{M}_{n\times m}(k)$ a matrix. Let q be the rank of M, that is, the dimension of the linear subspace of k^n spanned by the columns of M. Then there exist two square invertible matrices P, Q such that $M = PDQ$ with $d_{ii} = 1$ if $i \le q$ and $d_{ij} = 0$ in all other cases.*

Theorem 9.3 can be proved more directly, arguing with bases of $R(M)$ and of $\ker M$.

- The matrices P or Q involved in the first three steps of the proof of Theorem 9.2 are 2×2 matrices, up to identity blocks. When A is Euclidean, they are thus products of elementary matrices. Whence the following theorem.

Theorem 9.4 *If A is Euclidean, one passes from $M \in \mathbf{M}_{n\times m}(A)$ to an equivalent quasi-diagonal matrix with*

$$d_1|d_2,\ldots,d_i|d_{i+1},\ldots$$

by multiplying at right and left by elementary matrices.

9.3 Similarity Invariants and Jordan Reduction

From now on, k denotes a field and $A = k[X]$ the ring of polynomials over k. This ring is Euclidean, hence a principal ideal domain. In the sequel, the results are *effective*, in the sense that the normal forms that we define are obtained by means of an algorithm that uses right or left multiplications by elementary matrices of $\mathbf{M}_n(A)$, the computations being based upon the Euclidean division of polynomials.

Given a matrix $B \in \mathbf{M}_n(k)$, a square matrix with scalar entries, we consider the matrix $XI_n - B \in \mathbf{M}_n(A)$, where X is the indeterminate in A.

Definition 9.5 *If $B \in \mathbf{M}_n(k)$, the invariant factors of $M := XI_n - B$ are called invariant polynomials of B, or similarity invariants of B.*

This definition is motivated by the following statement.

Theorem 9.5 *Two matrices in $\mathbf{M}_n(k)$ are similar if and only if they have the same list of invariant polynomials (counted with their multiplicities).*

This theorem is a particular case of a more general one:

Theorem 9.6 *Let A_0, A_1, B_0, B_1 be matrices in $\mathbf{M}_n(k)$, with $A_0, A_1 \in \mathbf{GL}_n(k)$. Then the matrices $XA_0 + B_0$ and $XA_1 + B_1$ are equivalent (in $\mathbf{M}_n(A)$) if and only if there exist $G, H \in \mathbf{GL}_n(k)$ such that*

$$GA_0 = A_1H, \quad GB_0 = B_1H.$$

When $A_0 = A_1 = I_n$, Theorem 9.6 tells us that $XI_n - B_0$ and $XI_n - B_1$ are equivalent, namely that they have the same invariant polynomials, if there exists $P \in \mathbf{GL}_n(k)$ such that $PB_0 = B_1P$. Whence Theorem 9.5.

Proof. We prove Theorem 9.6. The condition is clearly sufficient.

Conversely, if $XA_0 + B_0$ and $XA_1 + B_1$ are equivalent, there exist matrices $P, Q \in \mathbf{GL}_n(A)$, such that $P(XA_0 + B_0) = (XA_1 + B_1)Q$. Because A_1 is invertible, one may perform Euclidean division[1] of P by $XA_1 + B_1$ on the right:

$$P = (XA_1 + B_1)P_1 + G,$$

where G is a matrix whose entries are scalars. We warn the reader that inasmuch as $\mathbf{M}_n(k)$ is not commutative, Euclidean division in $\mathbf{M}_n(k)[X]$ may be done either on the right or on the left, with distinct quotients and distinct remainders. It is always required that the coefficient of highest degree (here A_1) in the divisor (here $XA_1 + B_1$) be an invertible matrix. Likewise, we have $Q = Q_1(XA_0 + B_0) + H$ with $H \in \mathbf{M}_n(k)$. Let us then write

$$(XA_1 + B_1)(P_1 - Q_1)(XA_0 + B_0) = (XA_1 + B_1)H - G(XA_0 + B_0).$$

The left-hand side of this equality has degree (the degree is defined as the supremum of the degrees of the entries of the matrix) $2 + \deg(P_1 - Q_1)$, again because A_0, A_1 are nonsingular. Meanwhile, the right-hand side has degree less than or equal to one. The two sides, being equal, must vanish, and we conclude that

$$GA_0 = A_1H, \quad GB_0 = B_1H.$$

There remains to show that G and H are invertible. To do so, let us define $R \in \mathbf{M}_n(A)$ as the inverse matrix of P (which exists by assumption). We still have

$$R = (XA_0 + B_0)R_1 + K, \quad K \in \mathbf{M}_n(k).$$

Combining the equalities stated above, we obtain

$$I_n - GK = (XA_1 + B_1)(QR_1 + P_1K).$$

Inasmuch as the left-hand side is constant and the right-hand side has degree $1 + \deg(QR_1 + P_1K)$, we must have $I_n = GK$, so that G is invertible. Likewise, H is invertible. $\square$

We conclude this paragraph with a remarkable statement:

Theorem 9.7 *If $B \in \mathbf{M}_n(k)$, then B and B^T are similar.*

Indeed, $XI_n - B$ and $XI_n - B^T$ are transposes of each other, and hence have the same list of minors, and the same invariant factors.

---

[1] The fact that A_1 is invertible is essential, because the ring $\mathbf{M}_n(A)$ is not an integral domain.

9.3.1 Example: The Companion Matrix of a Polynomial

Given a polynomial

$$p(X) = X^n + a_1 X^{n-1} + \cdots + a_n,$$

there exists a matrix $B \in \mathbf{M}_n(k)$ such that the list of invariant factors of the matrix $XI_n - B$ is $(1, \ldots, 1, p)$. We may take the *companion matrix* associated with p to be

$$B_p := \begin{pmatrix} 0 \cdots & \cdots & 0 & -a_n \\ 1 & \ddots & & \vdots & \vdots \\ 0 & \ddots & \ddots & \vdots & \vdots \\ \vdots & \ddots & \ddots & 0 & \vdots \\ 0 \cdots & & 0 & 1 & -a_1 \end{pmatrix}.$$

Naturally, any matrix similar to B_p would do as well, because if $B = Q^{-1} B_p Q$, then $XI_n - B$ is similar, hence equivalent, to $XI_n - B_p$. In order to show that the invariant factors of B_p are the polynomials $(1, \ldots, 1, p)$, we observe that $XI_n - B_p$ possesses a minor of order $n - 1$ that is invertible, namely, the determinant of the submatrix

$$\begin{pmatrix} -1 & X & 0 & \cdots & 0 \\ 0 & \ddots & \ddots & \ddots & \vdots \\ \vdots & \ddots & \ddots & \ddots & 0 \\ \vdots & & \ddots & \ddots & X \\ 0 & \cdots & \cdots & 0 & -1 \end{pmatrix}$$

obtained by deleting the first row and the last column. We thus have $D_{n-1}(XI_n - B_p) = 1$, so that the invariant factors $d_1, \ldots, d_{n-1}$ are all equal to 1. Hence $d_n = D_n(XI_n - B_p) = \det(XI_n - B_p)$, the characteristic polynomial of B_p, namely p.

In this example p is also the minimal polynomial of B_p. In fact, if Q is a polynomial of degree less than or equal to $n - 1$,

$$Q(X) = b_0 X^{n-1} + \cdots + b_{n-1},$$

the vector $Q(A)\mathbf{e}^1$ reads

$$b_0 \mathbf{e}^n + \cdots + b_{n-1} \mathbf{e}^1.$$

Hence $Q(A) = 0$ and $\deg Q \le n - 1$ imply $Q = 0$. The minimal polynomial is therefore of degree at least n. It is thus equal to the characteristic polynomial.

9.3.2 First Canonical Form of a Square Matrix

Let $M \in \mathbf{M}_n(k)$ be a square matrix and $p_1, \ldots, p_n \in k[X]$ its similarity invariants. The sum of their degrees n_j ($1 \leq j \leq n$) is n. Let us denote by $M^{(j)} \in \mathbf{M}_{n_j}(k)$ the companion matrix of the polynomial p_j. Let us form the matrix M', block-diagonal, whose diagonal blocks are the $M^{(j)}$s. The few first polynomials p_j are generally constant (we show below that the only case where p_1 is not constant corresponds to $M = \alpha I_n$), and the corresponding blocks are empty, as are their bordering rows and columns. The actual number m of diagonal blocks is equal to the number of nonconstant similarity invariants.

The matrix $XI_{n_j} - M^{(j)}$ is equivalent to the matrix $N^{(j)} = \mathrm{diag}(1, \ldots, 1, p_j)$, therefore we have

$$XI_{n_j} - M^{(j)} = P^{(j)} N^{(j)} Q^{(j)},$$

where $P^{(j)}, Q^{(j)} \in \mathbf{GL}_{n_j}(k[X])$. Let us form matrices $P, Q \in \mathbf{GL}_n(k[X])$ by

$$P = \mathrm{diag}(P^{(1)}, \ldots, P^{(n)}), \quad Q = \mathrm{diag}(Q^{(1)}, \ldots, Q^{(n)}).$$

We obtain

$$XI_n - M' = PNQ, \quad N = \mathrm{diag}(N^{(1)}, \ldots, N^{(n)}).$$

Here N is a diagonal matrix, whose diagonal entries are the similarity invariants of M, up to the order. In fact, each nonconstant p_j appears in the corresponding block $N^{(j)}$. The other diagonal terms are the constant 1, which occurs $n - m$ times; these are the polynomials $p_1, \ldots, p_{n-m}$, as expected. Conjugating by a permutation matrix, we obtain that $XI_n - M'$ is equivalent to the matrix $\mathrm{diag}(p_1, \ldots, p_n)$. Hence $XI_n - M'$ is equivalent to $XI_n - M$. From Theorem 9.5, M and M' are similar.

Theorem 9.8 *Let k be a field, $M \in \mathbf{M}_n(k)$ a square matrix, and $p_1, \ldots, p_n$ its similarity invariants. Then M is similar to the block-diagonal matrix M' whose jth diagonal block is the companion matrix of p_j.*

The matrix M' is called the first canonical form *of M, or the* Frobenius canonical form *of M.*

Remark

If L is an extension of k (namely, a field containing k) and $M \in \mathbf{M}_n(k)$, then $M \in \mathbf{M}_n(L)$. Let $p_1, \ldots, p_n$ be the similarity invariants of M as a matrix with entries in k. Then $XI_n - M = P\mathrm{diag}(p_1, \ldots, p_n)Q$, where $P, Q \in \mathbf{GL}_n(k[X])$. Because P, Q, their inverses, and the diagonal matrix also belong to $\mathbf{M}_n(L[X])$, $p_1, \ldots, p_n$ are the similarity invariants of M as a matrix with entries in L. In other words, the similarity invariants depend on M but not on the field k. To compute them, it is enough to place ourselves in the smallest possible field, namely that spanned by the entries of M. The same remark holds true for the first canonical form. As we show in the next section, this is no longer true for the second canonical form, which is therefore less canonical.

9.3.2.1 The Minimal Polynomial

Theorem 9.9 *Let k be a field, $M \in \mathbf{M}_n(k)$ a square matrix, and $p_1,\ldots,p_n$ its similarity invariants. Then p_n is the minimal polynomial of M. In particular, the minimal polynomial does not depend on the field under consideration, as long as it contains the entries of M.*

Proof. We use the first canonical form M' of M. Because M' and M are similar, they have the same minimal polynomial. One thus can assume that M is in the canonical form $M = \mathrm{diag}(M_1,\ldots,M_n)$, where M_j is the companion matrix of p_j. Inasmuch as $p_j(M_j) = 0$ (Cayley–Hamilton theorem) and $p_j|p_n$, we have $p_n(M_j) = 0$ and thus $p_n(M) = 0_n$. Hence, the minimal polynomial π_M divides p_n. Conversely, $\pi_M(M) = 0_n$ implies $\pi_M(M_n) = 0$. Because M_n is the companion matrix of p_n, we already know that p_n is the minimal polynomial of M_n. Thus p_n divides π_M and finally $p_n = \pi_M$.

The similarity invariants do not depend on the choice of the field, therefore in particular p_n does not.   □

Warning: One may draw an incorrect conclusion if one applies Theorem 9.9 carelessly. Given a matrix $M \in \mathbf{M}_n(\mathbb{Z}) \subset \mathbf{M}_n(\mathbb{Q})$, one can define a matrix $M_{(p)}$ in $\mathbf{M}_n(\mathbb{Z}/q\mathbb{Z})$ by reduction modulo q (q a prime number). But the minimal polynomial of $M_{(q)}$ is not necessarily the reduction modulo q of π_M. Here is an example: Let us take $n = 2$ and

$$M = \begin{pmatrix} 2 & 2 \\ 0 & 2 \end{pmatrix}.$$

Then π_M divides $P_M = (X - 2)^2$, but $\pi_M \neq X - 2$, because $M \neq 2I_2$. Hence $\pi_M = (X - 2)^2$. On the other hand, $M_{(2)} = 0_2$, whose minimal polynomial is X, which is different from X^2, the reduction modulo 2 of π_M.

The explanation of this phenomenon is the following. The matrices M and $M_{(2)}$ are composed of scalars of different natures. There is no field L containing $\mathbb{Z}$ and $\mathbb{Z}/2\mathbb{Z}$ simultaneously. There is thus no context in which Theorem 9.9 could be applied.

9.3.3 Second Canonical Form of a Square Matrix

We now decompose the similarity invariants of M into products of irreducible polynomials. This decomposition depends, of course, on the choice of the field of scalars. Denoting by $q_1,\ldots,q_t$ the list of distinct irreducible (in $k[X]$) factors of p_n, we have

$$p_j = \prod_{k=1}^{t} q_k^{\alpha(j,k)}, \quad 1 \le j \le n$$

(because p_j divides p_n), where the $\alpha(j,k)$ are nondecreasing with respect to j, inasmuch as p_j divides p_{j+1}.

Definition 9.6 *The* elementary divisors *of the matrix $M \in \mathbf{M}_n(k)$ are the polynomials $q_k^{\alpha(j,k)}$ for which the exponent $\alpha(j,k)$ is nonzero. The* multiplicity *of an elementary divisor q_k^m is the number of solutions j of the equation $\alpha(j,k) = m$. The* list *of elementary divisors is the sequence of these polynomials, repeated with their multiplicities.*

Let us begin with the case of the companion matrix N of some polynomial p. Its similarity invariants are $(1,\ldots,1,p)$ (see above). Let $r_1,\ldots,r_t$ be its elementary divisors (we observe that each has multiplicity one). We then have $p = r_1 \cdots r_t$, and the r_ls are pairwise coprime. With each r_l we associate its companion matrix N_l, and we form a block-diagonal matrix $N' := \mathrm{diag}(N_1,\ldots,N_t)$. Each $N_l - XI_l$ is equivalent to a diagonal matrix

$$\begin{pmatrix} 1 & & & \\ & \ddots & & \\ & & 1 & \\ & & & r_l \end{pmatrix}$$

in $\mathbf{M}_{n(l)}(k[X])$, therefore the whole matrix $N' - XI_n$ is equivalent to

$$Q := \begin{pmatrix} 1 & & & & & \\ & \ddots & & & O & \\ & & 1 & & & \\ & & & r_1 & & \\ & O & & & \ddots & \\ & & & & & r_t \end{pmatrix}.$$

Let us now compute the similarity invariants of N', that is, the invariant factors of Q. It is enough to compute the greatest common divisor D_{n-1} of the minors of size $n-1$. Taking into account the principal minors of Q, we see that D_{n-1} must divide every product of the form

$$\prod_{l \ne k} r_l, \quad 1 \le k \le t.$$

Because the r_ls are pairwise coprime, this implies that $D_{n-1} = 1$. This means that the list of similarity invariants of N' has the form $(1,\ldots,1,\cdot)$, where the last polynomial must be the characteristic polynomial of N'. This polynomial is the product of the characteristic polynomials of the N_ls. These being equal to the r_ls, the characteristic polynomial of N' is p. Finally, N and N' have the same similarity invariants and are therefore similar.

Now let M be a general matrix in $\mathbf{M}_n(k)$. We apply the former reduction to every diagonal block M_j of its Frobenius canonical form. Each M_j is similar to a block-diagonal matrix whose diagonal blocks are companion matrices corresponding to the elementary divisors of M entering into the factorization of the jth invariant polynomial of M. We have thus proved the following statement.

Theorem 9.10 *Let $r_1, \ldots, r_s$ be the elementary divisors of $M \in \mathbf{M}_n(k)$. Then M is similar to a block-diagonal matrix M' whose diagonal blocks are companion matrices of the $r_l s$.*

The matrix M' is called the second canonical form of M.

Remark

The exact computation of the second canonical form of a given matrix is impossible in general, in contrast to the case of the first form. Indeed, if there were an algorithmic construction, it would provide an algorithm for factorizing polynomials into irreducible factors via the formation of the companion matrix, a task known to be impossible if $k = \mathbb{R}$ or $\mathbb{C}$. Recall that one of the most important results in Galois theory, known as Abel's theorem, states the impossibility of solving a general polynomial equation of degree at least five with complex coefficients, using only the basic operations and the extraction of roots of any order.

9.3.4 Jordan Form of a Matrix

If the characteristic polynomial splits over k, which holds for instance when the field k is algebraically closed, the elementary divisors have the form $(X - a)^r$ for $a \in k$ and $r \geq 1$. In that case, the second canonical form can be greatly simplified by replacing the companion matrix of the monomial $(X - a)^r$ by its *Jordan block*

$$
J(a;r) := \begin{pmatrix} a & 1 & 0 & \cdots & 0 \\ 0 & \ddots & \ddots & \ddots & \vdots \\ \vdots & \ddots & \ddots & \ddots & 0 \\ \vdots & & \ddots & \ddots & 1 \\ 0 & \cdots & \cdots & 0 & a \end{pmatrix}.
$$

The characteristic polynomial of $J(a;r)$ (an $r \times r$ triangular matrix) is $(X - a)^r$, whereas the matrix $XI_r - J(a;r)$ possesses an invertible minor of order $r - 1$, namely

$$\begin{pmatrix} -1 & 0 & \cdots & 0 \\ X-a & \ddots & \ddots & \vdots \\ & \ddots & \ddots & 0 \\ & & X-a & -1 \end{pmatrix},$$

which we obtain by deleting the first column and the last row. Again, this shows that $D_{n-1}(XI_r - J) = 1$, so that the invariant factors $d_1, \ldots, d_{r-1}$ are equal to 1. Hence $d_r = D_r(XI_r - J) = \det(XI_r - J) = (X - a)^r$. Its invariant factors are thus $1, \ldots, 1, (X - a)^r$, as required, hence the following theorem.

Theorem 9.11 *When an elementary divisor of M is $(X - a)^r$, one may, in the second canonical form of M, replace its companion matrix by the Jordan block $J(a;r)$.*

Corollary 9.1 *If the characteristic polynomial of M splits over k, then M is similar to a block-diagonal matrix whose jth diagonal block is a Jordan block $J(a_j;r_j)$. This form is unique, up to the order of blocks.*

Corollary 9.2 *If k is algebraically closed (e.g., if $k = \mathbb{C}$), then every square matrix M is similar to a block-diagonal matrix whose jth diagonal block is a Jordan block $J(a_j;r_j)$. This form is unique, up to the order of blocks.*

Exercises

1. Show that every principal ideal domain is a unique factorization domain.
2. Verify that the characteristic polynomial of the companion matrix of a polynomial p is equal to p.
3. Let k be a field and $M \in \mathbf{M}_n(k)$. Show that M, M^T have the same rank and that in general, the rank of $M^T M$ is less than or equal to that of M. Show that the equality of these ranks always holds if $k = \mathbb{R}$, but that strict inequality is possible, for example with $k = \mathbb{C}$.
4. Compute the elementary divisors of the matrices

$$\begin{pmatrix} 22 & 23 & 10 & -98 \\ 12 & 18 & 16 & -38 \\ -15 & -19 & -13 & 58 \\ 6 & 7 & 4 & -25 \end{pmatrix}, \quad \begin{pmatrix} 0 & -21 & -56 & -96 \\ 18 & 36 & 52 & -8 \\ -12 & -17 & -16 & 38 \\ 3 & 2 & -2 & -20 \end{pmatrix}$$

and

$$\begin{pmatrix} 44 & 89 & 120 & -32 \\ 0 & -12 & -32 & -56 \\ -14 & -20 & -16 & 49 \\ 8 & 14 & 16 & -16 \end{pmatrix}$$

in $\mathbf{M}_n(\mathbb{C})$. What are their Jordan reductions?

5. (Lagrange's theorem.)
   Let K be a field and $A \in \mathbf{M}_n(K)$. Let $X, Y \in K^n$ be vectors such that $X^T A Y \neq 0$.
   We normalize by $X^T A Y = 1$ and define

   $$B := A - (AY)(X^T A).$$

   Show that in the factorization

   $$PAQ = \begin{pmatrix} I_r & 0 \\ 0 & 0_{n-r} \end{pmatrix}, \quad P, Q \in \mathbf{GL}_n(K),$$

   one can choose Y as the first column of Q and X^T as the first row of P. Deduce
   that $\mathrm{rk}\, B = \mathrm{rk}\, A - 1$.
   More generally, show that if $X, Y \in \mathbf{M}_{n \times m}(K)$ are such that $X^T A Y \in \mathbf{GL}_m(K)$,
   then the rank of

   $$B := A - (AY)(X^T A Y)^{-1}(X^T A),$$

   equals $\mathrm{rk}\, A - m$.
   If $A \in \mathbf{Sym}_n(\mathbb{R})$ and if A is positive-semidefinite, and if $X = Y$, show that B is
   also positive-semidefinite.

6. For $A \in \mathbf{M}_n(\mathbb{C})$, consider the linear differential equation in $\mathbb{C}^n$

   $$\frac{dx}{dt} = Ax. \tag{9.1}$$

   a. Let $P \in \mathbf{GL}_n(\mathbb{C})$ and let $t \mapsto x(t)$ be a solution of (9.1). What is the differ-
      ential equation satisfied by $t \mapsto Px(t)$?
   b. Let $(X - a)^m$ be an elementary divisor of A. Show that for every $k = 0, \ldots, m - 1$, (9.1) possesses solutions of the form $e^{at} Q_k(t)$, where Q_k is
      a complex-valued polynomial map of degree k.

7. Consider the following differential equation of order n in $\mathbb{C}$:

   $$x^{(n)}(t) + a_1 x^{(n-1)}(t) + \cdots + a_n x(t) = 0. \tag{9.2}$$

   a. Define $p(X) = X^n + a_1 X^{n-1} + \cdots + a_n$ and let M be the companion matrix
      of p. Let

      $$p(X) = \prod_{a \in A}(X - a)^{n_a}$$

      be the factorization of p into irreducible factors. Compute the Jordan form
      of M.
   b. Using either the previous exercise or arguing directly, show that the set of
      solutions of (9.2) is spanned by the solutions of the form

      $$t \mapsto e^{at} R(t), \quad R \in \mathbb{C}[X], \quad \deg R < n_a.$$

8. Consider a linear recursion of order n in a field K

$$u_{m+n} + a_1 u_{m+n-1} + \cdots + a_n u_m = 0, \quad m \in \mathbb{N}. \tag{9.3}$$

We assume that p splits over K and use the notation of the previous exercise. Show that the set of solutions of (9.3) is spanned by the solutions of the form

$$(a^m R(m))_{m \in \mathbb{N}}, \quad R \in K[X], \quad \deg R < n_a.$$

9. Let $n \geq 2$ and let $M \in \mathbf{M}_n(\mathbb{Z})$ be the matrix defined by $m_{ij} = i + j - 1$:

$$M = \begin{pmatrix} 1 & 2 & \cdots & & n \\ 2 & & & & \vdots \\ \vdots & & & & \vdots \\ n & \cdots & \cdots & & 2n-1 \end{pmatrix}.$$

a. Show that M has rank 2 (you may look for two vectors $x, y \in \mathbb{Z}^n$ such that $m_{ij} = x_i x_j - y_i y_j$).
b. Compute the invariant factors of M in $\mathbf{M}_n(\mathbb{Z})$ (the equivalent diagonal form is obtained after five elementary operations).

10. The ground field is $\mathbb{C}$.

a. Define

$$N = J(0;n), \quad B = \begin{pmatrix} & \cdots & 0 & 1 \\ \vdots & & & 0 \\ 0 & & & \vdots \\ 1 & 0 & \cdots & \end{pmatrix}.$$

Compute NB, BN, and BNB. Show that $S := \dfrac{1}{\sqrt{2}}(I + iB)$ is unitary.

b. Deduce that N is similar to

$$\frac{1}{2}\begin{pmatrix} 0 & 1 & 0 & \dots & 0 \\ 1 & & & & \vdots \\ 0 & & & & 0 \\ \vdots & & & & 1 \\ 0 & \dots & 0 & 1 & 0 \end{pmatrix} + \frac{i}{2}\begin{pmatrix} 0 & \dots & 0 & -1 & 0 \\ \vdots & & & & 1 \\ 0 & & & & 0 \\ -1 & & & & \vdots \\ 0 & 1 & 0 & \dots & 0 \end{pmatrix}.$$

c. Deduce that every matrix $M \in \mathbf{M}_n(\mathbb{C})$ is similar to a complex symmetric matrix. Compare with the real case.

11. Let k be a field and $A \in \mathbf{M}_n(k)$, $B \in \mathbf{M}_m(k)$, $C \in \mathbf{M}_{n \times m}(k)$ be given matrices. If the equation $AX - XB = C$ is solvable, prove that the following $(n+m) \times (n+m)$ matrices are similar,

$$D := \begin{pmatrix} A & 0 \\ 0 & B \end{pmatrix}, \qquad T := \begin{pmatrix} A & C \\ 0 & B \end{pmatrix}.$$

We now prove the converse, following Flanders and Wimmer. Thus we assume that D and T are similar.

a. We define two homomorphisms ϕ_j of $\mathbf{M}_{n+m}(k)$:

$$\phi_0(K) := DK - KD, \qquad \phi_1(K) := TK - KD.$$

Prove that the kernels of ϕ_1 and ϕ_0 are isomorphic, hence of equal dimensions. **Hint:** This is where we use the assumption.

b. Let E be the subspace of $\mathbf{M}_{m \times (n+m)}(k)$, made of matrices (R, S) such that

$$BR = RA, \qquad BS = SB.$$

Verify that if

$$K := \begin{pmatrix} P & Q \\ R & S \end{pmatrix} \in \ker \phi_j \quad (j = 0 \text{ or } 1),$$

then $(R, S) \in E$. This allows us to define the projections $\mu_j(K) := (R, S)$, from $\ker \phi_j$ to E.

c. Verify that $\ker \mu_0 = \ker \mu_1$, and therefore $R(\mu_0)$ and $R(\mu_1)$ have equal dimensions.

d. Deduce that μ_1 is onto.

e. Show that there exists a matrix in $\ker \phi_1$, of the form

$$\begin{pmatrix} P & X \\ 0 & -I_m \end{pmatrix}.$$

Conclude.

12. Let $m, n \geq 1$ and $A \in \mathbf{GL}_n(k)$, $B \in \mathbf{GL}_m(k)$, $G, H \in \mathbf{M}_{m \times n}(k)$ be given. Let us define $R := A - G^T BH$ and $S := B^{-1} - HA^{-1}G^T$.

a. Show that the following matrices are equivalent within $\mathbf{M}_{m+n}(k)$,

$$\begin{pmatrix} B^{-1} & H \\ G^T & A \end{pmatrix}, \qquad \begin{pmatrix} B^{-1} & 0 \\ 0 & R \end{pmatrix}, \qquad \begin{pmatrix} S & 0 \\ 0 & A \end{pmatrix}.$$

b. Deduce the equality

$$\mathrm{rk}R - \mathrm{rk}S = n - m.$$

13. Let ϕ denote the Euler indicator, $\phi(m)$ is the number of integers less than m that are prime to m. We recall the formula

$$\sum_{d|n} \phi(d) = n.$$

In the sequel, we define the $n \times n$ matrices G (for gcd), Φ and D (for *divisibility*) by

$$g_{ij} := \gcd(i,j), \qquad \Phi := \mathrm{diag}(\phi(1),\dots,\phi(n)), \qquad d_{ij} := \begin{cases} 1 & \text{if } i|j, \\ 0 & \text{else.} \end{cases}$$

a. Prove that $D^T \Phi D = G$.

b. Deduce the Smith determinant formula:

$$\det((\gcd(i,j)))_{1 \le i,j \le n} = \phi(1)\phi(2) \cdots \phi(n).$$

c. Compute the invariant factors of G as a matrix of $\mathbf{M}_n(\mathbb{Z})$, for small values of n. Say up to $n = 10$.

14. Prove Theorem 9.3 directly.

15. Theorem 9.2 tells us that $D_k(M)^2$ divides $D_{k-1}(M)D_{k+1}(M)$ and therefore divides every product of minors of respective sizes $k-1$ and $k+1$. This is not obvious a priori. We verify this directly for $k = 2$.

a. We compute modulo the ideal generated by $D_2(M)^2$. Assume that $n, m \ge 4$. Show that

$$m_{44}M \begin{pmatrix} 1 & 2 & 3 \\ 1 & 2 & 3 \end{pmatrix} \equiv m_{14}M \begin{pmatrix} 4 & 2 & 3 \\ 1 & 2 & 3 \end{pmatrix}.$$

b. Using the result above three times, together with

$$M \begin{pmatrix} i_1 & i_2 & i_3 \\ j_1 & j_2 & j_3 \end{pmatrix} = -M \begin{pmatrix} i_2 & i_1 & i_3 \\ j_1 & j_2 & j_3 \end{pmatrix},$$

prove that

$$m_{44}M \begin{pmatrix} 1 & 2 & 3 \\ 1 & 2 & 3 \end{pmatrix} \equiv -m_{44}M \begin{pmatrix} 1 & 2 & 3 \\ 1 & 2 & 3 \end{pmatrix}.$$

c. Conclude.

Chapter 10
Exponential of a Matrix, Polar Decomposition, and Classical Groups

Polar decomposition and exponentiation are fundamental tools in the theory of finite-dimensional Lie groups and Lie algebras. We do not consider these notions here in their full generality, but restrict attention to their matricial aspects.

10.1 The Polar Decomposition

The polar decomposition of matrices is defined by analogy with that in the complex plane: if $z \in \mathbb{C}^*$, there exists a unique pair $(r, q) \in (0, +\infty) \times S^1$ (S^1 denotes the unit circle, the set of complex numbers of modulus 1) such that $z = rq$. If z acts on $\mathbb{C}$ (or on $\mathbb{C}^*$) by multiplication, this action can be decomposed as the product of a rotation of angle θ (where $q = \exp(i\theta)$) with a homothety of ratio $r > 0$. The fact that these two actions commute is a consequence of the commutativity of the multiplicative group $\mathbb{C}^*$; this commutation is false for the polar decomposition in $\mathbf{GL}_n(k)$, $k = \mathbb{R}$ or $\mathbb{C}$, because the general linear group is not commutative.

The factors $(0, +\infty)$ and S^1 are replaced by $\mathbf{HPD}_n$, the open cone of matrices of $\mathbf{M}_n(\mathbb{C})$ that are Hermitian positive-definite, and the unitary group $\mathbf{U}_n$. In $\mathbf{M}_n(\mathbb{R})$, we play instead with $\mathbf{SPD}_n$, the set of symmetric positive-definite matrices, and the orthogonal group $\mathbf{O}_n$. The groups $\mathbf{U}_n$ and $\mathbf{O}_n$ are compact, because they are closed and bounded in $\mathbf{M}_n(K = \mathbb{R}, \mathbb{C})$. The columns of unitary matrices are unit vectors, so that $\mathbf{U}_n$ is bounded. On the other hand, $\mathbf{U}_n$ is defined by an equation $U^*U = I_n$, where the map $U \mapsto U^*U$ is continuous; hence $\mathbf{U}_n$ is closed. By the same arguments, $\mathbf{O}_n$ is compact.

Theorem 10.1 *For every $M \in \mathbf{GL}_n(\mathbb{C})$, there exists a unique pair*

$$(H, Q) \in \mathbf{HPD}_n \times \mathbf{U}_n$$

such that $M = HQ$. If $M \in \mathbf{GL}_n(\mathbb{R})$, then $(H, Q) \in \mathbf{SPD}_n \times \mathbf{O}_n$.

The map $M \mapsto (H,Q)$, called the polar decomposition *of M, is a homeomorphism (i.e., a bicontinuous bijection) between $\mathbf{GL}_n(\mathbb{C})$ and $\mathbf{HPD}_n \times \mathbf{U}_n$ (respectively, between $\mathbf{GL}_n(\mathbb{R})$ and $\mathbf{SPD}_n \times \mathbf{O}_n$).*

Proof. Existence.   Because $MM^* \in \mathbf{HPD}_n$, we can set $H := \sqrt{MM^*}$ (the square root was defined in Section 6.1). Then $Q := H^{-1}M$ satisfies $Q^*Q = M^*H^{-2}M = M^*(MM^*)^{-1}M = I_n$, hence $Q \in \mathbf{U}_n$.

In the real case ($M \in \mathbf{GL}_n(\mathbb{R})$), MM^* is real symmetric. In fact, H is real symmetric. Hence Q is real orthogonal.

Uniqueness.   Let $M = HQ$ be a polar decomposition, then $MM^* = HQQ^*H = H^2$. Because of the uniqueness of the positive-definite square root, we have $H = \sqrt{MM^*}$ and thus $Q = H^{-1}M$.

Smoothness.   The map $(H,Q) \mapsto HQ$ is polynomial, hence continuous. Conversely, it is enough to prove that $M \mapsto (H,Q)$ is sequentially continuous, because $\mathbf{GL}_n(\mathbb{C})$ is a metric space. Let $(M_k)_{k \in \mathbb{N}}$ be a convergent sequence in $\mathbf{GL}_n(\mathbb{C})$, with limit M. Let us denote by $M_k = H_kQ_k$ and $M = HQ$ their respective polar decompositions. Because $\mathbf{U}_n$ is compact, the sequence $(Q_k)_{k \in \mathbb{N}}$ admits a cluster point R, that is, a limit of some subsequence $(Q_{k_\ell})_{l \in \mathbb{N}}$, with $k_\ell \to +\infty$. Then $H_{k_\ell} = M_{k_\ell}Q_{k_\ell}^*$ converges to $S := MR^*$. The matrix S is Hermitian positive-semidefinite (because it is the limit of the H_{k_ℓ}s) and invertible (because it is the product of M and R^*). It is thus positive-definite. Hence, SR is a polar decomposition of M. The uniqueness part ensures that $R = Q$ and $S = H$. The sequence $(Q_k)_{k \in \mathbb{N}}$, which is relatively compact and has at most one cluster point (namely Q), converges to Q. Finally, $H_k = M_kQ_k^*$ converges to $MQ^* = H$.

□

Remark

There is as well a polar decomposition $M = QH$ with the same properties. We may speak of left- and right-polar decomposition. We use one or the other depending on the context. We warn the reader that for a given matrix, the H-factors in both decompositions do not coincide. For example, in $M = HQ$, H is the square root of MM^*, although in $M = QH$, it is the square root of M^*M. However, the Q-factors coincide.

10.2 Exponential of a Matrix

The ground field is here $k = \mathbb{C}$. By restriction, we can also treat the case $k = \mathbb{R}$. Because $z \mapsto \exp z$ is holomorphic over $\mathbb{C}$, we may define $\exp(A)$ for every matrix through the functional calculus developed in Section 5.5. However, it is more efficient to give an explicit definition of $\exp A$ by using the exponential series. Both ways are, of course, equivalent.

For A in $\mathbf{M}_n(\mathbb{C})$, the series

$$\sum_{k=0}^{\infty} \frac{1}{k!} A^k$$

converges normally (which means that the series of norms is convergent), because for any matrix norm, we have

$$\sum_{k=0}^{\infty} \left\| \frac{1}{k!} A^k \right\| \leq \sum_{k=0}^{\infty} \frac{1}{k!} \|A\|^k = \exp \|A\|.$$

Because $\mathbf{M}_n(\mathbb{C})$ is complete, the series is convergent, and the estimation above shows that it converges uniformly on every bounded set. Its sum, denoted by $\exp A$, thus defines a continuous map $\exp : \mathbf{M}_n(\mathbb{C}) \to \mathbf{M}_n(\mathbb{C})$, called the *exponential*. When $A \in \mathbf{M}_n(\mathbb{R})$, we have $\exp A \in \mathbf{M}_n(\mathbb{R})$.

Given two matrices A and B in the general position, the binomial formula is not valid: $(A + B)^k$ does not necessarily coincide with

$$\sum_{j=0}^{j=k} \binom{k}{j} A^j B^{k-j}.$$

It thus follows that $\exp(A + B)$ differs in general from $\exp A \cdot \exp B$. A correct statement is the following.

Proposition 10.1 *Let $A, B \in \mathbf{M}_n(\mathbb{C})$ be commuting matrices; that is, $AB = BA$. Then $\exp(A + B) = (\exp A)(\exp B)$.*

Proof. The proof is exactly the same as for the exponential of complex numbers. We observe that because the series defining the exponential of a matrix is normally convergent, we may compute the product $(\exp A)(\exp B)$ by multiplying term by term the series

$$(\exp A)(\exp B) = \sum_{j,k=0}^{\infty} \frac{1}{j!k!} A^j B^k.$$

In other words,

$$(\exp A)(\exp B) = \sum_{\ell=0}^{\infty} \frac{1}{\ell!} C_\ell,$$

where

$$C_\ell := \sum_{j+k=\ell} \frac{\ell!}{j!k!} A^j B^k.$$

From the assumption $AB = BA$, we know that the binomial formula holds. Therefore, $C_\ell = (A + B)^\ell$, which proves the proposition. $\square$

Noting that $\exp 0_n = I_n$ and that A and $-A$ commute, we derive the following consequence.

Corollary 10.1 *For every $A \in \mathbf{M}_n(\mathbb{C})$, $\exp A$ is invertible, and its inverse is $\exp(-A)$.*

Given two conjugate matrices $B = P^{-1}AP$, we have $B^k = P^{-1}A^kP$ for each integer k and thus

$$\exp(P^{-1}AP) = P^{-1}(\exp A)P. \tag{10.1}$$

If $D = \mathrm{diag}(d_1, \ldots, d_n)$ is diagonal, we have $\exp D = \mathrm{diag}(\exp d_1, \ldots, \exp d_n)$. Of course, this formula, or more generally (10.1), can be combined with Jordan reduction in order to compute the exponential of a given matrix. Let us keep in mind, however, that Jordan reduction cannot be carried out explicitly.

Let us introduce a real parameter t and define a function g by $g(t) = \exp tA$. From Proposition 10.1, we see that g satisfies the functional equation

$$g(s+t) = g(s)g(t). \tag{10.2}$$

We have $g(0) = I_n$ and

$$\frac{g(t) - g(0)}{t} - A = \sum_{k=2}^{\infty} \frac{t^{k-1}}{k!} A^k.$$

Using any matrix norm, we deduce that

$$\left\| \frac{g(t) - g(0)}{t} - A \right\| \leq \frac{e^{\|tA\|} - 1 - \|tA\|}{|t|},$$

from which we obtain

$$\lim_{t \to 0} \frac{g(t) - g(0)}{t} = A.$$

We conclude that g has a derivative at $t = 0$, with $g'(0) = A$. Using the functional equation (10.2), we then obtain that g is differentiable everywhere, with

$$g'(t) = \lim_{s \to 0} \frac{g(t)g(s) - g(t)}{s} = g(t)A.$$

We observe that we also have

$$g'(t) = \lim_{s \to 0} \frac{g(s)g(t) - g(t)}{s} = Ag(t).$$

From either of these differential equations we see that g is actually infinitely differentiable. We retain the formula

$$\frac{d}{dt} \exp tA = A \exp tA = (\exp tA)A. \tag{10.3}$$

This differential equation is sometimes the most practical way to compute the exponential of a matrix. This is of particular relevance when A has real entries and has at least one nonreal eigenvalue, if one wishes to avoid the use of complex numbers.

Proposition 10.2 *For every $A \in \mathbf{M}_n(\mathbb{C})$,*

$$\det \exp A = \exp \operatorname{Tr} A. \tag{10.4}$$

Proof. We begin with a reduction of A of the form $A = P^{-1}TP$, where T is upper-triangular. Because T^k is still triangular, with diagonal entries equal to t_{jj}^k, $\exp T$ is triangular too, with diagonal entries equal to $\exp t_{jj}$. Hence

$$\det \exp T = \prod_j \exp t_{jj} = \exp \sum_j t_{jj} = \exp \operatorname{Tr} T.$$

This is the expected formula, inasmuch as $\exp A = P^{-1}(\exp T)P$ and $\operatorname{Tr} A = \operatorname{Tr} T$. □

Because $(M^*)^k = (M^k)^*$, we have $(\exp M)^* = \exp(M^*)$. In particular, the exponential of a skew-Hermitian matrix is unitary, for then

$$(\exp M)^* \exp M = \exp(M^*) \exp M = \exp(-M) \exp M = I_n.$$

Likewise, the exponential of an Hermitian matrix is Hermitian positive-definite, because

$$\exp M = \left(\exp \frac{1}{2}M \right)^2 = \left(\exp \frac{1}{2}M \right)^* \exp \frac{1}{2}M$$

and the fact that $\exp(M/2)$ is nonsingular. This calculation also shows that if M is Hermitian, then

$$\sqrt{\exp M} = \exp \frac{1}{2}M.$$

We have the following more accurate statement.

Proposition 10.3 *The map* $\exp : \mathbf{H}_n \to \mathbf{HPD}_n$ *is a homeomorphism.*

Proof. Injectivity: Let $A, B \in \mathbf{H}_n$ with $\exp A = \exp B =: H$. Then

$$\exp \frac{1}{2}A = \sqrt{H} = \exp \frac{1}{2}B.$$

By induction, we have

$$\exp 2^{-m}A = \exp 2^{-m}B, \quad m \in \mathbb{Z}.$$

Subtracting I_n, multiplying by 2^m, and passing to the limit as $m \to +\infty$, we obtain

$$\left. \frac{d}{dt} \right|_{t=0} \exp tA = \left. \frac{d}{dt} \right|_{t=0} \exp tB;$$

that is, $A = B$.

Surjectivity: Let $H \in \mathbf{HPD}_n$ be given. Then $H = U^* \operatorname{diag}(d_1, \ldots, d_n)U$, where U is unitary and $d_j \in (0, +\infty)$. From above, we know that $H = \exp M$ for

$$M := U^* \operatorname{diag}(\log d_1, \ldots, \log d_n)U,$$

which is Hermitian.

Continuity: The continuity of exp has already been proved. Let us investigate the continuity of the reciprocal map. Let $(H^\ell)_{\ell \in \mathbb{N}}$ be a sequence in $\mathbf{HPD}_n$ that converges to $H \in \mathbf{HPD}_n$. We denote by $M^\ell, M \in \mathbf{H}_n$, the Hermitian matrices whose exponentials are H^ℓ and H. The continuity of the spectral radius gives

$$\lim_{\ell \to +\infty} \rho(H^\ell) = \rho(H), \quad \lim_{\ell \to +\infty} \rho\left((H^\ell)^{-1}\right) = \rho\left(H^{-1}\right). \tag{10.5}$$

Because $\mathrm{Sp}(M^\ell) = \log \mathrm{Sp}(H^\ell)$, we have

$$\rho(M^\ell) = \log \max \left\{ \rho(H^\ell), \rho\left((H^\ell)^{-1}\right) \right\}. \tag{10.6}$$

Keeping in mind that the restriction to $\mathbf{H}_n$ of the induced norm $\|\cdot\|_2$ coincides with that of the spectral radius ρ, we deduce from (10.5) and (10.6) that the sequence $(M^\ell)_{\ell \in \mathbb{N}}$ is bounded. If N is a cluster point of the sequence, the continuity of the exponential implies $\exp N = H$. But the injectivity shown above implies $N = M$. The sequence $(M^\ell)_{\ell \in \mathbb{N}}$, bounded with a unique cluster point, is convergent.
□

10.3 Structure of Classical Groups

Proposition 10.4 *Let G be a subgroup of $\mathbf{GL}_n(\mathbb{C})$. We assume that G is stable under the map $M \mapsto M^*$ and that for every $M \in G \cap \mathbf{HPD}_n$, the square root $\sqrt{M}$ is an element of G. Then G is stable under polar decomposition. Furthermore, polar decomposition is a homeomorphism between G and*

$$(G \cap \mathbf{U}_n) \times (G \cap \mathbf{HPD}_n).$$

This proposition applies in particular to subgroups of $\mathbf{GL}_n(\mathbb{R})$ that are stable under transposition and under extraction of square roots in $\mathbf{SPD}_n$. One has then

$$G \overset{\text{homeo}}{\sim} (G \cap \mathbf{O}_n) \times (G \cap \mathbf{SPD}_n).$$

Proof. Let $M \in G$ be given and let HQ be its polar decomposition. Because $MM^* \in G \cdot G = G$, we have $H^2 \in G$, hence $H \in G$ by assumption. Finally, we have $Q = H^{-1}M \in G^{-1} \cdot G = G$. An application of Theorem 10.1 finishes the proof. □

We apply this general result to the classical groups $\mathbf{U}(p,q)$, $\mathbf{O}(p,q)$ (where $n = p + q$) and $\mathbf{Sp}_m$ (where $n = 2m$). These are, respectively, the *unitary* group of the Hermitian form $|z_1|^2 + \cdots + |z_p|^2 - |z_{p+1}|^2 - \cdots - |z_n|^2$, the orthogonal group of the quadratic form $x_1^2 + \cdots + x_p^2 - x_{p+1}^2 - \cdots - x_n^2$, and the symplectic group. They are defined by $G = \{M \in \mathbf{M}_n(k) \mid M^*JM = J\}$, with $k = \mathbb{C}$ for $\mathbf{U}(p,q)$, $k = \mathbb{R}$ otherwise.

The matrix J equals

$$\begin{pmatrix} I_p & 0_{p\times q} \\ 0_{q\times p} & -I_q \end{pmatrix},$$

for $\mathbf{U}(p,q)$ and $\mathbf{O}(p,q)$, and

$$\begin{pmatrix} 0_m & I_m \\ -I_m & 0_m \end{pmatrix},$$

for $\mathbf{Sp}_m$. In each case, $J^2 = \pm I_n$.

Proposition 10.5 *Let J be a complex $n \times n$ matrix satisfying $J^2 = \pm I_n$. The subgroup G of $\mathbf{M}_n(\mathbb{C})$ defined by the equation $M^* J M = J$ is invariant under polar decomposition. If $M \in G$, then $|\det M| = 1$.*

Proof. Let $M \in G$. Then $\det J = \det M^* \det M \det J$; that is, $|\det M|^2 = 1$. In particular M is nonsingular. Then we have

$$M^{-*} J M^{-1} = M^{-*}(M^* J M) M^{-1} = J,$$

thus $M^{-1} \in G$. If $M, N \in G$, we also have

$$(MN)^* J(MN) = N^*(M^* J M)N = N^* J N = J$$

and again $MN \in G$. Thus G is a group.

For stability under adjunction, let us write, for $M \in G$,

$$M^* J M (J M^*) = J^2 M^* = \pm M^* = M^* J^2.$$

Simplifying by $M^* J$ on the left, there remains $MJM^* = J$; that is, $M^* \in G$.

Because G is a group, $M \in G$ implies $M^k \in G$; that is, $(M^*)^k J = JM^{-k}$ for every $k \in \mathbb{N}$. By linearity, it follows that $p(M^*)J = Jp(M^{-1})$ holds for every polynomial $p \in \mathbb{R}[X]$. Let us assume in addition that $M \in \mathbf{HPD}_n$. We then have $M = U^* \operatorname{diag}(d_1, \ldots, d_n)U$, where U is unitary and the d_js are positive real numbers. Let A be the set formed by the numbers d_j and $1/d_j$. There exists a polynomial p with real entries such that $p(a) = \sqrt{a}$ for every $a \in A$. Then we have $p(M) = \sqrt{M}$ and $p(M^{-1}) = \sqrt{M}^{-1}$. Because $M^* = M$, we also have $p(M)J = Jp(M^{-1})$; that is, $\sqrt{M}J = J\sqrt{M}^{-1}$. Hence $\sqrt{M} \in G$. From Proposition 10.4, G is stable under polar decomposition. $\square$

This leads us to our main result.

Theorem 10.2 *Under the hypotheses of Proposition 10.5, the group G is homeomorphic to $(G \cap \mathbf{U}_n) \times \mathbb{R}^d$, for a suitable integer d.*

Comments

- Of course, if $G = \mathbf{O}(p,q)$ or $\mathbf{Sp}_m$, the subgroup $G \cap \mathbf{U}_n$ can also be written as $G \cap \mathbf{O}_n$.

- We call $G \cap \mathbf{U}_n$ a *maximal compact subgroup* of G, because one can prove that it is not a proper subgroup of a compact subgroup of G. Another deep result, which is beyond the scope of this book, is that every maximal compact subgroup of G is a conjugate of $G \cap \mathbf{U}_n$. In the sequel, when speaking about the maximal compact subgroup of G, we always have in mind $G \cap \mathbf{U}_n$.
- In practice, the maximal compact subgroup is both homeo- and iso-morphic to some simpler classical group.

10.3.1 Calculation Trick

According to Theorem 10.2, one important step in the study of a classical group G is the calculation of $G \cap \mathbf{U}_n$. Both the definitions of G and of $\mathbf{U}_n$ are quadratic, thus nonlinear. However, the calculation can be done linearly, in part: If $M \in G \cap \mathbf{U}_n$, then we have $M^* J M = J$ and $M^* = M^{-1}$, whence the linear equation

$$JM = MJ. \tag{10.7}$$

The elements of $G \cap \mathbf{U}_n$ are precisely the unitary matrices satisfying (10.7). This equation gives easy information about the blocks of M. There remains to describe the unitary matrices that have a rather simple prescribed block form.

Proof. (of Theorem 10.2.)

According to Proposition 10.4, the proof amounts to showing that the factor $G \cap \mathbf{HPD}_n$ is homeomorphic to some $\mathbb{R}^d$. To do this, we define

$$\mathscr{G} := \{N \in \mathbf{M}_n(k) \,|\, \exp tN \in G, \forall t \in \mathbb{R}\}.$$

Lemma 14. *The set $\mathscr{G}$ defined above satisfies*

$$\mathscr{G} = \{N \in \mathbf{M}_n(k) \,|\, N^* J + J N = 0_n\}.$$

Proof. If $N^* J + J N = 0_n$, let us set $M(t) = \exp tN$. Then $M(0) = I_n$ and, thanks to (10.3)

$$\frac{d}{dt} M(t)^* J M(t) = M^*(t)(N^* J + J N) M(t) = 0_n,$$

so that $M(t)^* J M(t) \equiv J$. We thus have $N \in \mathscr{G}$. Conversely, if $M(t) := \exp tN \in G$ for every t, then the derivative at $t = 0$ of $M^*(t) J M(t) = J$ gives $N^* J + J N = 0_n$. □

Lemma 15. *The map* $\exp : \mathscr{G} \cap \mathbf{H}_n \to G \cap \mathbf{HPD}_n$ *is a homeomorphism.*

Proof. We must show that $\exp : \mathscr{G} \cap \mathbf{H}_n \to G \cap \mathbf{HPD}_n$ is onto. Let $M \in G \cap \mathbf{HPD}_n$ and let N be the Hermitian matrix such that $\exp N = M$. Let $p \in \mathbb{R}[X]$ be a polynomial with real entries such that for every $\lambda \in \mathrm{Sp} M \cup \mathrm{Sp} M^{-1}$, we have $p(\lambda) = \log \lambda$. Such a polynomial exists, because the numbers λ are real and positive.

Let $N = U^* D U$ be a unitary diagonalization of N. We have

$$M = \exp N = U^*(\exp D)U \quad \text{and} \quad M^{-1} = \exp(-N) = U^*\exp(-D)U.$$

Hence, $p(M) = N$ and $p(M^{-1}) = -N$. However, $M \in G$ implies $MJ = JM^{-1}$, and therefore $q(M)J = Jq(M^{-1})$ for every $q \in \mathbb{R}[X]$. With $q = p$, we obtain $NJ = -JN$. $\square$

These two lemmas complete the proof of the theorem, because $\mathscr{G} \cap \mathbf{H}_n$ is an $\mathbb{R}$-vector space. The integer d mentioned in the theorem is its dimension. $\square$

We warn the reader that neither $\mathscr{G}$ nor $\mathbf{H}_n$ is a $\mathbb{C}$-vector space. We present examples in the next section which show that $\mathscr{G} \cap \mathbf{H}_n$ can be naturally $\mathbb{R}$-isomorphic to a $\mathbb{C}$-vector space, which is a source of confusion. One therefore must be cautious when computing d.

The reader eager to learn more about the theory of classical groups is advised to have a look at the book of Mneimné and Testard [30] or the one by Knapp [26].

10.4 The Groups $\mathbf{U}(\mathbf{p}, \mathbf{q})$

Let us begin with the study of the maximal compact subgroup of $\mathbf{U}(p,q)$. If $M \in \mathbf{U}(p,q) \cap \mathbf{U}_n$, let us write M blockwise:

$$M = \begin{pmatrix} A & B \\ C & D \end{pmatrix},$$

where $A \in \mathbf{M}_p(\mathbb{C})$, and so on. As discussed above, we have (10.7); here

$$\begin{pmatrix} I_p & 0 \\ 0 & -I_q \end{pmatrix} \begin{pmatrix} A & B \\ C & D \end{pmatrix} = \begin{pmatrix} A & B \\ C & D \end{pmatrix} \begin{pmatrix} I_p & 0 \\ 0 & -I_q \end{pmatrix},$$

which yields $B = 0$ and $C = 0 : M$ is block diagonal. Then M is unitary if and only if A and D are so. This shows that the maximal compact subgroup of $\mathbf{U}(p,q)$ is isomorphic (not only homeomorphic) to $\mathbf{U}_p \times \mathbf{U}_q$.

Next, $\mathscr{G} \cap \mathbf{H}_n$ is the set of matrices

$$N = \begin{pmatrix} A & B \\ B^* & D \end{pmatrix},$$

where $A \in \mathbf{H}_p$, $D \in \mathbf{H}_q$, which satisfy $NJ + JN = 0_n$; that is, $A = 0_p$, $D = 0_q$. Hence $\mathscr{G} \cap \mathbf{H}_n$ is $\mathbb{R}$-isomorphic to $\mathbf{M}_{p \times q}(\mathbb{C})$. One therefore has $d = 2pq$.

Proposition 10.6 *The unitary group $\mathbf{U}(p,q)$ is homeomorphic to $\mathbf{U}_p \times \mathbf{U}_q \times \mathbb{R}^{2pq}$. In particular, $\mathbf{U}(p,q)$ is connected.*

There remains to show connectedness. It is a straightforward consequence of the following lemma.

Lemma 16. *The unitary group $\mathbf{U}_n$ is connected.*

Inasmuch as $\mathbf{GL}_n(\mathbb{C})$ is homeomorphic to $\mathbf{U}_n \times \mathbf{HPD}_n$ (via polar decomposition), hence to $\mathbf{U}_n \times \mathbf{H}_n$ via the exponential), it is equivalent to the following statement.

Lemma 17. *The linear group* $\mathbf{GL}_n(\mathbb{C})$ *is connected.*

Proof. Let $M \in \mathbf{GL}_n(\mathbb{C})$ be given. Define $A := \mathbb{C} \setminus \{(1-\lambda)^{-1} | \lambda \in \mathrm{Sp}(M)\}$, which is arcwise connected because its complement is finite. The set A contains the origin and the point $z = 1$, because $0 \notin \mathrm{Sp}(M)$. There exists a path γ joining 0 to 1 in $A : \gamma \in \mathscr{C}([0,1];A)$, $\gamma(0) = 0$, and $\gamma(1) = 1$. Let us define $M(t) := \gamma(t)M + (1 - \gamma(t))I_n$. By construction, $M(t)$ is invertible for every t, and $M(0) = I_n, M(1) = M$. The connected component of I_n is thus all of $\mathbf{GL}_n(\mathbb{C})$. $\square$

10.5 The Orthogonal Groups O(p, q)

The analysis of the maximal compact subgroup and of $\mathscr{G} \cap \mathbf{H}_n$ for the group $\mathbf{O}(p,q)$ is identical to that for $\mathbf{U}(p,q)$. On the one hand, $\mathbf{O}(p,q) \cap \mathbf{O}_n$ is isomorphic to $\mathbf{O}_p \times \mathbf{O}_q$. However, $\mathscr{G} \cap \mathbf{H}_n$ is isomorphic to $\mathbf{M}_{p \times q}(\mathbb{R})$, which is of dimension $d = pq$.

Proposition 10.7 *Let* $n \geq 1$. *The group* $\mathbf{O}(p,q)$ *is homeomorphic to* $\mathbf{O}_p \times \mathbf{O}_q \times \mathbb{R}^{pq}$. *The number of its connected components is two if* p *or* q *is zero, four otherwise.*

Proof. We must show that $\mathbf{O}_n$ has two connected components. It is the disjoint union of $\mathbf{SO}_n$ (matrices of determinant $+1$) and of $\mathbf{O}_n^-$ (matrices of determinant -1). Because $\mathbf{O}_n^- = M \cdot \mathbf{SO}_n$ for any matrix $M \in \mathbf{O}_n^-$ (e.g., a hyperplane symmetry), there remains to show that the special orthogonal group $\mathbf{SO}_n$ is connected, in fact arcwise connected. We use the following property:

Lemma 18. *Given* $M \in \mathbf{O}_n$, *there exists* $Q \in \mathbf{O}_n$ *such that the matrix* $Q^{-1}MQ$ *has the form*

$$\begin{pmatrix} (\cdot) & 0 & \cdots & 0 \\ 0 & \ddots & \ddots & \vdots \\ \vdots & \ddots & \ddots & 0 \\ 0 & \cdots & 0 & (\cdot) \end{pmatrix}, \tag{10.8}$$

where the diagonal blocks are of size 1×1 *or* 2×2 *and are orthogonal. The* 1×1 *blocks are* (± 1), *whereas those of size* 2×2 *are rotation matrices:*

$$\begin{pmatrix} \cos\theta & \sin\theta \\ -\sin\theta & \cos\theta \end{pmatrix}. \tag{10.9}$$

Let us apply Lemma 18 to $M \in \mathbf{SO}_n$. The determinant of M, which is the product of the determinants of the diagonal blocks, equals $(-1)^m$, m being the multiplicity of the eigenvalue -1. Because $\det M = 1$, m is even, and we can assemble the diagonal -1s pairwise in order to form matrices of the form (10.9), with $\theta = \pi$. Finally, there exists $Q \in \mathbf{O}_n$ such that

$$M = Q^T \begin{pmatrix} \begin{pmatrix} \cos\theta_1 & \sin\theta_1 \\ -\sin\theta_1 & \cos\theta_1 \end{pmatrix} & 0 & \cdots & & \cdots \cdots & 0 \\ 0 & \ddots & & \ddots & & \vdots \\ \vdots & & \ddots & \begin{pmatrix} \cos\theta_r & \sin\theta_r \\ -\sin\theta_r & \cos\theta_r \end{pmatrix} & \ddots & 0 \\ \vdots & & & \ddots & 1 & \ddots & \vdots \\ \vdots & & & & & \ddots & \ddots & 0 \\ 0 & \cdots & & \cdots & & \cdots & 0 & 1 \end{pmatrix} Q.$$

Let us now define a matrix $M(t)$ by the same formula, in which we replace the angles θ_j by $t\theta_j$. We thus obtain a path in $\mathbf{SO}_n$, from $M(0) = I_n$ to $M(1) = M$. The connected component of I_n is thus the whole of $\mathbf{SO}_n$. $\quad\square$

We now prove Lemma 18. As an orthogonal matrix, M is normal. From Theorem 5.5, it decomposes into a matrix of the form (10.8), the 1×1 diagonal blocks being the real eigenvalues. These eigenvalues are ± 1, inasmuch as M is orthogonal. The diagonal blocks 2×2 are direct similitude matrices. However, they are isometries, because $Q^{-1}MQ$ is orthogonal. Hence they are rotation matrices.

10.5.1 Notable Subgroups of $\mathbf{O(p,q)}$

We describe here the four connected components of $\mathbf{O}(p,q)$ when $p,q \geq 1$.

Let us write $M \in \mathbf{O}(p,q)$ blockwise

$$M = \begin{pmatrix} A & B \\ C & D \end{pmatrix},$$

where $A \in \mathbf{M}_p(\mathbb{R})$, and so on. When writing $M^T J M = J$, we find in particular $A^T A = C^T C + I_p$, as well as $D^T D = B^T B + I_q$. From the former identity, $A^T A$ is larger than I_p as a symmetric matrix, hence $\det A$ cannot vanish. More precisely, $|\det A| \geq 1$. Likewise, the latter shows that $\det D$ does not vanish. The continuous map $M \mapsto (\det A, \det D)$ thus sends $\mathbf{O}(p,q)$ to $((-\infty,-1] \cup [1,+\infty))^2$. The sign map from $(-\infty,-1] \cup [1,+\infty)$ to $\{-,+\}$ is continuous, therefore we define a continuous function

$$\mathbf{O}(p,q) \xrightarrow{\;\sigma\;} \{-,+\}^2 \sim (\mathbb{Z}/2\mathbb{Z})^2,$$
$$M \mapsto (\operatorname{sgn}\det A, \operatorname{sgn}\det D).$$

The diagonal matrices whose diagonal entries are ± 1 belong to $\mathbf{O}(p,q)$. It follows that σ is onto. Because σ is continuous, the preimage G_α of an element α of $\{-,+\}^2$ is the union of some connected components of $\mathbf{O}(p,q)$. Let $n(\alpha)$ be the number of these components. Then $n(\alpha) \geq 1$ (σ being onto), and $\sum_\alpha n(\alpha)$ equals 4, the number of connected components of $\mathbf{O}(p,q)$. There are four terms in this sum, therefore we obtain $n(\alpha) = 1$ for every α. Finally, the connected components of $\mathbf{O}(p,q)$ are the G_αs, where $\alpha \in \{-,+\}^2$.

The left multiplication by an element M of $\mathbf{O}(p,q)$ is continuous, bijective, and its inverse (another multiplication) is continuous. It thus induces a permutation of the set π_0 of connected components of $\mathbf{O}(p,q)$. Because σ induces a bijection between π_0 and $\{-,+\}^2$, there exists a permutation q_M of $\{-,+\}^2$ such that $\sigma(MM') = q_M(\sigma(M'))$. Likewise, the multiplication at the right by M' is a homeomorphism, whence another permutation $p_{M'}$ of $\{-,+\}^2$ such that $\sigma(MM') = p_{M'}(\sigma(M))$. The equality

$$p_{M'}(\sigma(M)) = q_M(\sigma(M'))$$

shows that p_M and q_M actually depend only on $\sigma(M)$. In other words, $\sigma(MM')$ depends only on $\sigma(M)$ and $\sigma(M')$. A direct evaluation in the special case of matrices in $\mathbf{O}(p,q) \cap \mathbf{O}_n(\mathbb{R})$ leads to the following conclusion.

Proposition 10.8 ($p,q \geq 1$) *The connected components of $G = \mathbf{O}(p,q)$ are the sets $G_\alpha := \sigma^{-1}(\alpha)$, defined by $\alpha_1 \det A > 0$ and $\alpha_2 \det D > 0$, when a matrix M is written blockwise as above. The map $\sigma : \mathbf{O}(p,q) \to \{-,+\}^2$ is a surjective group homomorphism; that is, $\sigma(MM') = \sigma(M)\sigma(M')$. In particular:*

1. $G_\alpha^{-1} = G_\alpha$.
2. $G_\alpha \cdot G_{\alpha'} = G_{\alpha\alpha'}$.

Remark

The map σ admits a right inverse, namely

$$\alpha \mapsto M^\alpha := \mathrm{diag}(\alpha_1, 1, \ldots, 1, \alpha_2).$$

The group $\mathbf{O}(p,q)$ is therefore the semidirect product of G_{++} with $(\mathbb{Z}/2\mathbb{Z})^2$.

We deduce immediately from the proposition that $\mathbf{O}(p,q)$ possesses five open and closed normal subgroups, the preimages of the five subgroups of $(\mathbb{Z}/2\mathbb{Z})^2$:

- $\mathbf{O}(p,q)$ itself.
- G_{++}, which we also denote by G_0 (see Exercise 15), the connected component of the unit element I_n.
- $G_{++} \cup G_\alpha$, for the three other choices of an element α.

One of these groups, namely $G_{++} \cup G_{--}$ is equal to the kernel $\mathbf{SO}(p,q)$ of the homomorphism $M \mapsto \det M$. In fact, this kernel is open and closed, thus is the union of connected components of $\mathbf{O}(p,q)$. However, the sign of $\det M$ for $M \in G_\alpha$ is that of $\alpha_1 \alpha_2$, which can be seen directly from the case of diagonal matrices M^α.

10.5.2 *The Lorentz Group* $\mathbf{O}(1,3)$

If $p = 1$ and $q = 3$, the group $\mathbf{O}(1,3)$ is isomorphic to the orthogonal group of the Lorentz quadratic form $dt^2 - dx_1^2 - dx_2^2 - dx_3^2$, which defines the space–time distance in special relativity.[1] Each element M of $\mathbf{O}(1,3)$ corresponds to the transformation

$$\begin{pmatrix} t \\ x \end{pmatrix} \mapsto M \begin{pmatrix} t \\ x \end{pmatrix},$$

which we still denote by M, by abuse of notation. This transformation preserves the light cone of equation $t^2 - x_1^2 - x_2^2 - x_3^2 = 0$. Because it is a homeomorphism of $\mathbb{R}^4$, it permutes the connected components of the complement $\mathscr{C}$ of that cone. There are three such components (see Figure 10.1):

- The convex set $C_+ := \{(t,x) \mid \|x\| < t\}$;
- The convex set $C_- := \{(t,x) \mid t < -\|x\|\}$;
- The "ring" $\mathscr{A} := \{(t,x) \mid |t| < \|x\|\}$.

Clearly, C_+ and C_- are homeomorphic. For example, they are so via the time reversal $t \mapsto -t$. However, they are not homeomorphic to $\mathscr{A}$, because the latter is homeomorphic to $S^2 \times \mathbb{R}^2$ (here, S^2 denotes the unit sphere), which is not contractible, whereas a convex set is always contractible. Because M is a homeomorphism, one deduces that necessarily, $M\mathscr{A} = \mathscr{A}$, and $MC_+ = C_\pm$, $MC_- = C_\mp$.

The transformations that preserve C_+ (we say that they preserve the *time arrow*), and therefore every connected component of $\mathscr{C}$, form the *orthochronous Lorentz group*, denoted $\mathbf{O}^+(1,3)$. Its elements are those that send the vector $\mathbf{e}_0 := (1,0,0,0)^T$ to C_+; that is, those for which the first component of $M\mathbf{e}_0$ is positive. Because this component is A (here a scalar), this group must be $G_{++} \cup G_{+-}$. The transformations that preserve the time arrow and the orientation form the group $G_{++} =: \mathbf{SO}^+(1,3)$.

10.6 The Symplectic Group $\mathbf{Sp_n}$

To begin with, we describe the maximal compact subgroup $\mathbf{Sp}_n \cap \mathbf{O}_{2n}$. If

$$M = \begin{pmatrix} A & B \\ C & D \end{pmatrix} \in \mathbf{Sp}_n \cap \mathbf{O}_{2n},$$

with blocks of size $n \times n$, then M satisfies (10.7); that is,

$$C = -B, \qquad D = A.$$

Hence

---

[1] We have selected a system of units in which the speed of light equals one.

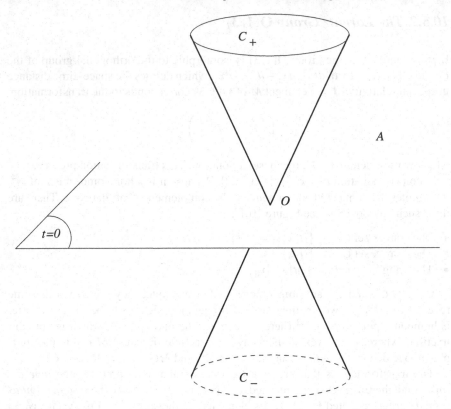

Fig. 10.1 The Lorentz cone. The spatial dimension has been reduced from 3 to 2 for the sake of clarity.

$$M = \begin{pmatrix} A & B \\ -B & A \end{pmatrix}.$$

There remains to write that M is orthogonal, which gives

$$A^T A + B^T B = I_n, \quad A^T B = B^T A.$$

This amounts to saying that $A + iB$ is unitary. One immediately checks that the map $M \mapsto A + iB$ is an isomorphism from $\mathbf{Sp}_n \cap \mathbf{O}_{2n}$ onto $\mathbf{U}_n$.
   Next, if

$$N \doteq \begin{pmatrix} A & B \\ B^T & D \end{pmatrix}$$

is symmetric and $NJ + JN = 0_{2n}$, we have, in fact,

$$N = \begin{pmatrix} A & B \\ B & -A \end{pmatrix},$$

where A and B are symmetric. Hence $\mathscr{G} \cap \mathbf{Sym}_{2n}$ is isomorphic to $\mathbf{Sym}_n \times \mathbf{Sym}_n$, that is, to $\mathbb{R}^{n(n+1)}$.

Proposition 10.9 *The symplectic group* $\mathbf{Sp}_n$ *is homeomorphic to* $\mathbf{U}_n \times \mathbb{R}^{n(n+1)}$.

Corollary 10.2 *In particular, every symplectic matrix has determinant* $+1$.

Indeed, Proposition 10.9 implies that $\mathbf{Sp}_n$ is connected. Because the determinant is continuous, with values in $\{-1,1\}$, it is constant, equal to $+1$.

Remark

Corollary 10.2 follows as well, and for every scalar field, from the formula (3.21).

Exercises

1. In the left- and right-polar decompositions of $M \in \mathbf{GL}_n(\mathbb{C})$, the unitary factors equal respectively $(MM^*)^{-1/2}M$ and $M(M^*M)^{-1/2}$. Deduce that they are equal to each other. **Hint:** More generally, $f(MM^*)M = Mf(M^*M)$ for every polynomial.
   Show that M is normal if and only if the Hermitian factor is the same in the left- and right-polar decompositions.
2. Let $M \in \mathbf{M}_n(k)$ be given, with $k = \mathbb{R}$ or $\mathbb{C}$. Show that there exists a polynomial $P \in k(X)$, of degree at most $n - 1$, such that $P(M) = \exp M$. However, show that this polynomial cannot be chosen independently of the matrix.
   Compute this polynomial when M is nilpotent.
3. For $t \in \mathbb{R}$, define *Pascal's matrix* $P(t)$ by $p_{ij}(t) = 0$ if $i < j$ (the matrix is lower-triangular) and
$$p_{ij}(t) = t^{i-j}\binom{i-1}{j-1}$$
   otherwise. Let us emphasize that for just this once in this book, P is an *infinite* matrix, meaning that its indices range over the infinite set $\mathbb{N}^*$. Compute $P'(t)$ and deduce that there exists a matrix L such that $P(t) = \exp(tL)$. Calculate L explicitly.
4. We use *Schur's norm* $\|A\| = (\mathrm{Tr}\,A^*A)^{1/2}$.

   a. If $A \in \mathbf{M}_n(\mathbb{C})$, show that there exists $Q \in \mathbf{U}_n$ such that $\|A - Q\| \leq \|A - U\|$ for every $U \in \mathbf{U}_n$. We define $S := Q^{-1}A$. We therefore have $\|S - I_n\| \leq \|S - U\|$ for every $U \in \mathbf{U}_n$.
   b. Let $H \in \mathbf{H}_n$ be an Hermitian matrix. Show that $\exp(itH) \in \mathbf{U}_n$ for every $t \in \mathbb{R}$. Compute the derivative at $t = 0$ of
$$t \mapsto \|S - \exp(itH)\|^2$$

and deduce that $S \in \mathbf{H}_n$.

c. Let D be a diagonal matrix, unitarily similar to S. Show that $\|D - I_n\| \leq \|DU - I_n\|$ for every $U \in \mathbf{U}_n$. By selecting a suitable U, deduce that $S \geq 0_n$.

d. If $A \in \mathbf{GL}_n(\mathbb{C})$, show that QS is the polar decomposition of A.

e. Deduce that if $H \in \mathbf{HPD}_n$ and if $U \in \mathbf{U}_n, U \neq I_n$, then $\|H - I_n\| < \|H - U\|$.

f. Finally, show that if $H \in \mathbf{H}_n, H \geq 0_n$ and $U \in \mathbf{U}_n$, then $\|H - I_n\| \leq \|H - U\|$.

5. Let $A \in \mathbf{M}_n(\mathbb{C})$ and $h \in \mathbb{C}$. Show that $I_n - hA$ is invertible as soon as $|h| < 1/\rho(A)$. One then denotes its inverse by $R(h;A)$ (the *resolvant* of A).

a. Let $r \in (0, 1/\rho(A))$. Show that there exists a $c_0 > 0$ such that for every $h \in \mathbb{C}$ with $|h| \leq r$, we have

$$\|R(h;A) - e^{hA}\| \leq c_0 |h|^2.$$

b. Verify the formula

$$C^m - B^m = (C - B)C^{m-1} + \cdots + B^{\ell-1}(C - B)C^{m-\ell} + \cdots + B^{m-1}(C - B),$$

and deduce the bound

$$\|R(h;A)^m - e^{mhA}\| \leq c_0 m |h|^2 e^{c_2 m |h|},$$

when $|h| \leq r$ and $m \in \mathbb{N}$.

c. Show that for every $t \in \mathbb{C}$,

$$\lim_{m \to +\infty} R(t/m;A)^m = e^{tA}.$$

This is the convergence of the implicit Euler difference scheme for the differential equation

$$\frac{dx}{dt} = Ax. \tag{10.10}$$

6. a. Let $J(a;r)$ be a Jordan block of size r, with $a \in \mathbb{C}^*$. Let $b \in \mathbb{C}$ be such that $a = e^b$. Show that there exists a nilpotent $N \in \mathbf{M}_r(\mathbb{C})$ such that $J(a;r) = \exp(bI_r + N)$.

b. Show that $\exp : \mathbf{M}_n(\mathbb{C}) \to \mathbf{GL}_n(\mathbb{C})$ is onto, but that it is not one-to-one. Deduce that $X \mapsto X^2$ is onto $\mathbf{GL}_n(\mathbb{C})$. Verify that it is not onto $\mathbf{M}_n(\mathbb{C})$.

7. a. Show that the matrix

$$J_2 = \begin{pmatrix} -1 & 1 \\ 0 & -1 \end{pmatrix}$$

is not the square of any matrix of $\mathbf{M}_2(\mathbb{R})$.

b. Show, however, that the matrix $J_4 := \mathrm{diag}(J_2, J_2)$ is the square of a matrix of $\mathbf{M}_4(\mathbb{R})$. Show also that the matrix

$$J_3 = \begin{pmatrix} J_2 & I_2 \\ 0_2 & J_2 \end{pmatrix}$$

is not the square of a matrix of $\mathbf{M}_4(\mathbb{R})$.

   c. Show that J_2 is not the exponential of any matrix of $\mathbf{M}_2(\mathbb{R})$. Compare with the previous exercise.

   d. Show that J_4 is the exponential of a matrix of $\mathbf{M}_4(\mathbb{R})$, but that J_3 is not.

8. Let $\mathbf{A}_n(\mathbb{C})$ be the set of skew-Hermitian matrices of size n. Show that $\exp : \mathbf{A}_n(\mathbb{C}) \to \mathbf{U}_n$ is onto. **Hint:** If U is unitary, diagonalize it.

9.   a. Let $\theta \in \mathbb{R}$ be given. Compute $\exp B$, where

$$B = \begin{pmatrix} 0 & \theta \\ -\theta & 0 \end{pmatrix}.$$

   b. Let $\mathbf{A}_n(\mathbb{R})$ be the set of real skew-symmetric matrices of size n. Show that $\exp : \mathbf{A}_n(\mathbb{R}) \to \mathbf{SO}_n$ is onto. **Hint:** Use the reduction of direct orthogonal matrices.

10. (J. Duncan.) We denote $\langle x, y \rangle$ the usual sesquilinear product in $\mathbb{C}^n$. Let $M \in \mathbf{GL}_n(\mathbb{C})$ be given. Thanks to Exercise 1, the left- and right-polar decompositions of M write $M = UH = KU$, with $H = \sqrt{M^*M}$ and $K = \sqrt{MM^*}$.

   a. Prove that $U\sqrt{H} = \sqrt{K}U$.

   b. Check that

$$\langle Mx, y \rangle = \langle \sqrt{H}x, U^*\sqrt{K}y \rangle, \qquad \forall x, y \in \mathbb{C}^n.$$

   Deduce that

$$|\langle Mx, y \rangle|^2 \le \langle Hx, x \rangle \langle Ky, y \rangle.$$

   c. More generally, let a rectangular matrix $A \in \mathbf{M}_{n \times m}(\mathbb{C})$ be given. Prove the generalized Cauchy–Schwarz inequality

$$|\langle Ax, y \rangle|^2 \le \langle \sqrt{A^*A}x, x \rangle \langle \sqrt{AA^*}y, y \rangle, \qquad \forall x, y \in \mathbb{C}^n.$$

   Hint: Use the decompositions

$$\mathbb{C}^m = \ker A \oplus^\perp R(A^*), \qquad \mathbb{C}^n = \ker A^* \oplus^\perp R(A),$$

   then apply the case above to the restriction of A from $R(A^*)$ to $R(A)$.

11. Let $\phi : \mathbf{M}_n(\mathbb{R}) \to \mathbb{R}$ be a nonnull map satisfying $\phi(AB) = \phi(A)\phi(B)$ for every $A, B \in \mathbf{M}_n(\mathbb{R})$. If $\alpha \in \mathbb{R}$, we set $\delta(\alpha) = |\phi(\alpha I_n)|^{1/n}$. We have seen, in Exercise 5 of Chapter 5, that $|\phi(M)| = \delta(\det M)$ for every $M \in \mathbf{M}_n(\mathbb{R})$.

   a. Show that on the range of $M \mapsto M^2$ and on that of $M \mapsto \exp M$, $\phi \equiv \delta \circ \det$.

   b. Deduce that $\phi \equiv \delta \circ \det$ on $\mathbf{SO}_n$ (use Exercise 9) and on $\mathbf{SPD}_n$.

   c. Show that either $\phi \equiv \delta \circ \det$ or $\phi \equiv (\text{sgn}(\det))\delta \circ \det$.

12. Let A be a Banach algebra ($K = \mathbb{R}$ or $\mathbb{C}$) with a unit denoted by e. If $x \in A$, define $x^0 := e$.

   a. Given $x \in A$, show that the series

$$\sum_{m \in \mathbb{N}} \frac{1}{m!} x^m$$

converges normally, hence converges in A. Its sum is denoted by $\exp x$.

b. If $x, y \in A$, $[x, y] = xy - yx$ is called the "commutator" of x and y. Verify that $[x, y] = 0$ implies

$$\exp(x + y) = (\exp x)(\exp y), \quad [x, \exp y] = 0.$$

c. Show that the map $t \mapsto \exp tx$ is differentiable on $\mathbb{R}$, with

$$\frac{d}{dt} \exp tx = x \exp tx = (\exp tx)x.$$

d. Let $x, y \in A$ be given. Assume all along this part that $[x, y]$ commutes with x and y.
   i. Show that $(\exp -tx)xy(\exp tx) = xy + t[y, x]x$.
   ii. Deduce that $[\exp -tx, y] = t[y, x] \exp -tx$.
   iii. Compute the derivative of $t \mapsto (\exp -ty)(\exp -tx) \exp t(x + y)$. Finally, prove the Campbell–Hausdorff formula

$$\exp(x + y) = (\exp x)(\exp y) \left(\exp \frac{1}{2}[y, x] \right).$$

e. In $A = \mathbf{M}_3(\mathbb{R})$, construct an example that satisfies the above hypothesis ($[x, y]$ commutes with x and y), where $[x, y]$ is nonzero.

13. Show that the map

$$H \mapsto f(H) := (iI_n + H)(iI_n - H)^{-1}$$

induces a homeomorphism from $\mathbf{H}_n$ onto the set of matrices of $\mathbf{U}_n$ whose spectrum does not contain -1. Find an equivalent of $f(tH) - \exp(-2itH)$ as $t \to 0$.

14. Let G be a group satisfying the hypotheses of Proposition 10.5.

   a. Show that $\mathscr{G}$ is a *Lie algebra*, meaning that it is stable under the bilinear map $(A, B) \mapsto [A, B] := AB - BA$.
   b. Show that for $t \to 0+$,

$$\exp(tA)\exp(tB)\exp(-tA)\exp(-tB) = I_n + t^2[A, B] + O(t^3).$$

   Deduce another proof of the stability of $\mathscr{G}$ by $[\cdot, \cdot]$.
   c. Show that the map $M \mapsto [A, M]$ is a derivation, meaning that the Jacobi identity

$$[A, [B, C]] = [[A, B], C] + [B, [A, C]]$$

   holds true.

15. A *topological group* is a group G endowed with a topology for which the maps $(g,h) \mapsto gh$ and $g \mapsto g^{-1}$ are continuous. Show that in a topological group, the connected component of the unit element is a normal subgroup. Show also that the open subgroups are closed. Illustrate this result with $G = \mathbf{O}(p,q)$. Give an example of a closed subgroup that is not open.

16. One identifies $\mathbb{R}^{2n}$ with $\mathbb{C}^n$ by the map

$$\begin{pmatrix} x \\ y \end{pmatrix} \mapsto x + iy.$$

Therefore, every matrix $M \in \mathbf{M}_{2n}(\mathbb{R})$ defines an $\mathbb{R}$-linear map $\tilde{M}$ from $\mathbb{C}^n$ into itself.

  a. Let

$$M = \begin{pmatrix} A & B \\ C & D \end{pmatrix} \in \mathbf{M}_{2n}(\mathbb{R})$$

  be given. Under what condition on the blocks A, B, C, D is the map $\tilde{M}$ $\mathbb{C}$-linear?

  b. Show that $M \mapsto \tilde{M}$ is an isomorphism from $\mathbf{Sp}_n \cap \mathbf{O}_{2n}$ onto $\mathbf{U}_n$.

17. Let k be a field and

$$P = \begin{pmatrix} A & B \\ C & D \end{pmatrix}$$

be an orthogonal matrix, with A and D square.
Prove that

$$\det D = \pm \det A.$$

Hint: multiply P by

$$\begin{pmatrix} A^T & C^T \\ 0 & I \end{pmatrix}.$$

Extend this result to elements P of a group $\mathbf{O}(p,q)$.

18. Let $A \in \mathbf{M}_n(\mathbb{C})$ be given, and $U(t) := \exp(tA)$.

  a. Show that $\|U(t)\| \le \exp(t\|A\|)$ for $t \ge 0$ and any matrix norm. Deduce that the integral

$$\int_0^{+\infty} e^{-2\gamma t} U(t)^* U(t)\, dt$$

  converges for every $\gamma > \|A\|$.

  b. Denote H_γ the value of this integral, when it is defined. Computing the derivative at $h = 0$ of $h \mapsto U(h)^* H_\gamma U(h)$, by two different methods, deduce that H_γ is a solution of

$$A^* X + XA = 2\gamma X - I_n, \quad X \in \mathbf{HPD}_n. \tag{10.11}$$

    c. Let γ be larger than the supremum of the real parts of eigenvalues of A. Show that Equation (10.11) admits a unique solution in $\mathbf{HPD}_n$, and that the above integral converges.

    d. In particular, if the spectrum of M has positive real part, and if $K \in \mathbf{HPD}_n$ is given, then the Lyapunov equation

$$M^*H + HM = K, \qquad H \in \mathbf{HPD}_n$$

admits a unique solution.

Let $x(t)$ be a solution of the differential equation $\dot{x} + Mx = 0$, show that $t \mapsto x^*Hx$ decays, and strictly if $x \neq 0$.

19. This exercise shows that a matrix $M \in \mathbf{GL}_n(\mathbb{R})$ is the exponential of a real matrix if and only if it is the square of another real matrix.

    a. Show that, in $\mathbf{M}_n(\mathbb{R})$, every exponential is a square.

    b. Given a matrix $A \in \mathbf{M}_n(\mathbb{C})$, we denote $\mathscr{A}$ the $\mathbb{C}$-algebra spanned by A, that is, the set of matrices $P(A)$ as P runs over $\mathbb{C}[X]$.

       i. Check that $\mathscr{A}$ is commutative, and that the exponential map is a homomorphism from $(\mathscr{A}, +)$ to $(\mathscr{A}^*, \times)$, where $\mathscr{A}^*$ denotes the subset of invertible matrices (a multiplicative group.)

      ii. Show that $\mathscr{A}^*$ is an open and connected subset of $\mathscr{A}$.

      iii. Let E denote $\exp(\mathscr{A})$, so that E is a subgroup of $\mathscr{A}^*$. Show that E is a neighbourhood of the identity. **Hint:** Use the implicit function theorem.

      iv. Deduce that E is closed in $\mathscr{A}^*$. **Hint:** The complement F of E in $\mathscr{A}$ satisfies $F = F \cdot E$ and thus is open. Conclude that $E = \mathscr{A}^*$.

      v. Finally, show that every matrix $B \in \mathbf{GL}_n(\mathbb{C})$ reads $B = \exp(P(B))$ for some polynomial $P \in \mathbb{C}[X]$.

    c. Let $B \in \mathbf{GL}_n(\mathbb{R})$ and $P \in \mathbb{C}[X]$ be as above. Show that

$$B^2 = \exp(P(B) + \bar{P}(B)).$$

    Conclusion?

20. (See also [41].)
Let M belong to $\mathbf{Sp}_n(\mathbb{R})$. We recall that $M^T J M = J$, $J^2 = -I_{2n}$, and $J^T = -J$.

    a. Show that the characteristic polynomial is reciprocal:

$$P_M(X) = X^{2n} P_M\left(\frac{1}{X}\right).$$

    Deduce a classification of the eigenvalues of M.

    b. Define the quadratic form

$$q(x) := 2x^T J M x.$$

    Verify that M is a q-isometry.

c. Let $(e^{-i\theta}, e^{i\theta})$ be a pair of *simple* eigenvalues of M on the unit circle. Let Π be the corresponding invariant subspace:

$$\Pi := \ker(M^2 - 2(\cos\theta)M + I_{2n}).$$

   i. Show that $J\Pi^\perp$ is invariant under M.

   ii. Using the formula (5.3), show that $e^{\pm i\theta}$ are not eigenvalues of $M|_{J\Pi^\perp}$.

   iii. Deduce that $\mathbb{R}^{2n} = \Pi \oplus J\Pi^\perp$.

d. (Continued.)

   i. Show that q does not vanish on $\Pi \setminus \{0\}$. Hence q defines a Euclidean structure on Π.

   ii. Check that $M|_\Pi$ is direct (its determinant is positive).

   iii. Show that $M|_\Pi$ is a rotation with respect to the Euclidean structure defined by q, whose angle is either θ or $-\theta$.

e. More generally, assume that a plane Π is invariant under a symplectic matrix M, with corresponding eigenvalues $e^{\pm i\theta}$, and that Π is not Lagrangian: $(x, y) \mapsto y^T J x$ is not identically zero on Π. Show that $M|_\Pi$ acts as rotation of angle $\pm\theta$. In particular, if $M = J$, show that $\theta = +\pi/2$.

f. Let H be an invariant subspace of M, on which the form q is either positive or negative-definite. Prove that the spectrum of $M|_H$ lies in the unit circle and that $M|_H$ is semisimple (the Jordan form is diagonal).

g. Equivalently, let λ be an eigenvalue of M (say a simple one) with $\lambda \notin \mathbb{R}$ and $|\lambda| \neq 1$. Let H be the invariant subspace associated with the eigenvalues $(\lambda, \bar\lambda, 1/\lambda, 1/\bar\lambda)$. Show that the restriction of the form q to H is neither positive nor negative-definite. Show that the invariant subspace K associated with the eigenvalues λ and $\bar\lambda$ is q-isotropic. Thus, if $q|_H$ is non-degenerate, its signature is $(2, 2)$.

21. This is a sequel of Exercise 23, Chapter 7. Let Σ denote the unit sphere of $\mathbf{M}_2(\mathbb{R})$ for the induced norm $\|\cdot\|_2$. Recall that Σ is the union of the segments $[r, s]$ where $r \in \mathscr{R} := \mathbf{SO}_2(\mathbb{R})$ and $s \in \mathscr{S}$, the set of orthogonal symmetries. Both $\mathscr{R}$ and $\mathscr{S}$ are circles. Finally, two distinct segments may intersect only at an extremity.

a. Show that there is a well-defined map $\rho : \Sigma \setminus \mathscr{S} \to \mathscr{R}$, such that M belongs to some segment $[\rho(M), s)$ with $s \in \mathscr{S}$. For which M is the other extremity s unique?

b. Show that the map ρ above is continuous, and that ρ coincides with the identity over $\mathscr{R}$. We say that ρ is a *retraction* from $\Sigma \setminus \mathscr{S}$ onto $\mathscr{R}$.

c. Let $f : D \to \Sigma$ be a continuous function, where D is the unit disk of the complex plane, such that $f(\exp(i\theta))$ is the rotation of angle θ. Show that $f(D)$ contains an element of $\mathscr{S}$.

Hint: Otherwise, there would be a retraction of D onto the unit circle, which is impossible (an equivalent statement to Brouwer fixed point theorem).

Meaning. Likewise, one finds that if a disk D' is immersed in Σ, with boundary $\mathscr{S}$, then it contains an element of $\mathscr{R}$. We say that the circles $\mathscr{R}$ and $\mathscr{S}$ of Σ are *linked*.

This result tells us that $\mathscr{R}$ and $\mathscr{S}$ are linked within Σ.

22. In $\mathbb{R}^{1+m}$ we denote the generic point by $(t,x)^T$, with $t \in \mathbb{R}$ and $x \in \mathbb{R}^m$. Let $\mathscr{C}^+$ be the cone defined by $t > \|x\|$. Recall that those matrices of $\mathbf{O}(1,m)$ that preserve $\mathscr{C}^+$ form the subgroup $\mathbf{O}^+(1,m)$. The Hermitian form $(t,x) \mapsto \|x\|^2 - |t|^2$ is denoted by q.
   Let M belong to $\mathbf{O}^+(1,m)$.

   a. Given a point x in the unit closed ball B of $\mathbb{R}^m$, let $(t,y)^T$ be the image of $(1,x)^T$ under M. Define $f(x) := y/t$. Prove that f is a continous map from B into itself. Deduce that it has a fixed point. Deduce that M has at least one real positive eigenvalue, associated with an eigenvector in the closure of $\mathscr{C}^+$. **Note:** If m is odd, one can prove that this eigenvector can be taken in the light cone $t = \|x\|$.
   b. If $Mv = \lambda v$ with $v \in \mathbb{C}^{1+m}$ and $q(v)) \neq 0$, show that $|\lambda| = 1$.
   c. Let $v = (t,x)$ and $w = (s,y)$ be *light* vectors (i.e., $q(v) = q(w) = 0$), linearly independent. Show that $v^*Jw \neq 0$.
   d. Assume that M admits an eigenvalue λ of modulus different from 1, v being an eigenvector. Show that $1/\lambda$ is also an eigenvector. Denote by w a corresponding eigenvector. Let $< v,w >^\circ$ be the orthogonal of v and w with respect to q. Using the previous question, show that the restriction q_1 of q to $< v,w >^\circ$ is positive-definite. Show that $< v,w >^\circ$ is invariant under M and deduce that the remaining eigenvalues have unit modulus. In particular, λ is real.
   e. Show that, for every $M \in \mathbf{O}^+(1,m)$, $\rho(M)$ is an eigenvalue of M.

23. We endow $\mathbf{M}_n(\mathbb{C})$ with the induced norm $\|\cdot\|_2$. Let G be a subgroup of $\mathbf{GL}_n(\mathbb{C})$ that is contained in the open ball $B(I_n; r)$ for some $r < 2$.

   a. Show that for every $M \in G$, there exists an integer $p \geq 1$ such that $M^p = I_n$.
   b. Let $A, B \in G$ be such that $\mathrm{Tr}(AM) = \mathrm{Tr}(BM)$ for every $M \in G$. Prove that $A = B$.
   c. Deduce that G is a finite group.
   d. Conversely, let R be a rotation in the plane ($n = 2$) of angle $\theta \notin \pi\mathbb{Q}$. Prove that the subgroup spanned by R is infinite and is contained in $B(I_2; 2)$.

24. Let $m \in \mathbb{N}^*$ be given. We denote $P_m : A \mapsto A^m$ the mth power in $\mathbf{M}_n(\mathbb{C})$. Show that the differential of P_m at A is given by

$$dP_m(A) \cdot B = \sum_{j=0}^{m-1} A^j B A^{m-1-j}.$$

Deduce the formula

$$\mathrm{D}\exp(A)\cdot B = \int_0^1 e^{(1-t)A}Be^{tA}dt.$$

25. Let $A \in \mathbf{M}_n(\mathbb{R})$ be a matrix satisfying $a_{ij} \geq 0$ for every pair (i,j) of distinct indices.

    a. Using the Exercise 3 of Chapter 8, show that

    $$R(h;A) := (I_n - hA)^{-1} \geq 0,$$

    for $h > 0$ small enough.
    b. Deduce that $\exp(tA) \geq 0$ for every $t > 0$. **Hint:** Use Exercise 5.
    c. Deduce that if $x(0) \geq 0$, then the solution of (10.10) is nonnegative for every nonnegative t.
    d. Deduce also that

    $$\sigma := \sup\{\Re\lambda \,|\, \lambda \in \mathrm{Sp}\, A\}$$

    is an eigenvalue of A.

26. We use the scalar product over $\mathbf{M}_n(\mathbb{C})$, given by $\langle M, N \rangle = \mathrm{Tr}(M^*N)$. We recall that the corresponding norm is the Schur–Frobenius norm $\|\cdot\|_F$. If $T \in \mathbf{GL}_n(\mathbb{C})$, we denote $T = U|T|$ the polar decomposition, with $|T| := \sqrt{T^*T}$ and $U \in \mathbf{U}_n$. The Aluthge transform $\Delta(T)$ is defined by

    $$\Delta(T) := |T|^{1/2}U|T|^{1/2}.$$

    a. Check that $\Delta(T)$ is similar to T.
    b. If T is normal, show that $\Delta(T) = T$.
    c. Show that $\|\Delta(T)\|_F \leq \|T\|_F$, with equality if and only if T is normal.
    d. We define Δ^n by induction, with $\Delta^n(T) := \Delta(\Delta^{n-1}(T))$.
        i. Given $T \in \mathbf{GL}_n(\mathbb{C})$, show that the sequence $(\Delta^k(T))_{k\in\mathbb{N}}$ is bounded.
        ii. Show that its limit points are normal matrices with the same characteristic polynomial as T (Jung, Ko and Pearcy, or Ando).
        iii. Deduce that when T has only one eigenvalue μ, then the sequence converges towards μI_n.
    e. If T is not diagonalizable, show that these limit points are not similar to T.

Chapter 11
Matrix Factorizations and Their Applications

The techniques described below are often called *direct solving methods*.

The direct solution (by Cramer's method, see Section 3.3.2) of a linear system $Mx = b$, when $M \in \mathbf{GL}_n(k)$ ($b \in k^n$) is computationally expensive, especially if one wishes to solve the system many times with various values of b. In the next chapter we study iterative methods for the case $k = \mathbb{R}$ or $\mathbb{C}$. Here we concentrate on a simple idea: to decompose M as a product PQ in such a way that the resolution of the intermediate systems $Py = b$ and $Qx = y$ is "cheap". In general, at least one of the matrices is triangular. For example, if P is lower-triangular ($p_{ij} = 0$ if $i < j$), then its diagonal entries p_{ii} are nonzero, and one may solve the system $Py = b$ step by step:

$$y_1 = \frac{b_1}{p_{11}},$$

$$\vdots$$

$$y_i = \frac{b_i - p_{i1}y_1 - \cdots - p_{i,i-1}y_{i-1}}{p_{ii}},$$

$$\vdots$$

$$y_n = \frac{b_n - p_{n1}y_1 - \cdots - p_{n,n-1}y_{n-1}}{p_{nn}}.$$

The computation of y_i needs $2i - 1$ operations and the final result is obtained in n^2 operations. This is not expensive if one notices that even computing the product $x = M^{-1}b$ (assuming that M^{-1} is computed once and for all, an expensive task) needs $2n^2 - n$ operations in general, and still n^2 in the triangular case.

Another example of easily invertible matrices is that of orthogonal matrices: if $Q \in \mathbf{O}_n$ (or $Q \in \mathbf{U}_n$), then $Qx = y$ is equivalent to $x = Q^T y$ (or $x = Q^* y$), which provides x in $O(n^2)$ operations.

11.1 The *LU* Factorization

Let k be a field.

Definition 11.1 *Let $M \in \mathbf{GL}_n(k)$ be given. We say that M admits an LU factorization if there exist in $\mathbf{GL}_n(k)$ two matrices L (lower-triangular with 1s on the diagonal) and U (upper-triangular) such that $M = LU$.*

Remarks

- The diagonal entries of U are not equal to 1 in general. The *LU* factorization is thus asymmetric with respect to L and U.
- The letters L and U recall the shape of the matrices: L for *lower* and U for *upper*.
- If there exists an *LU* factorization (which is unique, as we show below), then it can be computed by induction on the size of the matrix. The algorithm is provided in the proof of the next theorem. Indeed, if $N^{(p)}$ denotes the matrix extracted from N by keeping only the first p rows and columns, we have easily

$$M^{(p)} = L^{(p)} U^{(p)},$$

  which is nothing but the *LU* factorization of $M^{(p)}$.

Definition 11.2 *The* leading principal minors *of M are the determinants of the matrices $M^{(p)}$ defined above, for $1 \le p \le n$.*

Theorem 11.1 *The matrix $M \in \mathbf{GL}_n(k)$ admits an LU factorization if and only if its leading principal minors are nonzero. When this condition is fulfilled, the LU factorization is unique.*

Proof. Let us begin with uniqueness: if $LU = L'U'$, then $(L')^{-1}L = U'U^{-1}$, which reads $L'' = U''$, where L'' and U'' are triangular of opposite types, the diagonal entries of L'' being 1s. We deduce $L'' = U'' = I_n$; that is, $L' = L$, $U' = U$.

We next prove the necessity. Let us assume that M admits an *LU* factorization. Then $\det M^{(p)} = \det L^{(p)} \det U^{(p)} = \prod_{1 \le j \le p} u_{jj}$, which is nonzero because U is invertible.

Finally, we prove the sufficiency, that is, the existence of an *LU* factorization. We proceed by induction on the size of the matrices. It is clear if $n = 1$. Otherwise, let us assume that the statement is true up to the order $n - 1$ and let $M \in \mathbf{GL}_n(k)$ be given, with nonzero leading principal minors. We look for L and U in the blockwise form

$$L = \begin{pmatrix} L' & 0 \\ X^T & 1 \end{pmatrix}, \quad U = \begin{pmatrix} U' & Y \\ 0 & u \end{pmatrix},$$

with $L', U' \in \mathbf{M}_{n-1}(k)$, and so on. We likewise obtain the description

$$M = \begin{pmatrix} M' & R \\ S^T & m \end{pmatrix}.$$

Multiplying blockwise, we obtain the equations

$$L'U' = M', \quad L'Y = R, \quad (U')^T X = S, \quad u = m - X^T Y.$$

By assumption, the leading principal minors of M' are nonzero. The induction hypothesis guarantees the existence of the factorization $M' = L'U'$. Then Y and X are the unique solutions of (triangular) Cramer systems. Finally, u is explicitly given. □

Let us evaluate the number of operations needed in the computation of L and U. We pass from a factorization in $\mathbf{GL}_{n-1}(k)$ to a factorization in $\mathbf{GL}_n(k)$ by means of the computations of X (in $(n-1)(n-2)$ operations), Y (in $(n-1)^2$ operations) and u (in $2(n-1)$ operations), for a total of $(n-1)(2n-1)$ operations. Finally, the computation *ex nihilo* of an *LU* factorization costs

$$P(n) = 3 + 10 + \cdots + (n-1)(2n-1) = \frac{2}{3}n^3 + O(n^2)$$

operations.

Proposition 11.1 *The LU factorization is computable in $\frac{2}{3}n^3 + O(n^2)$ operations.*

One says that the *complexity* of the *LU* factorization is $\frac{2}{3}n^3$.

Remark

When all leading principal minors but the last (the determinant of M) are nonzero, the proof above furnishes a factorization $M = LU$, in which U is not invertible; that is, $u_{nn} = 0$.

11.1.1 Block Factorization

One can likewise perform a *blockwise LU* factorization. If $n = p_1 + \cdots + p_r$ with $p_j \geq 1$, the matrices L and U are block-triangular. The diagonal blocks are square, of respective sizes $p_1, \ldots, p_r$. Those of L are of the form I_{p_j}, whereas those of U are invertible. A necessary and sufficient condition for such a factorization to exist is that the leading principal minors of M, of orders $p_1 + \cdots + p_j$ $(j \leq r)$, be nonzero. As above, we may allow that $\det M \neq 0$, with the price that the last diagonal block of U be singular.

Such a factorization is useful for the resolution of the linear system $MX = b$ if the diagonal blocks of U are easily inverted, for instance if their sizes are small enough (say $p_j \approx \sqrt{n}$). Another favorable situation is when most of the diagonal blocks are equal to each other, because then one has to invert only a few blocks.

We have performed this blockwise factorization in Section 3.3.1 when $r = 2$. Recall that if

$$M = \begin{pmatrix} A & B \\ C & D \end{pmatrix}, \tag{11.1}$$

where the diagonal blocks are square and A is invertible, then

$$M = LU \qquad \text{with} \qquad L = \begin{pmatrix} I & 0 \\ CA^{-1} & I \end{pmatrix}, \quad U = \begin{pmatrix} A & B \\ 0 & D - CA^{-1}B \end{pmatrix}. \tag{11.2}$$

From this, we see that if M is nonsingular too, then

$$M^{-1} = U^{-1}L^{-1} = \begin{pmatrix} A^{-1} & \cdot \\ 0 & (D - CA^{-1}B)^{-1} \end{pmatrix} \cdot \begin{pmatrix} I & 0 \\ \cdot & I \end{pmatrix} = \begin{pmatrix} \cdot & \cdot \\ \cdot & (D - CA^{-1}B)^{-1} \end{pmatrix}.$$

When all the blocks have the same size, a similar analysis yields Banachiewicz' formula

Corollary 11.1 *Let $M \in \mathbf{GL}_n(k)$, with $n = 2m$, read blockwise*

$$M = \begin{pmatrix} A & B \\ C & D \end{pmatrix}, \quad A, B, C, D \in \mathbf{GL}_m(k).$$

Then

$$M^{-1} = \begin{pmatrix} (A - BD^{-1}C)^{-1} & (C - DB^{-1}A)^{-1} \\ (B - AC^{-1}D)^{-1} & (D - CA^{-1}B)^{-1} \end{pmatrix}.$$

Proof. We can verify the formula by multiplying by M. The only point to show is that the inverses are meaningful: $A - BD^{-1}C, \ldots$ are invertible. Because of the symmetry of the formulæ, it is enough to check it for a single term, namely $D - CA^{-1}B$. Schur's complement formula gives $\det(D - CA^{-1}B) = \det M / \det A$, which is nonzero by assumption. □

11.1.2 Complexity of Matrix Inversion

We can now show that the complexity of inverting a matrix is not higher than that of matrix multiplication, at equivalent sizes. This fact is due independently to Boltz, Banachiewicz, and to Strassen We assume here that $k = \mathbb{R}$ or $\mathbb{C}$.

Notation 11.1 *We denote by J_n the number of operations in k used in the inversion of a typical $n \times n$ matrix, and by P_n the number of operations (in k) used in the product of two $n \times n$ matrices.*

Of course, the number J_n must be understood for generic matrices, that is, for matrices within a dense open subset of $\mathbf{M}_n(k)$. More important, J_n, P_n also depend on the algorithm chosen for inversion or for multiplication. In the sequel we wish to adapt the inversion algorithm to the one used for multiplication.

Let us examine first of all the matrices whose size n has the form 2^k.

We decompose the matrices $M \in \mathbf{GL}_n(k)$ blockwise as in (11.1), with blocks of size $n/2 \times n/2$. The condition $A \in \mathbf{GL}_{n/2}(k)$ defines a dense open set, because $M \mapsto \det A$ is a nonzero polynomial. Suppose that we are given an inversion algorithm for generic matrices in $\mathbf{GL}_{n/2}(k)$ in $j_{k-1} = J_{2^{k-1}}$ operations. Then blockwise *LU* factorization and the formula $M^{-1} = U^{-1}L^{-1}$, where

$$L^{-1} = \begin{pmatrix} I & 0 \\ -CA^{-1} & I \end{pmatrix}, \quad U = \begin{pmatrix} A^{-1} & -A^{-1}B(D-CA^{-1}B)^{-1} \\ 0 & (D-CA^{-1}B)^{-1} \end{pmatrix},$$

furnish an inversion algorithm for generic matrices in $\mathbf{GL}_n(k)$. We can then bound j_k by means of j_{k-1} and the number $\pi_{k-1} = P_{2^{k-1}}$ of operations used in the computation of the product of two matrices of size $2^{k-1} \times 2^{k-1}$. We also denote by $\sigma_k = 2^{2k}$ the number of operations involved in the computation of the sum of matrices in $\mathbf{M}_{2^k}(k)$.

To compute M^{-1}, we first compute A^{-1}, then CA^{-1}, which gives us L^{-1} in $j_{k-1} + \pi_{k-1}$ operations. Then we compute $(D-CA^{-1}B)^{-1}$ (this amounts to $\sigma_{k-1} + \pi_{k-1} + j_{k-1}$ operations) and $A^{-1}B(D-CA^{-1}B)^{-1}$ (at cost $2\pi_{k-1}$), which furnishes U^{-1}. The computation of $U^{-1}L^{-1}$ is done with the expense of $\sigma_{k-1} + 2\pi_{k-1}$ operations. Finally,

$$j_k \le 2j_{k-1} + 2\sigma_{k-1} + 6\pi_{k-1}.$$

In other words,

$$2^{-k}j_k - 2^{1-k}j_{k-1} \le 2^{k-1} + 3 \cdot 2^{1-k}\pi_{k-1}. \qquad (11.3)$$

The complexity of the product in $\mathbf{M}_n(k)$ obeys the inequalities

$$n^2 \le P_n \le n^2(2n-1).$$

The first inequality is due to the number of data $(2n^2)$ and the fact that each operation involves only two of them. The second is given by the naive algorithm that consists in computing n^2 scalar products.

Lemma 19. *If $P_n \le c_\alpha n^\alpha$ (with $2 \le \alpha \le 3$), then $j_\ell \le C_\alpha \pi_\ell$, where $C_\alpha = 1 + 3c_\alpha/(2^{\alpha-1} - 1)$.*

Proof. It is enough to sum (11.3) from $k = 1$ to l and use the inequality $1 + q + \cdots + q^{l-1} \le q^l/(q-1)$ for $q > 1$. $\square$

When n is not a power of 2, we obtain M^{-1} by computing the inverse of a block-diagonal matrix $\mathrm{diag}(M,I)$, whose size N satisfies $n \le N = 2^\ell < 2n$. We obtain $J_n \le j_\ell \le C_\alpha \pi_\ell$. This is the first part of the following result.

Proposition 11.2 *If the complexity P_n of the product in $\mathbf{M}_n(\mathbb{C})$ is bounded by $c_\alpha n^\alpha$, then the complexity J_n of inversion in $\mathbf{GL}_n(\mathbb{C})$ is bounded by $d_\alpha n^\alpha$, where*

$$d_\alpha = \left(1 + \frac{3c_\alpha}{2^{\alpha-1} - 1}\right) 2^\alpha.$$

Conversely, if the complexity of inversion in $\mathbf{GL}_n(\mathbb{C})$ is bounded by $\delta_\alpha n^\alpha$, then the complexity of the product in $\mathbf{M}_n(\mathbb{C})$ is bounded by $\gamma_\alpha n^\alpha$, where

$$\gamma_\alpha = 3^\alpha \delta_\alpha.$$

That can be summarized as follows:

Those who know how to multiply also know how to invert.

Proof. There remains to prove the second part. We notice that if $A, B \in \mathbf{M}_n(\mathbb{C})$ are given, then the matrix

$$M = \begin{pmatrix} I_n & -A & 0_n \\ 0_n & I_n & -B \\ 0_n & 0_n & I_n \end{pmatrix} \in \mathbf{M}_{3n}(\mathbb{C})$$

is invertible, with inverse

$$M^{-1} = \begin{pmatrix} I_n & A & AB \\ 0_n & I_n & B \\ 0_n & 0_n & I_n \end{pmatrix}.$$

Given A and B, we compute M^{-1}, thus AB, in $\delta_\alpha (3n)^\alpha$ operations at most (and certainly much less). $\square$

11.1.3 Complexity of the Matrix Product

The ideas that follow apply to the product of rectangular matrices, but for the sake of simplicity, we present only the case of square matrices.

As we have seen above, the complexity P_n of matrix multiplication in $M_n(k)$ is $O(n^3)$. However, better algorithms allow us to improve the exponent 3. The simplest and oldest one is Strassen's algorithm, which uses a recursion argument. It is based upon a way of computing the product of two 2×2 matrices by means of 7 multiplications and 18 additions. Two features of Strassen's formula are essential. First, the number of multiplications that it involves is strictly less than that (eight) of the naive algorithm. The second is that the method is valid when the matrices have entries in a *noncommutative* ring, and so it can be employed for two matrices $M, N \in \mathbf{M}_n(k)$, considered as elements of $\mathbf{M}_2(A)$, with $A := \mathbf{M}_{n/2}(k)$. This trick yields

$$P_n \le 7 P_{n/2} + 18 \left(\frac{n}{2} \right)^2.$$

For $n = 2^\ell$, we infer

$$7^{-\ell} \pi_\ell - 7^{1-\ell} \pi_{\ell-1} \le \frac{9}{2} \left(\frac{4}{7} \right)^\ell,$$

which, after summation from $k = 0$ to ℓ, gives

$$7^{-\ell} \pi_\ell \le \frac{9}{2} \times \frac{1}{1 - 4/7},$$

because of $\frac{4}{7} < 1$. Finally,

$$\pi_\ell \le \frac{21}{2} 7^\ell.$$

When n is not a power of two, one chooses ℓ such that $n \le 2^\ell < 2n$ and we obtain the following result.

Proposition 11.3 *The complexity of the multiplication of $n \times n$ matrices is $O(n^\alpha)$, with $\alpha = \log 7/\log 2 = 2.807\ldots$ More precisely,*

$$P_n \le \frac{147}{2} n^{\log 7/\log 2}.$$

The exponent α can be improved, at the cost of greater complication and a larger constant c_α. The best exponent known in 2009, due to Coppersmith and Winograd [11], is $\alpha = 2.376\ldots$ It is fifteen years old, whereas Strassen's is forty years old. A rather complete analysis can be found in the book by Bürgisser, Clausen, and Shokrollahi [7].

Here is Strassen's formula [37]. Let $M, N \in \mathbf{M}_2(A)$, with

$$M = \begin{pmatrix} a & b \\ c & d \end{pmatrix}, \quad N = \begin{pmatrix} x & y \\ z & t \end{pmatrix}.$$

One first forms the expressions $x_1 = (a+d)(x+t)$, $x_2 = (c+d)x$, $x_3 = a(y-t)$, $x_4 = d(z-x)$, $x_5 = (a+b)t$, $x_6 = (c-a)(x+y)$, $x_7 = (b-d)(z+t)$. Each one involves one multiplication and either one or two addition(s). Then the product is given by eight more additions:

$$MN = \begin{pmatrix} x_1 + x_4 - x_5 + x_7 & x_3 + x_5 \\ x_2 + x_4 & x_1 - x_2 + x_3 + x_6 \end{pmatrix}.$$

Remark

The use of a fast method for matrix multiplication does reduce the complexity of many algorithms. Let us consider for instance the calculation of the characteristic polynomial $P(A)$ in the form improved by Preparata and Sawarte (see Section 3.10.2). If matrix multiplication is done in $O(n^\alpha)$ operations, then P_A is obtained in $O(n^\beta)$ operations, with $\beta = \max\{\alpha + \frac{1}{2}, 3\}$. If one has a not too cumbersome method with some $\alpha \le 2.5$, it is thus useless to try to reduce α.

11.2 Choleski Factorization

In this section $k = \mathbb{R}$, and we consider symmetric positive-definite matrices.

Theorem 11.2 *Let $M \in \mathbf{SPD}_n$. Then there exists a unique lower-triangular matrix $L \in \mathbf{M}_n(\mathbb{R})$, with strictly positive diagonal entries, satisfying $M = LL^T$.*

We warn the reader that, because of the symmetry between the lower- and upper-triangular factors, the diagonal entries of the matrix L are not units in general.

Proof. Uniqueness.   If L_1 and L_2 have the properties stated above, then $I_n = LL^T$, for $L = L_2^{-1}L_1$, which still has the same form. In other words, $L = L^{-T}$, where both sides are triangular matrices, but of opposite types (lower and upper). This equality shows that L is actually diagonal, with $L^2 = I_n$. Because its diagonal is positive, we obtain $L = I_n$; that is, $L_2 = L_1$.

We give two constructions of L.

First method.   The matrix $M^{(p)}$ is positive-definite (test the quadratic form induced by M on the linear subspace $\mathbb{R}^p \times \{0\}$). The leading principal minors of M are thus nonzero and there exists an LU factorization $M = L_0 U_0$. Let D be the diagonal of U_0, which is invertible. Then $U_0 = DU_1$, where the diagonal entries of U_1 equal 1. By transposition, we have $M = U_1^T D_0 L_0^T$. From uniqueness of the LU factorization, we deduce $U_1 = L_0^T$ and $M = L_0 DL_0^T$. Then $L = \sqrt{D}L_0$ satisfies the conditions of the theorem. Observe that $D > 0$ because $D = PMP^T$ with $P = L_0^{-1}$, and thus D is positive-definite.

Second method.   We proceed by induction over n. The statement is clear if $n = 1$. Otherwise, we seek an L of the form

$$L = \begin{pmatrix} L' & 0 \\ X^T & \ell \end{pmatrix},$$

knowing that

$$M = \begin{pmatrix} M' & R \\ R^T & m \end{pmatrix}.$$

The matrix L' is obtained by Choleski factorization of M', which belongs to $\mathbf{SPD}_{n-1}$. Then X is obtained by solving $L'X = R$. Finally, ℓ is a square root of $m - \|X\|^2$. Because $0 < \det M = (\ell \det L')^2$, we see that $m - \|X\|^2 > 0$; we thus choose $\ell = \sqrt{m - \|X\|^2}$. This method again shows uniqueness.
$\square$

Remark

Choleski factorization extends to Hermitian positive-definite matrices. In that case, L has complex entries, but its diagonal entries are still real and positive.

11.3 The *QR* Factorization

We turn to the situation where one factor is triangular, and the other one is unitary.

Proposition 11.4 *Let $M \in \mathbf{GL}_n(\mathbb{C})$ be given. Then there exist a unitary matrix Q and an upper-triangular matrix R; the diagonal entries of the latter real positive, such that $M = QR$. This factorization is unique.*

We observe that the condition on the numbers r_{jj} is essential for uniqueness. In fact, if D is diagonal with $|d_{jj}| = 1$ for every j, then $Q' := Q\bar{D}$ is unitary, $R' := DR$ is upper-triangular, and $M = Q'R'$, which gives an infinity of factorizations "QU". Even in the real case, where Q is orthogonal, there are 2^n "QU" factorizations.

Proof. Uniqueness. If (Q_1, R_1) and (Q_2, R_2) give two factorizations, then $Q = R$ with $Q := Q_2^{-1} Q_1$ and $R := R_2 R_1^{-1}$. Because Q is unitary, that is, $Q^* = Q^{-1}$, this implies $R^* = R^{-1}$. Because the inverse of a triangular matrix is a triangular matrix of the same type, whereas R^* is of opposite type, this tells us that R is diagonal. In additional, its diagonal part is strictly positive. Then $R^2 = R^*R = Q^*Q = I_n$ gives $R = I_n$. Finally, $R_2 = R_1$ and consequently, $Q_2 = Q_1$.

Existence.   It follows from that of Choleski factorization. If $M \in \mathbf{GL}_n(\mathbb{C})$, the matrix M^*M is Hermitian positive-definite, and hence it admits a Choleski factorization R^*R, where R is upper-triangular with real positive diagonal entries. Defining $Q := MR^{-1}$, we have

$$Q^*Q = R^{-*}M^*MR^{-1} = R^{-*}R^*RR^{-1} = I_n;$$

hence Q is unitary. Finally, $M = QR$ by construction.

$\square$

The method used above is unsatisfactory from a practical point of view, because one can compute Q and R directly, at a lower cost, without computing M^*M or its Choleski factorization. Moreover, the direct method, which we present now, is based on a theoretical observation: the QR factorization is nothing but the Gram–Schmidt orthonormalization procedure in $\mathbb{C}^n$, with respect to the canonical scalar product $\langle \cdot, \cdot \rangle$. In fact, giving M in $\mathbf{GL}_n(\mathbb{C})$ amounts to giving a basis $\{V^1, \ldots, V^n\}$ of $\mathbb{C}^n$, where $V^1, \ldots, V^n$ are the column vectors of M. If $Y^1, \ldots, Y^n$ denote the column vectors of Q, then $\{Y^1, \ldots, Y^n\}$ is an orthonormal basis. If $M = QR$, then

$$V^k = \sum_{j=1}^{k} r_{jk} Y^j.$$

Denoting by E_k the linear subspace spanned by $Y^1, \ldots, Y^k$, of dimension k, one sees that $V^1, \ldots, V^k$ are in E_k. Hence $\{V^1, \ldots, V^k\}$ is a basis of E_k. The columns of Q are therefore obtained by the Gram–Schmidt procedure, applied to the columns of M : Y^k is a unitary vector in E_k, orthogonal to E_{k-1}, where $E_k := \mathrm{Span}(V^1, \ldots, V^k)$.

The practical computation of Q and R is done by induction on k. If $k = 1$, then

$$r_{11} = \|V^1\|, \quad Y^1 = \frac{1}{r_{11}} V^1.$$

If $k > 1$, and if $Y^1, \ldots, Y^{k-1}$ are already known, one looks for Y^k and the entries r_{jk} $(j \leq k)$. For $j < k$, we have

$$r_{jk} = \langle V^k, Y^j \rangle.$$

Then

$$r_{kk} = \|Z_k\|, \quad Y^k = \frac{1}{r_{kk}} Z^k,$$

where

$$Z^k := V^k - \sum_{j=1}^{k-1} r_{jk} Y^j.$$

Let us examine the complexity of the procedure described above. To pass from the step $k - 1$ to the step k, one computes $k - 1$ scalar products, then Z^k, its norm, and finally Y^k. This requires $(4n - 1)k + 3n$ operations. Summing from $k = 1$ to n yields $2n^3 + O(n^2)$ operations. This method is not optimal, as we show in Section 13.3.3.

The interest of this construction lies also in giving a more complete statement than Proposition 11.4.

Theorem 11.3 *Let $M \in \mathbf{M}_n(\mathbb{C})$ be a matrix of rank p. There exists $Q \in \mathbf{U}_n$ and an upper-triangular matrix R, with $r_{\ell\ell} \in \mathbb{R}^+$ for every ℓ and $r_{jk} = 0$ for $j > p$, such that $M = QR$.*

Remark

The QR factorization of a singular matrix (i.e., a noninvertible one) is not unique. There exists, in fact, a QR factorization for rectangular matrices in which R is a "quasi-triangular" matrix.

11.4 Singular Value Decomposition

As we show in Exercise 14 below (see also Exercise 11 of Chapter 7), the eigenvalues of the matrix $H = \sqrt{M^*M}$, the Hermitian factor in the polar decomposition of a nonsingular matrix $M \in \mathbf{M}_n(\mathbb{C})$, are of some practical importance. They are called the *singular values* of M. These are the square roots of the eigenvalues of M^*M, thus one may even speak of the singular values of an arbitrary matrix, neither an invertible, nor even a square one. Recalling that (see Exercise 14 in Chapter 3) when M is $n \times m$, M^*M and MM^* have the same nonzero eigenvalues, counting them with multiplicities, one may even speak of the singular values of a rectangular matrix, up to an ambiguity concerning the multiplicity of the eigenvalue 0.

The main result of this section is the following.

Theorem 11.4 *Let $M \in \mathbf{M}_{n \times m}(\mathbb{C})$ be given. There exist two unitary matrices $U \in \mathbf{U}_n$, $V \in \mathbf{U}_m$ and a quasi-diagonal matrix*

$$D = \begin{pmatrix} s_1 & & & & \\ & \ddots & & & \\ & & s_r & & \\ & & & 0 & \\ & & & & \ddots \end{pmatrix},$$

with $s_1, \ldots, s_r \in (0, +\infty)$ and $s_1 \geq \cdots \geq s_r$, such that $M = UDV$. The numbers r and $s_1, \ldots, s_r$ are uniquely defined; they are respectively the rank and the nonzero singular values of M.

If $M \in \mathbf{M}_{n \times m}(\mathbb{R})$, one may choose U, V to be real orthogonal.

We notice that although D is uniquely defined, the other factors U and V are not unique. For instance, $M = I_n$ yields $D = I_n$ and $V = U^*$, where U can be an arbitrary unitary matrix.

Proof. To begin with, let us recall the following facts. We have

$$\mathbb{C}^n = R(M) \oplus^\perp \ker M^*, \qquad \ker MM^* = \ker M^*, \qquad R(M) = R(MM^*), \quad (11.4)$$

and on the other hand

$$\mathbb{C}^m = \ker(M) \oplus^\perp R(M^*), \qquad R(M^*) = R(M^*M), \qquad \ker M = \ker M^*M. \quad (11.5)$$

Inasmuch as MM^* is positive-semidefinite, we may write its eigenvalues as

$$s_1^2, \ldots, s_r^2, 0, \ldots,$$

where the s_js, the singular values of M, are positive real numbers arranged in decreasing order. The spectrum of M^*M has the same form, except for the multiplicity of 0. The index r is the rank of MM^*, that is, that of M, or as well that of M^*. The multiplicities of 0 as an eigenvalue of M^*M and MM^*, respectively, differ by $n - m$, whereas the multiplicities of other eigenvalues are the same for both matrices. We set $S = \text{diag}(s_1, \ldots, s_r)$.

Because MM^* is hermitian, there exists an orthonormal basis $\{\mathbf{u}_1, \ldots, \mathbf{u}_n\}$ of $\mathbb{C}^n$ consisting of eigenvectors associated with the s_j^2s, followed by vectors of $\ker M^*$ (because of (11.4)). Let us form the unitary matrix

$$U = (\mathbf{u}_1 | \ldots | \mathbf{u}_n).$$

Written blockwise, we have $U = (U_R, U_K)$, where

$$MM^*U_R = U_R S^2, \qquad M^*U_K = 0.$$

Let us define $V_R := M^*U_R S^{-1}$. From above, we have

$$V_R^*V_R = S^{-1}U_R^*MM^*U_R S^{-1} = I_r.$$

This means that the columns $\mathbf{v}_1, \ldots, \mathbf{v}_r$ of V_R constitute an orthonormal family. Obviously, it is included in $R(M^*)$.

Because $\dim R(M^*) = r$, $\{\mathbf{v}_1, \ldots, \mathbf{v}_r\}$ form an orthonormal basis of this space and can be extended to an orthonormal basis $\{\mathbf{v}_1, \ldots, \mathbf{v}_m\}$ of $\mathbb{C}^m$, where $\mathbf{v}_{r+1}, \ldots, \mathbf{v}_m$ belong to $\ker M$ (because of (11.5)). Let $V =: (V_R, V_K)$ be the unitary matrix whose columns are $\mathbf{v}_1, \ldots, \mathbf{v}_m$.

We compute blockwise the product U^*MV. From $MV_K = 0$ and $M^*U_K^* = 0$, we get

$$U^*MV = \begin{pmatrix} U_R^*MV_R & 0 \\ 0 & 0 \end{pmatrix}.$$

Finally, we obtain

$$U_R^*MV_R = U_R^*MM^*U_R S^{-1} = U_R^*U_R S = S.$$

$\square$

11.5 The Moore–Penrose Generalized Inverse

The resolution of a general linear system $Ax = b$, where A may be singular and may even not be square, is a delicate question, whose treatment is made much simpler by the use of the Moore–Penrose generalized inverse.

Theorem 11.5 *Let $A \in M_{n \times m}(\mathbb{C})$ be given. There exists a unique matrix $A^\dagger \in M_{m \times n}(\mathbb{C})$, called the Moore–Penrose generalized inverse, satisfying the following four properties.*

1. $AA^\dagger A = A$.
2. $A^\dagger AA^\dagger = A^\dagger$.
3. $AA^\dagger \in H_n$.
4. $A^\dagger A \in H_m$.

Finally, if A has real entries, then so has $A^\dagger$.

When $A \in \mathbf{GL}_n(\mathbb{C})$, $A^\dagger$ coincides with the standard inverse A^{-1}, because the latter obviously satisfies the four properties. More generally, if A is onto, then Property 1 shows that $AA^\dagger = I_n$, (i.e. $A^\dagger$ is a right inverse of A). Likewise, if A is one-to-one, then $A^\dagger A = I_m$, (i.e. $A^\dagger$ is a left inverse of A).

Proof. We first remark that if X satisfies these four properties, and if $U \in \mathbf{U}_n$, $V \in \mathbf{U}_m$, then V^*XU^* is a generalized inverse of UAV. Therefore, existence and uniqueness need to be proved for only a single representative D of the equivalence class of A modulo unitary multiplications on the right and the left. From Theorem 11.4, we may choose a quasi-diagonal matrix D, with given $s_1, \ldots, s_r$, the nonzero singular values of A.

Let $D^\dagger$ be any generalized inverse of D, which we write blockwise as

$$D^\dagger = \begin{pmatrix} G & H \\ J & K \end{pmatrix}$$

with $G \in \mathbf{M}_r(\mathbb{C})$. From Property 1, we obtain $S = SGS$, where $S := \mathrm{diag}(s_1, \ldots, s_r)$. Inasmuch as S is nonsingular, we obtain $G = S^{-1}$. Next, Property 3 implies $SH = 0$, that is, $H = 0$. Likewise, Property 4 gives $JS = 0$, that is, $J = 0$. Finally, Property 2 yields $K = JSH = 0$. We see, then, that $D^\dagger$ must equal (uniqueness)

$$\begin{pmatrix} S^{-1} & 0 \\ 0 & 0 \end{pmatrix}.$$

One easily checks that this matrix solves our problem (existence). $\square$

Some obvious properties are stated in the following proposition. We warn the reader that, contrary to what happens for the standard inverse, the generalized inverse of AB does not equal $B^\dagger A^\dagger$ in general.

Proposition 11.5 *The following equalities hold for the generalized inverse:*

$$(\lambda A)^\dagger = \frac{1}{\lambda} A^\dagger \quad (\lambda \neq 0), \quad (A^\dagger)^\dagger = A, \quad (A^\dagger)^* = (A^*)^\dagger.$$

If $A \in GL_n(\mathbb{C})$, then $A^\dagger = A^{-1}$.

Because $(AA^\dagger)^2 = AA^\dagger$, the matrix $AA^\dagger$ is a projector, which can therefore be described in terms of its range and kernel. Because $AA^\dagger$ is Hermitian, these subspaces are orthogonal to each other. Obviously, $R(AA^\dagger) \subset R(A)$. But because $AA^\dagger A = A$, the reverse inclusion holds too. Finally, we have

$$R(AA^\dagger) = R(A),$$

and $AA^\dagger$ is the orthogonal projector onto $R(A)$. Likewise, $A^\dagger A$ is an orthogonal projector. Obviously, $\ker A \subset \ker A^\dagger A$, and the identity $AA^\dagger A = A$ implies the reverse inclusion, so that

$$\ker A^\dagger A = \ker A.$$

Finally, $A^\dagger A$ is the orthogonal projector onto $(\ker A)^\perp$.

11.5.1 Solutions of the General Linear System

Given a matrix $M \in M_{n \times m}(\mathbb{C})$ and a vector $b \in \mathbb{C}^n$, let us consider the linear system

$$Mx = b. \tag{11.6}$$

In (11.6), the matrix M need not be square, even not of full rank. From Property 1, a necessary condition for the solvability of (11.6) is $MM^\dagger b = b$. Obviously, this is also sufficient, because it ensures that $x_0 := M^\dagger b$ is a solution. Hence, the generalized inverse plays one of the roles of the standard inverse, namely to provide one solution of (11.6) when it is solvable. To catch every solution of that system, it remains to solve the homogeneous problem $My = 0$. From the analysis done in the previous section, $\ker M$ is nothing but the range of $I_m - M^\dagger M$. Therefore, we may state the following proposition:

Proposition 11.6 *The system* (11.6) *is solvable if and only if* $b = MM^\dagger b$. *When it is solvable, its general solution is* $x = M^\dagger b + (I_m - M^\dagger M)z$, *where* z *ranges* $\mathbb{C}^m$. *Finally, the special solution* $x_0 := M^\dagger b$ *is the one of least Hermitian norm.*

There remains to prove that x_0 has the smallest norm among the solutions. That comes from the Pythagorean theorem and from the fact that $R(M^\dagger) = R(M^\dagger M) = (\ker M)^\perp$.

Exercises

1. Assume that there exists an algorithm for multiplying two $N \times N$ matrices with entries in a noncommutative ring by means of K multiplications and L additions. Show that the complexity of the multiplication in $\mathbf{M}_n(k)$ is $O(n^\alpha)$, with $\alpha = \log K / \log N$.
2. What is the complexity of Choleski factorization?
3. Let $M \in \mathbf{SPD}_n$ be also tridiagonal. What is the structure of L in the Choleski factorization? More generally, what is the structure of L when $m_{ij} = 0$ for $|i - j| > r$? (When $r \ll n$ we say that M is a *band matrix*.)
4. (Continuation of Exercise 3)
   For $i \leq n$, denote by $\phi(i)$ the smallest index j such that $m_{ij} \neq 0$. In Choleski factorization, show that $l_{ij} = 0$ for every pair (i, j) such that $j < \phi(i)$.
5. In the QR factorization, show that the map $M \mapsto (Q, R)$ is continuous on $\mathbf{GL}_n(\mathbb{C})$.
6. Let H be an $n \times n$ Hermitian matrix, that blockwise reads

$$H = \begin{pmatrix} A & B^* \\ B & C \end{pmatrix}.$$

Assume that $A \in \mathbf{HPD}_{n-k}$ $(1 \leq k \leq n - 1)$.
Find a matrix T of the form

$$T = \begin{pmatrix} I_{n-k} & 0 \\ \cdot & I_k \end{pmatrix}$$

such that THT^* is block-diagonal. Deduce that if $W \in \mathbf{H}_k$, then

$$H - \begin{pmatrix} 0 & 0 \\ 0 & W \end{pmatrix}$$

is positive-(semi)definite if and only if $S - W$ is so, where S is the Schur complement of A in H.

7. (Continuation of Exercise 6) Fix the size k. We keep $A \in \mathbf{HPD}_{n-k}$ constant and let vary B and C. We denote by $S(H)$ the Schur complement of A. Using the previous exercise, show that if $\lambda \in [0,1]$:

   a. $S(\lambda H + (1-\lambda)H') - \lambda S(H) - (1-\lambda)S(H')$ is positive-semidefinite.
   b. If $H - H'$ is positive-semidefinite, then so is $S(H) - S(H')$.

   In other words, $H \mapsto S$ is "concave nondecreasing" from the affine subspace formed of those matrices of $\mathbf{H}_n$ with prescribed $A \in \mathbf{HPD}_{n-k}$, into the ordered set $\mathbf{H}_k$.

8. In Proposition 11.4, find an alternative proof of the uniqueness part, by inspection of the spectrum of the matrix $Q := Q_2^{-1} Q_1 = R_2 R_1^{-1}$.

9. Identify the generalized inverse of row matrices and column matrices.

10. What is the generalized inverse of an orthogonal projector, that is, an Hermitian matrix P satisfying $P^2 = P$? Deduce that the description of $AA^\dagger$ and $A^\dagger A$ as orthogonal projectors does not characterize $A^\dagger$ uniquely.

11. Given a matrix $B \in \mathbf{M}_{p \times q}(\mathbb{C})$ and a vector $a \in \mathbb{C}^p$, let us form the matrix $A := (B,a) \in \mathbf{M}_{p \times (q+1)}(\mathbb{C})$.

   a. Let us define $d := B^\dagger a$, $c := a - Bd$, and

   $$b := \begin{cases} c^\dagger, & \text{if } c \neq 0, \\ (1 + |d|^2)^{-1} d^* B^\dagger, & \text{if } c = 0. \end{cases}$$

   Prove that

   $$A^\dagger = \begin{pmatrix} B^\dagger - db \\ b \end{pmatrix}.$$

   b. Deduce an algorithm (Greville's algorithm) in $O(pq^2)$ operations for the computation of the generalized inverse of a $p \times q$ matrix. **Hint:** To get started with the algorithm, use Exercise 9.

12. Let $A \in \mathbf{M}_n(\mathbb{C})$ be given, with eigenvalues λ_j and singular values σ_j, $1 \le j \le n$ (we include zeroes in this list if A is singular). We choose the decreasing orders:

   $$|\lambda_1| \ge |\lambda_2| \ge \cdots \ge |\lambda_n|, \quad \sigma_1 \ge \sigma_2 \ge \cdots \ge \sigma_n.$$

   Recall that the σ_js are the square roots of the eigenvalues of A^*A. We wish to prove the inequality

   $$\prod_{j=1}^{k} |\lambda_j| \le \prod_{j=1}^{k} \sigma_j, \quad 1 \le k \le n.$$

   a. Directly prove the case $k = 1$. Show the equality in the case $k = n$.
   b. Working within the exterior algebra (see Chapter 4), we define an endomorphism $A^{\wedge p}$ over $\Lambda^p(\mathbb{C}^n)$ by

$$A^{\wedge p}(x_1 \wedge \cdots \wedge x_p) := (Ax_1) \wedge \cdots \wedge (Ax_p), \qquad \forall x_1, \ldots, x_p \in \mathbb{C}^n.$$

Prove that the eigenvalues of $A^{\wedge p}$ are the products of p terms λ_j with pairwise distinct indices. Deduce the value of the spectral radius.

  c. Let $\{e^{i_1}, \ldots, e^{i_n}\}$ be the canonical basis of $\mathbb{C}^n$. We endow $\Lambda^p(\mathbb{C}^n)$ with the natural Hermitian norm in which the canonical basis made of $e^{i_1} \wedge \cdots \wedge e^{i_p}$ with $i_1 < \cdots < i_p$, is orthonormal. We denote by $\langle \cdot, \cdot \rangle$ the scalar product in $\Lambda^p(\mathbb{C}^n)$.

      i. If $x_1, \ldots, x_p, y_1, \ldots, y_p \in \mathbb{C}^n$, prove that

$$\langle x_1 \wedge \cdots \wedge x_p, y_1 \wedge \cdots \wedge y_p \rangle = \det(x_i^* y_j)_{1 \le i, j \le p}.$$

      ii. For $M \in \mathbf{M}_n(\mathbb{C})$, show that the Hermitian adjoint of $M^{\wedge p}$ is $(M^*)^{\wedge p}$.

      iii. If $U \in \mathbf{U}_n$, show that $U^{\wedge p}$ is unitary.

      iv. Deduce that the norm of $A^{\wedge p}$ equals $\sigma_1 \cdots \sigma_p$.

  d. Conclude.

13. Let A, B, C be complex matrices of respective sizes $n \times r$, $s \times m$, and $n \times m$. Prove that the equation

$$AXB = C$$

is solvable if and only if

$$AA^\dagger C B^\dagger B = C.$$

In this case, verify that every solution is of the form

$$A^\dagger C B^\dagger + Y - A^\dagger A Y B B^\dagger,$$

where Y is an arbitrary $r \times s$ matrix. We recall that $M^\dagger$ is the Moore–Penrose inverse of M.

14. The deformation of an elastic body is represented at each point by a square matrix $F \in \mathbf{GL}_3^+(\mathbb{R})$ (the sign $+$ expresses that $\det F > 0$). More generally, $F \in \mathbf{GL}_n^+(\mathbb{R})$ in space dimension n. The density of elastic energy is given by a function $F \mapsto W(F) \in \mathbb{R}^+$.

  a. The principle of frame indifference says that $W(QF) = W(F)$ for every $F \in \mathbf{GL}_n^+(\mathbb{R})$ and every rotation Q. Show that there exists a map $w : \mathbf{SPD}_n \to \mathbb{R}^+$ such that $W(F) = w(H)$, where $F = QH$ is the polar decomposition.

  b. When the body is isotropic, we also have $W(FQ) = W(F)$, for every $F \in \mathbf{GL}_n^+(\mathbb{R})$ and every rotation Q. Show that there exists a map $\phi : \mathbb{R}^n \to \mathbb{R}^+$ such that $W(F) = \phi(h_1, \ldots, h_n)$, where the h_j are the entries of the characteristic polynomial of H. In other words, $W(F)$ depends only on the singular values of F.

15. A matrix $A \in \mathbf{M}_n(\mathbb{R})$ is called a *totally positive* matrix when all minors

$$A \begin{pmatrix} i_1 & i_2 & \cdots & i_p \\ j_1 & j_2 & \cdots & j_p \end{pmatrix}$$

with $1 \leq p \leq n$, $1 \leq i_1 < \cdots < i_p \leq n$ and $1 \leq j_1 < \cdots < j_p \leq n$ are positive.

a. Prove that the product of totally positive matrices is totally positive.

b. Prove that a totally positive matrix admits an LU factorization and that every "nontrivial" minor of L and U is positive. Here, "nontrivial" means

$$L \begin{pmatrix} i_1 & i_2 & \cdots & i_p \\ j_1 & j_2 & \cdots & j_p \end{pmatrix}$$

with $1 \leq p \leq n$, $1 \leq i_1 < \cdots < i_p \leq n$, $1 \leq j_1 < \cdots < j_p \leq l$, and $i_s \geq j_s$ for every s, because every other minor vanishes trivially. For U, read $i_s \leq j_s$ instead. **Note**: One says that L and U are *triangular totally positive*.

c. Show that a Vandermonde matrix (see Exercise 17 of Chapter 3) is totally positive whenever $0 < a_1 < \cdots < a_n$.

Chapter 12
Iterative Methods for Linear Systems

In this chapter the field of scalars is $K = \mathbb{R}$ or $\mathbb{C}$.

We have seen in the previous chapter a few direct methods for solving a linear system $Ax = b$, when $A \in \mathbf{M}_n(K)$ is invertible. For example, if A admits an LU factorization, the successive resolution of $Ly = b$, $Ux = y$ is called the *Gauss method*. When a leading principal minor of A vanishes, a permutation of the columns allows us to return to the generic case. More generally, the Gauss method with pivoting consists in permuting the columns at each step of the factorization in such a way as to limit the magnitude of roundoff errors and that of the conditioning number of the matrices L, U.

The direct computation of the solution of a Cramer's linear system $Ax = b$, by the Gauss method or by any other direct method, is rather costly, on the order of n^3 operations. It also presents several inconveniences. It does not completely exploit the sparse shape of many matrices A; in numerical analysis it happens frequently that an $n \times n$ matrix has only $O(n)$ nonzero entries, instead of $O(n^2)$. These matrices are often bandshape with bandwidth an $O(n^\alpha)$ with $\alpha = \frac{1}{2}$ or $\frac{2}{3}$. On the other hand, the computation of an LU factorization is rather unstable, because the roundoff errors produced by the computer are amplified at each step of the computation.

For these reasons, one often prefers using an iterative method to compute an approximate solution x^m, instead of an exact solution(which after all is not exact because of roundoff errors). The iterative methods fully exploit the sparse structure of A. The number of operations is $O(am)$, where a is the number of nonzero entries in A. The choice of m depends on the accuracy that one requires a priori. It is, however, modest because the error $\|x^m - \bar{x}\|$ from the exact solution $\bar{x}$ is of order r^m up to a multiplicative constant, where $r < 1$ whenever the method converges. Typically, a dozen iterations give a rather good result, and then $O(10a) \ll O(n^3)$. Another advantage of the iterative methods is that the roundoff errors are damped by the subsequent iterations, instead of being amplified.

General Principle

We choose a decomposition of $A = M - N$ and rewrite the system, assuming that M is invertible:

$$x = M^{-1}(Nx + b).$$

Then choosing a starting point $x^0 \in K^n$, which is either an arbitrary vector or a rather coarse approximation of the solution, we construct a sequence $(x^m)_{m \in \mathbb{N}}$ by induction:

$$x^{m+1} = M^{-1}(Nx^m + b). \tag{12.1}$$

In practice, one does not compute M^{-1} explicitly but one solves the linear systems $Mx^{m+1} = \cdots$. It is thus important that this resolution be cheap. This is the case when M is triangular. In that case, the invertibility of M can be read from its diagonal, because it occurs precisely when the diagonal entries are nonzero.

12.1 A Convergence Criterion

Definition 12.1 *Let us assume that A and M are invertible, $A = M - N$. We say that an iterative method is* convergent *if for every pair $(x^0, b) \in K^n \times K^n$, we have*

$$\lim_{m \to +\infty} x^m = A^{-1}b.$$

Proposition 12.1 *An iterative method is convergent if and only if $\rho(M^{-1}N) < 1$.*

Proof. If the method is convergent, then for $b = 0$,

$$\lim_{m \to +\infty} (M^{-1}N)^m x^0 = 0,$$

for every $x^0 \in K^n$. In other words,

$$\lim_{m \to +\infty} (M^{-1}N)^m = 0.$$

From Proposition 7.8, this implies $\rho(M^{-1}N) < 1$.
    Conversely, if $\rho(M^{-1}N) < 1$, then by Proposition 7.8,

$$\lim_{m \to +\infty} (M^{-1}N)^m = 0,$$

and hence

$$x^m - A^{-1}b = (M^{-1}N)^m(x^0 - A^{-1}b) \to 0.$$

$\square$

To be more precise, if $\| \cdot \|$ is a norm on K^n, then

$$\|x^m - A^{-1}b\| \le \|(M^{-1}N)^m\| \, \|x^0 - A^{-1}b\|.$$

From Householder's theorem (Theorem 7.1), there exists for every $\varepsilon > 0$ a constant $C(\varepsilon) < \infty$ such that

$$\|x^m - A^{-1}b\| \leq C(\varepsilon)\|x^0 - \bar{x}\|(\rho(M^{-1}N) + \varepsilon)^m.$$

In most cases (in fact, when there exists an induced norm satisfying $\|M^{-1}N\| = \rho(M^{-1}N)$), one can choose $\varepsilon = 0$ in this inequality, thus obtaining

$$\|x^m - A^{-1}b\| = O(\rho(M^{-1}N)^m).$$

The choice of a vector x^0 such that $x^0 - A^{-1}b$ is an eigenvector associated with an eigenvalue of maximal modulus shows that this inequality cannot be improved in general. For this reason, we call the positive number

$$\tau := -\log\rho(M^{-1}N)$$

the *convergence rate* of the method. Given two convergent methods, we say that the first one converges faster than the second one if $\tau_1 > \tau_2$. For example, we say that it converges twice as fast if $\tau_1 = 2\tau_2$. In fact, with an error of order $\rho(M^{-1}N)^m = \exp(-m\tau)$, we see that the faster method needs only half as many iterations to reach the same accuracy.

12.2 Basic Methods

There are three basic iterative methods, of which the first has only an historical and theoretical interest. Each one uses the decomposition of A into three parts, a diagonal one D, a lower-triangular $-E$, and an upper-triangular one $-F$:

$$A = D - E - F = \begin{pmatrix} d_1 & & \\ & \ddots & -F \\ -E & & \ddots \\ & & d_n \end{pmatrix}.$$

In all cases, one assumes that D is invertible: the diagonal entries of A are nonzero.

Jacobi method: One chooses $M = D$ thus $N = E + F$. The iteration matrix is $J := D^{-1}(E + F)$. Knowing the vector x^m, one computes the components of the vector x^{m+1} by the formula

$$x_i^{m+1} = \frac{1}{a_{ii}}\left(b_i - \sum_{j \neq i} a_{ij}x_j^m\right).$$

Gauss–Seidel method: One chooses $M = D - E$, and thus $N = F$. The iteration matrix is $G := (D - E)^{-1}F$. As we show below, one never computes G explicitly. One computes the approximate solutions by a *double loop*, an outer one over m

and an inner one over $i \in \{1, \ldots, n\}$:

$$x_i^{m+1} = \frac{1}{a_{ii}} \left(b_i - \sum_{j=1}^{i-1} a_{ij} x_j^{m+1} - \sum_{j=i+1}^{j=n} a_{ij} x_j^m \right).$$

The difference between the two methods is that in Gauss–Seidel one always uses the latest computed values of each coordinate.

Relaxation method: It often happens that the Gauss–Seidel method converges exceedingly slowly, even if it is more efficient than Jacobi's method. We wish to improve the Gauss–Seidel method by looking for a "best" approximated value of the x_j (with $j < i$) when computing x_i^{m+1}. Instead of being simply x_j^m, as in the Jacobi method, or x_j^{m+1}, as in that of Gauss–Seidel, this best value is an interpolation of both (we show that it is merely an extrapolation). This justifies the choice of

$$M = \frac{1}{\omega} D - E, \quad N = \left(\frac{1}{\omega} - 1 \right) D + F,$$

where $\omega \in \mathbb{C}$ is a parameter. This parameter remains, in general, constant throughout the calculations. The method is called successive relaxation. The extrapolation case $\omega > 1$ bears the name *successive overrelaxation* (SOR). The iteration matrix is

$$\mathscr{L}_\omega := (D - \omega E)^{-1}((1 - \omega)D + \omega F).$$

The Gauss–Seidel method is a particular case of the relaxation method, with $\omega = 1$: $\mathscr{L}_1 = G$. Special attention is given to the choice of ω, in order to minimize $\rho(\mathscr{L}_\omega)$, that is, to maximize the convergence rate. The computation of the approximate solutions is done through a double loop:

$$x_i^{m+1} = \frac{\omega}{a_{ii}} \left(b_i - \sum_{j=1}^{i-1} a_{ij} x_j^{m+1} - \sum_{j=i+1}^{j=n} a_{ij} x_j^m + \left(\frac{1}{\omega} - 1 \right) a_{ii} x_i^m \right).$$

Without additional assumptions relative to the matrix A, the only result concerning the convergence is the following.

Proposition 12.2 *We have $\rho(\mathscr{L}_\omega) \geq |\omega - 1|$. In particular, if the relaxation method converges for a matrix $A \in \mathbf{M}_n(\mathbb{C})$ and a parameter $\omega \in \mathbb{C}$, then*

$$|\omega - 1| < 1.$$

In other words, it is necessary that ω belong to the disk for which $(0, 2)$ is a diameter.

Proof. If the method is convergent, we have $\rho(\mathscr{L}_\omega) < 1$. However,

$$\det \mathscr{L}_\omega = \frac{\det((1 - \omega)D + \omega F)}{\det(D - \omega E)} = \frac{\det((1 - \omega)D)}{\det D} = (1 - \omega)^n.$$

Hence

$$\rho(\mathscr{L}_\omega) \geq |\det \mathscr{L}_\omega|^{1/n} = |1 - \omega|.$$

□

12.3 Two Cases of Convergence

In this section and the following one we show that simple and natural hypotheses on A imply the convergence of the classical methods. We also compare their efficiencies.

12.3.1 The Diagonally Dominant Case

We assume here that one of the following two properties is satisfied:

1. A is strictly diagonally dominant.
2. A is irreducible and strongly diagonally dominant.

Proposition 12.3 *Under Hypothesis* (1) *or* (2), *the Jacobi method converges, as well as the relaxation method, for every* $\omega \in (0,1]$.

Proof. Jacobi method: The matrix $J = D^{-1}(E + F)$ is clearly irreducible when A is so. Furthermore,

$$\sum_{j=1}^{n} |J_{ij}| \leq 1, \quad i = 1, \ldots, n,$$

in which all inequalities are strict if (1) holds, and at least one inequality is strict under the hypothesis (2). Then either Gershgorin's theorem (Theorem 5.7) or its improvement, Proposition 5.13 for irreducible matrices, yields $\rho(J) < 1$.

Relaxation method: We assume that $\omega \in (0,1]$. Let $\lambda \in \mathbb{C}$ be a nonzero eigenvalue of $\mathscr{L}_\omega$. It is a root of

$$\det((1 - \omega - \lambda)D + \lambda \omega E + \omega F) = 0.$$

Hence, $\lambda + \omega - 1$ is an eigenvalue of $A' := \omega D^{-1}(\lambda E + F)$. This matrix is irreducible when A is so. Gershgorin's theorem (Theorem 5.7) shows that

$$|\lambda + \omega - 1| \leq \max \left\{ \frac{\omega}{|a_{ii}|} \left(|\lambda| \sum_{j<i} |a_{ij}| + \sum_{j>i} |a_{ij}| \right) ; 1 \leq i \leq n \right\}. \tag{12.2}$$

If $|\lambda| \geq 1$, we deduce that

$$|\lambda + \omega - 1| \leq \max \left\{ \frac{\omega |\lambda|}{|a_{ii}|} \sum_{j \neq i} |a_{ij}| ; 1 \leq i \leq n \right\}.$$

In case (1), this yields

$$|\lambda + \omega - 1| < \omega |\lambda|,$$

so that $|\lambda| \leq |\lambda + \omega - 1| + |1 - \omega| < |\lambda| \omega + 1 - \omega$; that is, $(|\lambda| - 1)(1 - \omega) < 0$, which is a contradiction. In case (2), Proposition 5.13 says that inequality (12.2) is strict. One concludes the proof the same way as in case (1).
    □

Of course, this result is not fully satisfactory, because $\omega \leq 1$ is not the hypothesis that we should like. Recall that in practice, one wants to use overrelaxation (i.e., $\omega > 1$), which turns out to be much more efficient than the Gauss–Seidel method for an appropriate choice of the parameter.

12.3.2 The Case of an Hermitian Positive-Definite Matrix

Let us begin with an intermediate result.

Lemma 20. *If A and $M^* + N$ are Hermitian positive-definite (in a decomposition $A = M - N$), then $\rho(M^{-1}N) < 1$.*

Proof. Let us remark first that $M^* + N = M^* + M - A$ is necessarily Hermitian when A is so.

It is therefore enough to show that $\|M^{-1}Nx\|_A < \|x\|_A$ for every nonzero $x \in \mathbb{C}^n$, where $\|\cdot\|_A$ denotes the norm associated with A:

$$\|x\|_A = \sqrt{x^*Ax}.$$

We have $M^{-1}Nx = x - y$ with $y = M^{-1}Ax$. Hence,

$$\|M^{-1}Nx\|_A^2 = \|x\|_A^2 - y^*Ax - x^*Ay + y^*Ay$$
$$= \|x\|_A^2 - y^*(M^* + N)y.$$

We conclude by observing that y is not zero; hence $y^*(M^* + N)y > 0$.   □

This proof gives a slightly more accurate result than what was claimed: by taking the supremum of $\|M^{-1}Nx\|_A$ on the unit ball, which is compact, we obtain $\|M^{-1}N\| < 1$ for the matrix norm induced by $\|\cdot\|_A$.

The main application of this lemma is the following theorem.

Theorem 12.1 *If A is Hermitian positive-definite, then the relaxation method converges if and only if $|\omega - 1| < 1$.*

Proof. We have seen in Proposition 12.2 that the convergence implies $|\omega - 1| < 1$. Let us see the converse. We have $E^* = F$ and $D^* = D$. Thus

$$M^* + N = \left(\frac{1}{\omega} + \frac{1}{\bar{\omega}} - 1\right)D = \frac{1 - |\omega - 1|^2}{|\omega|^2}D.$$

Because D is positive-definite, $M^* + N$ is positive-definite if and only if $|\omega - 1| < 1$.
□

However, Lemma 20 does not apply to the Jacobi method, inasmuch as the hypothesis (A positive-definite) does not imply that $M^* + N = D + E + F$ be positive-definite. We show in an exercise that this method diverges for certain matrices $A \in \mathbf{HPD}_n$, although it converges when $A \in \mathbf{HPD}_n$ is tridiagonal.

12.4 The Tridiagonal Case

We consider here the important case of tridiagonal matrices A, frequently encountered in the approximation of partial differential equations by finite differences or finite elements. The general structure of A is the following,

$$A = \begin{pmatrix} x & x' & 0 & \cdots & 0 \\ x'' & \ddots & \ddots & \ddots & \vdots \\ 0 & \ddots & \ddots & \ddots & 0 \\ \vdots & \ddots & \ddots & \ddots & y' \\ 0 & \cdots & 0 & y'' & y \end{pmatrix} .$$

In other words, the entries a_{ij} are zero as soon as $|j - i| \geq 2$.

Notice that in real cases, these matrices are blockwise tridiagonal instead of just tridiagonal. This means that the A_{ij}s are matrices, the diagonal ones being square matrices. The decomposition $A = D - E - F$ is done blockwise. The corresponding iterative methods, which are performed blockwise, need the inversion of the diagonal blocks, whose sizes are much smaller than n. These inversions are usually done by using a direct method. We do not detail here this extension of the classical methods. The analysis below can be adapted to this more realistic situation.

The structure of the matrix allows us to write a useful algebraic relation:

Lemma 21. *Let μ be a nonzero complex number and C a tridiagonal matrix, of diagonal C_0, of upper-triangular part C_+ and lower-triangular part C_-. Then*

$$\det C = \det \left(C_0 + \frac{1}{\mu} C_- + \mu C_+ \right) .$$

Proof. It is enough to observe that the matrix C is conjugate to

$$C_0 + \frac{1}{\mu} C_- + \mu C_+ ,$$

by the nonsingular matrix

$$Q_\mu = \begin{pmatrix} \mu & & & \\ & \mu^2 & & 0 \\ & & \ddots & \\ & 0 & & \ddots \\ & & & & \mu^n \end{pmatrix}.$$

□

Let us apply the lemma to the computation of the characteristic polynomial P_ω of $\mathscr{L}_\omega$. We have

$$\begin{aligned}
(\det D) P_\omega(\lambda) &= \det((D - \omega E)(\lambda I_n - \mathscr{L}_\omega)) \\
&= \det((\omega + \lambda - 1)D - \omega F - \lambda \omega E) \\
&= \det\left((\omega + \lambda - 1)D - \mu \omega F - \frac{\lambda \omega}{\mu} E\right),
\end{aligned}$$

for every nonzero μ. Let us choose for μ any square root of λ. We then have

$$\begin{aligned}
(\det D) P_\omega(\mu^2) &= \det((\omega + \mu^2 - 1)D - \mu\omega(E + F)) \\
&= (\det D) \det((\omega + \mu^2 - 1)I_n - \mu\omega J),
\end{aligned}$$

hence the following lemma.

Lemma 22. *If A is tridiagonal and D invertible, then*

$$P_\omega(\mu^2) = (\mu\omega)^n P_J\left(\frac{\mu^2 + \omega - 1}{\mu\omega}\right),$$

where P_J is the characteristic polynomial of the Jacobi matrix J.

Let us begin with the analysis of a simple case, that of the Gauss–Seidel method, for which $G = \mathscr{L}_1$.

Proposition 12.4 *If A is tridiagonal and D invertible, then:*

1. *$P_G(X^2) = X^n P_J(X)$, where P_G is the characteristic polynomial of the Gauss–Seidel matrix G,*
2. *$\rho(G) = \rho(J)^2$,*
3. *The Gauss–Seidel method converges if and only if the Jacobi method converges; in case of convergence, the Gauss–Seidel method converges twice as fast as the Jacobi method,*
4. *The spectrum of J is even: $\mathrm{Sp}\, J = -\mathrm{Sp}\, J$.*

Proof. Point 1 comes from Lemma 22. The spectrum of G is thus formed of $\lambda = 0$ (which is of multiplicity $[(n+1)/2]$ at least) and of squares of the eigenvalues of J, which proves Point 2. Point 3 follows immediately. Finally, if $\mu \in \mathrm{Sp}\, J$, then $P_J(\mu) = 0$, and also $P_G(\mu^2) = 0$, so that $(-\mu)^n P_J(-\mu) = 0$. Finally, either $P_J(-\mu) = 0$, or $\mu = 0 = -\mu$, in which case $P_J(-\mu)$ also vanishes. □

In fact, the comparison given in Point 3 of the proposition holds under various assumptions. For example (see Exercises 3 and 8), it holds true when D is positive and E, F are nonnegative.

We go back to the SOR, with an additional hypothesis: the spectrum of J is real, and the Jacobi method converges. This property is satisfied, for instance, when A is Hermitian positive-definite, inasmuch as Theorem 12.1 and Proposition 12.4 ensure the convergence of the Jacobi method, and on the other hand J is similar to the Hermitian matrix $D^{-1/2}(E+F)D^{-1/2}$.

We also select a real ω, that is, $\omega \in (0,2)$, taking into account Proposition 12.2. The spectrum of J is thus formed of the eigenvalues

$$-\lambda_r < \cdots < -\lambda_1 \leq \lambda_1 < \cdots < \lambda_r = \rho(J) < 1,$$

from Proposition 12.4. This notation does not mean that n is even: if n is odd, $\lambda_1 = 0$. Apart from the zero eigenvalue, which does not enter into the computation of the spectral radius, the eigenvalues of $\mathcal{L}_\omega$ are the squares of the roots of

$$\mu^2 + \omega - 1 = \mu \omega \lambda_a, \tag{12.3}$$

for $1 \leq a \leq r$. Indeed, taking $-\lambda_a$ instead of λ_a furnishes the same squares.

Let us denote $\Delta(\lambda) := \omega^2 \lambda^2 + 4(1 - \omega)$ the discriminant of (12.3). If $\Delta(\lambda_a)$ is nonpositive, both roots of (12.3) are complex conjugate, hence have modulus $|\omega - 1|^{1/2}$. If instead it is strictly positive, the roots are real and of distinct modulus. One of them, denoted by μ_a, satisfies $\mu_a^2 > |\omega - 1|$, the other one satisfying the opposite inequality.

From Proposition 12.2, $\rho(\mathcal{L}_\omega)$ is thus equal to one of the following.

- $|\omega - 1|$, if $\Delta(\lambda_a) \leq 0$ for every a, that is, if $\Delta(\rho(J)) \leq 0$.
- The maximum of the μ_a^2s defined above, otherwise.

The first case corresponds to the choice $\omega \in [\omega_J, 2)$, where

$$\omega_J = 2\frac{1 - \sqrt{1 - \rho(J)^2}}{\rho(J)^2} = \frac{2}{1 + \sqrt{1 - \rho(J)^2}} \in [1,2).$$

Then $\rho(\mathcal{L}_\omega) = \omega - 1$.

The second case is $\omega \in (0, \omega_J)$. If $\Delta(\lambda_a) > 0$, let us denote by $Q_a(X)$ the polynomial $X^2 + \omega - 1 - X\omega\lambda_a$. The sum of its roots being positive, μ_a is the largest one; it is thus positive. Moreover, $Q_a(1) = \omega(1 - \lambda_a) > 0$ shows that both roots belong to the same half-line of $\mathbb{R} \setminus \{1\}$. Their product has modulus less than or equal to one, therefore they are less than or equal to one. In particular,

$$|\omega - 1|^{1/2} < \mu_a < 1.$$

This shows that $\rho(\mathcal{L}_\omega) < 1$ holds for every $\omega \in (0,2)$. Under our hypotheses, the relaxation method is convergent.

If $\lambda_a \neq \rho(J)$, we have $Q_r(\mu_a) = \mu_a \omega(\lambda_a - \rho(J)) < 0$. Hence, μ_a lies between both roots of Q_r, so that $\mu_a < \mu_r$. Finally, the case $\Delta(\rho(J)) \geq 0$ furnishes $\rho(\mathscr{L}_\omega) = \mu_r^2$.

There remains to analyze the dependence of $\rho(\mathscr{L}_\omega)$ upon ω. We have

$$(2\mu_r - \omega\rho(J))\frac{d\mu_r}{d\omega} + 1 - \mu_r\rho(J) = 0.$$

Because $2\mu_r$ is larger than the sum $\omega\rho(J)$ of the roots and because $\mu_r, \rho(J) \in [0,1)$, one deduces that $\omega \mapsto \rho(\mathscr{L}_\omega)$ is nonincreasing over $(0, \omega_J)$.

We conclude that $\rho(\mathscr{L}_\omega)$ reaches its minimum at ω_J, that minimum being

$$\omega_J - 1 = \frac{1 - \sqrt{1 - \rho(J)^2}}{1 + \sqrt{1 - \rho(J)^2}}.$$

Theorem 12.2 *(See Figure* 12.1*) Suppose that A is tridiagonal, D is invertible, and that the eigenvalues of J are real and belong to* $(-1, 1)$*. Assume also that* $\omega \in \mathbb{R}$*.*

Then the relaxation method converges if and only if $\omega \in (0, 2)$*. Furthermore, the convergence rate is optimal for the parameter*

$$\omega_J := \frac{2}{1 + \sqrt{1 - \rho(J)^2}} \in [1, 2),$$

where the spectral radius of $\mathscr{L}_{\omega_J}$ *is*

$$(\omega_J - 1 =) \quad \frac{1 - \sqrt{1 - \rho(J)^2}}{1 + \sqrt{1 - \rho(J)^2}} = \left(\frac{1 - \sqrt{1 - \rho(J)^2}}{\rho(J)}\right)^2.$$

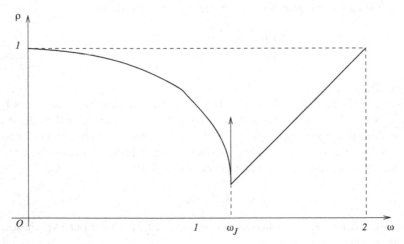

Fig. 12.1 $\rho(\mathscr{L}_\omega)$ in the tridiagonal case.

Remarks

- We show in Exercise 7 that Theorem 12.2 extends to complex values of ω: under the same assumptions, $\rho(\mathcal{L}_\omega)$ is minimal at ω_J, and the relaxation method converges if and only if $|\omega - 1| < 1$.
- The Gauss–Seidel method is not optimal in general; $\omega_J = 1$ holds only when $\rho(J) = 0$, although in practice $\rho(J)$ is close to 1. A typical example is the resolution of an elliptic PDE by the finite element method.

  For values of $\rho(J)$ that are not too close to 1, the relaxation method with optimal parameter ω_J, although improving the convergence rate, is not overwhelmingly more efficient than Gauss–Seidel. In fact,

$$\rho(G)/\rho(\mathcal{L}_{\omega_J}) = \left(1 + \sqrt{1 - \rho(J)^2}\right)^2$$

lies between 1 (for $\rho(J)$ close to 1) and 4 (for $\rho(J) = 0$), so that the ratio

$$\log \rho(\mathcal{L}_{\omega_J})/\log \rho(G)$$

remains moderate, as long as $\rho(J)$ keeps away from 1. However, in the realistic case where $\rho(J)$ is close to 1, we have

$$\log \rho(G)/\log \rho(\mathcal{L}_{\omega_J}) \approx \sqrt{\frac{1 - \rho(J)}{2}},$$

which is very small. The number of iterations needed for a prescribed accuracy is multiplied by that ratio when one replaces the Gauss–Seidel method by the relaxation method with the optimal parameter.

12.5 The Method of the Conjugate Gradient

We present here the *conjugate gradient* method in the most appropriate framework, namely that of systems $Ax = b$ where A is real symmetric positive-definite ($A \in \mathbf{SPD}_n$). As we show below, it is a *direct* method, in the sense that it furnishes the solution $\bar{x}$ after a finite number of iterations (at most n). However, the roundoff errors pollute the final result, and we would prefer to consider the conjugate gradient as an *iterative* method in which the number N of iterations, much less than n, gives a rather good approximation of $\bar{x}$. We show that the choice of N is linked to the *condition number* of the matrix A.

We denote by $\langle \cdot, \cdot \rangle$ the canonical scalar product on $\mathbb{R}^n$, and by $\|\cdot\|_A$ the norm associated with A :

$$\|x\|_A := \sqrt{\langle Ax, x \rangle}.$$

We also use the quadratic form E associated with $A : E(x) := \langle Ax, x \rangle = \|x\|_A^2$. It is the square of a norm of $\mathbb{R}^n$. The character $\perp_A$ indicates the orthogonality with respect to the scalar product defined by A :

$$(x \perp_A y) \iff (\langle Ax, y \rangle = 0).$$

We point out the importance in this paragraph of distinguishing between the canonical scalar product, and that associated with A. Both are used in the sequel.

When $A \in \mathbf{SPD}_n$ and $b \in \mathbb{R}^n$, the function

$$x \mapsto J(x) := \frac{1}{2}\langle Ax, x \rangle - \langle b, x \rangle$$

is strictly convex. By Cauchy–Schwarz, we have

$$J(x) \geq \frac{1}{2}\|x\|_A^2 - \|x\|_A \|b\|_A \overset{\|x\|_A \to +\infty}{\longrightarrow} +\infty.$$

Let m be the infimum of J. The nonempty set $K := \{x \mid J(x) < m+1\}$ is closed and bounded, thus compact. Hence J reaches its infimum over K, which is its infimum over $\mathbb{R}^n$. Because of the strict convexity, the infimum is taken at a unique point $\bar{x}$, which is the unique vector where the gradient of J vanishes. We denote by r (for *residue*) the gradient of J: $r(x) = Ax - b$. Therefore $\bar{x}$ is the unique solution of the linear system $Ax = b$.

The characterization of $\bar{x} = A^{-1}b$ as $\mathrm{argmin}(J)$ suggests employing a *descent method*. This motivates the method of the conjugate gradient.

12.5.1 A Theoretical Analysis

Let $x_0 \in \mathbb{R}^n$ be given. We define $e_0 = x_0 - \bar{x}$, $r_0 = r(x_0) = Ae_0$. We may assume that $e_0 \neq 0$; otherwise, we should already have got the solution.

For $k \geq 1$, let us define the vector space

$$\mathscr{H}_k := \{P(A)r_0 \mid P \in \mathbb{R}[X], \deg P \leq k-1\}, \quad \mathscr{H}_0 = \{0\}.$$

It can be obtained inductively, using either $\mathscr{H}_{k+1} = A\mathscr{H}_k + \mathscr{H}_k$ or $\mathscr{H}_{k+1} = A^k r_0 + \mathscr{H}_k$.

In $\mathscr{H}_{k+1}$, the subspace $\mathscr{H}_k$ is of codimension 0 or 1. In the first case, $\mathscr{H}_{k+1} = \mathscr{H}_k$, and it follows that $\mathscr{H}_{k+2} = A\mathscr{H}_{k+1} + \mathscr{H}_{k+1} = A\mathscr{H}_k + \mathscr{H}_k = \mathscr{H}_{k+1} = \mathscr{H}_k$. By induction, $\mathscr{H}_k = \mathscr{H}_m$ for every $m > k$. Let us denote by ℓ the smallest index such that $\mathscr{H}_\ell = \mathscr{H}_{\ell+1}$. For $k < \ell$, $\mathscr{H}_k$ is thus of codimension one in $\mathscr{H}_{k+1}$, whereas if $k \geq \ell$, then $\mathscr{H}_k = \mathscr{H}_{k+1}$. It follows that $\dim \mathscr{H}_k = k$ if $k \leq \ell$. In particular, $\ell \leq n$.

One can always find, by applying Gram–Schmidt orthonormalization, a basis $\{p_0, \ldots, p_{l-1}\}$ of $\mathscr{H}_\ell$ that is A-orthogonal (i.e., such that $\langle Ap_j, p_i \rangle = 0$ if $i \neq j$), and such that $\{p_0, \ldots, p_{k-1}\}$ is a basis of $\mathscr{H}_k$ when $k \leq l$. The vectors p_j, which are not necessarily unit vectors, are defined, up to a scalar multiple, by

$$p_k \in \mathscr{H}_{k+1}, \quad p_k \perp_A \mathscr{H}_k.$$

One says that the vectors p_j are pairwise *conjugate*. Of course, conjugation means A-orthogonality. This explains the name of the method.

The quadratic function J, strictly convex, reaches its infimum on the affine subspace $x_0 + \mathscr{H}_k$ at a unique vector, which we denote by x_k (same proof as above). The notation x_k is consistent for $k = 0$. If $x = y + \gamma p_k \in x_0 + \mathscr{H}_{k+1}$ with $y \in x_0 + \mathscr{H}_k$, then

$$
\begin{aligned}
J(x) &= J(\bar{x}) + \frac{1}{2} E(x - \bar{x}) \\
&= J(\bar{x}) + \frac{1}{2} E(y - \bar{x}) + \frac{1}{2} \gamma^2 E(p_k) + \gamma \langle A p_k, y - \bar{x} \rangle \\
&= J(y) + \frac{1}{2} \gamma^2 E(p_k) - \gamma \langle A p_k, e_0 \rangle,
\end{aligned}
$$

because $\langle A p_k, y - x_0 \rangle = 0$. Hence, minimizing J over $x_0 + \mathscr{H}_{k+1}$ amounts to minimizing J separately with respect to y, and to γ. The first minimization is that of $J(y)$ over $x_0 + \mathscr{H}_k$ and yields $y = x_k$, whereas the second one is that of $\gamma \mapsto \frac{1}{2} \gamma^2 E(p_k) - \gamma \langle p_k, r_0 \rangle$ over $\mathbb{R}$. We therefore have

$$x_{k+1} - x_k \in \mathbb{R} p_k. \tag{12.4}$$

By definition of ℓ there exists a nonzero polynomial P of degree ℓ such that $P(A) r_0 = 0$, that is, $A P(A) e_0 = 0$. Because A is invertible, $P(A) e_0 = 0$. If $P(0)$ vanished, then $P(X) = X Q(X)$ with $\deg Q = \ell - 1$. Therefore, $Q(A) r_0 = 0$: the map $S \mapsto S(A) r_0$ would not be one-to-one over the polynomials of degree less than or equal to $\ell - 1$. One should have $\dim \mathscr{H}_\ell < \ell$, a contradiction. Hence $P(0) \neq 0$, and we may normalize $P(0) = 1$. Then $P(X) = 1 - X R(X)$, where $\deg R = \ell - 1$. Thus $e_0 = R(A) r_0 \in \mathscr{H}_\ell$ or, equivalently, $\bar{x} \in x_0 + \mathscr{H}_\ell$. Conversely, if $k \leq \ell$ and $\bar{x} \in x_0 + \mathscr{H}_k$, then $e_0 \in \mathscr{H}_k$; that is, $e_0 = Q(A) r_0$, where $\deg Q \leq k - 1$. Then $Q_1(A) e_0 = 0$, with $Q_1(X) = 1 - X Q(X)$. Therefore, $Q_1(A) r_0 = 0$, $Q_1(0) \neq 0$, and $\deg Q_1 \leq k$. Hence $k \geq \ell$; that is, $k = \ell$. Summing up, we have $\bar{x} \in x_0 + \mathscr{H}_\ell$ but $\bar{x} \notin x_0 + \mathscr{H}_{\ell-1}$. Therefore, $x_\ell = \bar{x}$ and $x_k \neq \bar{x}$ if $k < \ell$.

We now set $r_k = r(x_k) = A(x_k - \bar{x})$. We have seen that $r_\ell = 0$ and that $r_k \neq 0$ if $k < l$. In fact, r_k is the gradient of J at x_k. The minimality of J at x_k over $x_0 + \mathscr{H}_k$ thus implies that $r_k \perp \mathscr{H}_k$ (for the usual scalar product). In other words, we have $\langle r_k, p_j \rangle = 0$ if $j < k$. However, $x_k - \bar{x} \in e_0 + \mathscr{H}_k$ can also be written as $x_k - \bar{x} = Q(A) e_0$ with $\deg Q \leq k$, which implies $r_k = Q(A) r_0$, so that $r_k \in \mathscr{H}_{k+1}$. If $k < l$, one therefore has $\mathscr{H}_{k+1} = \mathscr{H}_k \oplus \mathbb{R} r_k$.

We now normalize p_k (which was not done up to now) by

$$p_k - r_k \in \mathscr{H}_k.$$

In other words, p_k is the A-orthogonal projection of $r_k = r(x_k)$, parallel to $\mathscr{H}_k$. It is actually an element of $\mathscr{H}_{k+1}$, because $r_k \in \mathscr{H}_{k+1}$. It is also nonzero because $r_k \notin \mathscr{H}_k$. We note that r_k is orthogonal to $\mathscr{H}_k$ with respect to the usual scalar product, although p_k is orthogonal to $\mathscr{H}_k$ with respect to the A-scalar product; this explains why p_k and r_k are generally different.

If $j \leq k - 2$, we compute $\langle A(p_k - r_k), p_j \rangle = -\langle Ar_k, p_j \rangle = -\langle r_k, Ap_j \rangle = 0$. We have successively used the conjugation of the p_k, the symmetry of A, the fact that $Ap_j \in \mathscr{H}_{j+2}$, and the orthogonality of r_k and $\mathscr{H}_k$. We have therefore $p_k - r_k \perp_A \mathscr{H}_{k-1}$, so that

$$p_k = r_k + \delta_k p_{k-1} \tag{12.5}$$

for a suitable number δ_k.

12.5.1.1 Error Estimate

Lemma 23. *Let us denote by $\lambda_n \geq \cdots \geq \lambda_1 (> 0)$ the eigenvalues of A. If $k \leq \ell$, then*

$$E(x_k - \bar{x}) \leq E(e_0) \cdot \min_{\deg Q \leq k-1} \max_j |1 + \lambda_j Q(\lambda_j)|^2.$$

Proof. Let us compute

$$\begin{aligned}
E(x_k - \bar{x}) &= \min\{E(x - \bar{x}) \,|\, x \in x_0 + \mathscr{H}_k\} \\
&= \min\{E(e_0 + y) \,|\, y \in \mathscr{H}_k\} \\
&= \min\{E((I_n + AQ(A))e_0) \,|\, \deg Q \leq k - 1\} \\
&= \min\{\|(I_n + AQ(A))A^{1/2}e_0\|_2^2 \,|\, \deg Q \leq k - 1\},
\end{aligned}$$

where we have used the equality $\langle Aw, w \rangle = \|A^{1/2}w\|_2^2$. Hence

$$\begin{aligned}
E(x_k - \bar{x}) &\leq \min\{\|I_n + AQ(A)\|_2^2 \|A^{1/2}e_0\|_2^2 \,|\, \deg Q \leq k - 1\} \\
&= E(e_0) \min\{\rho(I_n + AQ(A))^2 \,|\, \deg Q \leq k - 1\},
\end{aligned}$$

because $\rho(S) = \|S\|_2$ holds true for every real symmetric matrix. $\square$

From Lemma 23, we deduce an estimate of the error $E(x_k - \bar{x})$ by bounding the right-hand side by

$$\min_{\deg Q \leq k-1} \max_{t \in [\lambda_1, \lambda_n]} |1 + tQ(t)|^2.$$

Classically, this minimum is reached for

$$1 + XQ(X) = \omega_k T_k \left(\frac{2X - \lambda_1 - \lambda_n}{\lambda_n - \lambda_1} \right),$$

where T_k is a Chebyshev polynomial:

$$T_k(t) = \begin{cases} \cos k \arccos t & \text{if } |t| \leq 1, \\ \cosh k \operatorname{arcosh} t & \text{if } t \geq 1, \\ (-1)^k \cosh k \operatorname{arcosh} |t| & \text{if } t \leq -1. \end{cases}$$

The number ω_k is that which furnishes the value 1 at $X = 0$, namely

$$\omega_k = \frac{(-1)^k}{T_k\left(\frac{\lambda_n + \lambda_1}{\lambda_n - \lambda_1}\right)}.$$

Then

$$\max_{[\lambda_1, \lambda_n]} |1 + t Q(t)| = |\omega_k| = \frac{1}{\cosh k \operatorname{arcosh} \frac{\lambda_n + \lambda_1}{\lambda_n - \lambda_1}}.$$

Hence $E(x_k - \bar{x}) \leq |\omega_k|^2 E(e_0)$. However, if

$$\theta := \operatorname{arcosh} \frac{\lambda_n + \lambda_1}{\lambda_n - \lambda_1},$$

then $|\omega_k| = (\cosh k\theta)^{-1} \leq 2\exp(-k\theta)$, and $\exp(-\theta)$ is the root, less than one, of the quadratic polynomial

$$T^2 - 2\frac{\lambda_n + \lambda_1}{\lambda_n - \lambda_1} T + 1.$$

Setting $\operatorname{cond}(A) := \|A\|_2 \|A^{-1}\|_2 = \lambda_n/\lambda_1$ the *condition number* of A, we obtain

$$e^{-\theta} = \frac{\lambda_n + \lambda_1}{\lambda_n - \lambda_1} - \sqrt{\left(\frac{\lambda_n + \lambda_1}{\lambda_n - \lambda_1}\right)^2 - 1} = \frac{\sqrt{\lambda_n} - \sqrt{\lambda_1}}{\sqrt{\lambda_n} + \sqrt{\lambda_1}} = \frac{\sqrt{\operatorname{cond}(A)} - 1}{\sqrt{\operatorname{cond}(A)} + 1}.$$

The final result is the following.

Theorem 12.3 *If $k \leq l$, then*

$$E(x_k - \bar{x}) \leq 4E(x_0 - \bar{x})\left(\frac{\sqrt{\operatorname{cond}(A)} - 1}{\sqrt{\operatorname{cond}(A)} + 1}\right)^{2k}. \tag{12.6}$$

12.5.2 Implementing the Conjugate Gradient

The main feature of the conjugate gradient is the simplicity of the computation of the vectors x_k. We proceed by induction. To begin with, we have $p_0 = r_0 = Ax_0 - b$, where x_0 is at our disposal.

Let us assume that x_k and p_{k-1} are known. Then $r_k = Ax_k - b$. If $r_k = 0$, we already have the solution. Otherwise, the formulæ (12.4) and (12.5) show that in fact, x_{k+1} minimizes J over the plane $x_k + \mathbb{R}r_k \oplus \mathbb{R}p_{k-1}$. We therefore have $x_{k+1} = x_k + \alpha_k r_k + \beta_k p_{k-1}$, where the entries α_k, β_k are obtained by solving the linear system of two equations

$$\alpha_k \langle Ar_k, r_k \rangle + \beta_k \langle Ar_k, p_{k-1} \rangle + \|r_k\|^2 = 0, \quad \alpha_k \langle Ar_k, p_{k-1} \rangle + \beta_k \langle Ap_{k-1}, p_{k-1} \rangle = 0$$

(we have used $\langle r_k, p_{k-1} \rangle = 0$). Then we have $\delta_k = \beta_k / \alpha_k$. Observe that α_k is nonzero, because otherwise β_k would vanish and r_k would too.

Summing up, the algorithm reads as follows

- Choose x_0. Define $p_0 = r_0 = r(x_0) := Ax_0 - b$.
- FOR $k = 0, \ldots, n$, DO

    – Compute $r_k := r(x_k) = Ax_k - b$. IF $r_k = 0$, THEN $\bar{x} := x_k$ and STOP.
    – ELSE, minimize $J(x_k + \alpha r_k + \beta p_{k-1})$, by computing α_k, β_k as above.
    – Define

    $$p_k = r_k + (\beta_k / \alpha_k) p_{k-1}, \quad x_{k+1} = x_k + \alpha_k p_k.$$

A priori, this computation furnishes the exact solution $\bar{x}$ in ℓ iterations. However, ℓ equals n in general, and the cost of each iteration is $O(n^2)$. The conjugate gradient, viewed as a direct method, is thus rather slow, or at least not faster than other methods. One often uses this method for sparse matrices, whose maximal number of nonzero elements m per rows is small compared to n. The complexity of an iteration is then $O(mn)$. However, that is still rather costly as a direct method ($O(mn^2)$ operations in all), because the complexity of iterative methods is also reduced for sparse matrices.

The advantage of the conjugate gradient is its *stability*, stated in Theorem 12.3. It explains why one prefers to consider the conjugate gradient as an *iterative* method, in which one makes only a few iterations $N \ll n$. Strictly speaking, Theorem 12.3 does not define a convergence rate τ, because one does not have, in general, an inequality of the form

$$\|x_{k+1} - \bar{x}\| \le e^{-\tau} \|x_k - \bar{x}\|.$$

In particular, one is not certain that $\|x_1 - \bar{x}\|$ be smaller than $\|x_0 - \bar{x}\|$. However, the inequality (12.6) is analogous to what we have for a classical iterative method, up to the factor 4. We therefore say that the conjugate gradient admits a *convergence rate* τ_{CG} that satisfies

$$\tau_{CG} \le \theta = -\log \frac{\sqrt{\mathrm{cond}(A)} - 1}{\sqrt{\mathrm{cond}(A)} + 1}. \tag{12.7}$$

This rate is equivalent to $2\mathrm{cond}(A)^{-1/2}$ when $\mathrm{cond}(A)$ is large. This method can be considered as an iterative method when $n\tau_{CG} \gg 1$ because then it is possible to choose $N \ll n$. Obviously, a sufficient condition is $\mathrm{cond}(A) \ll n^2$.

12.5.2.1 Application

Let us consider the Laplace equation in an open bounded set Ω of $\mathbb{R}^d$, with a Dirichlet boundary condition:

$$\Delta u = f \text{ in } \Omega, \qquad u = 0 \text{ on } \partial\Omega.$$

An approximate resolution by a finite elements method yields a finite-dimensional linear system $Ax = b$, where the matrix A is symmetric, reflecting the symmetry of the variational formulation

$$\int_\Omega (\nabla u \cdot \nabla v + fv)\, dx = 0, \quad \forall v \in H_0^1(\Omega).$$

If the diameter of the grid is h with $0 < h \ll 1$, and if that grid is regular enough, the number of degrees of freedom (the size of the matrix) n is of order C/h^d, where C is a constant. The matrix is sparse with $m = O(1)$. Each iteration thus needs $O(n)$ operations. Finally, the condition number of A is of order c/h^2. Hence, a number of iterations $N \gg 1/h$ is appropriate. This is worthwhile as soon as $d \geq 2$. The method becomes more useful as d gets larger, inasmuch as the threshold $1/h$ is independent of the dimension.

12.5.2.2 Preconditioning

Theorem 12.3 suggests that the performance of the method is improved by *preconditioning* the matrix A. The idea is to replace the system $Ax = b$ by $B^T ABy = B^T b$, where the inversion of B is easy; for example, B is block-triangular, or block-diagonal with small blocks, or block-triangular with small bandwidth. If BB^T is close enough to A^{-1}, the condition number of the new matrix $B^T AB$ is smaller, and the number of iterations is reduced. Actually, when the condition number reaches its infimum $K = 1$, we have $A = I_n$, and the solution $\bar{x} = b$ is obvious. The simplest preconditioning consists in choosing $B = D^{-1/2}$, D the diagonal of A. Its efficiency is clear in the (trivial) case where A is diagonal, because the matrix of the new system is I_n, and the condition number is lowered to 1. Observe that preconditioning is also used with SOR, because it allows us to diminish the value of $\rho(J)$, hence also the convergence rate. We show in Exercise 5 that, if $A \in \mathbf{SPD}_n$ is tridiagonal and if $D = dI_n$ (which corresponds to the preconditioning described above), the conjugate gradient method is twice as slow as the relaxation method with optimal parameter; that is,

$$\theta = \frac{1}{2}\tau_{\mathrm{RL}}.$$

This equality is obtained by computing θ and the optimal convergence rate τ_{RL} of the relaxation method in terms of $\rho(J)$. In the real world, in which A might not be tridiagonal, or be only blockwise tridiagonal, the map $\rho(J) \mapsto \theta$ remains the same, and τ_{RL} deteriorates. Then the conjugate gradient method becomes more efficient than the relaxation method. Another advantage is that it does not need the preliminary computation of $\rho(J)$, in contrast to the relaxation method with optimal parameter.

The reader may find a deeper analysis of the method of the conjugate gradient in the article of J.-F. Maître [1].

Exercises

1. Let A be a tridiagonal matrix with an invertible diagonal and let J be its Jacobi matrix. Show that J is conjugate to $-J$. Compare with Proposition 12.4.

2. We fix $n \geq 2$. Use Theorem 6.8 to construct a matrix $A \in \mathbf{SPD}_n$ for which the Jacobi method does not converge. Show in particular that

$$\sup\{\rho(J) \,|\, A \in \mathbf{SPD}_n, D = I_n\} = n - 1.$$

3. Let $A \in \mathbf{M}_n(\mathbb{R})$ satisfy $a_{ii} > 0$ for every index i, and $a_{ij} \leq 0$ whenever $j \neq i$. Using (several times) the weak form of the Perron–Frobenius theorem, prove that either $1 \leq \rho(J) \leq \rho(G)$ or $\rho(G) \leq \rho(J) \leq 1$. In particular, as in point 3 of Proposition 12.4, the Jacobi and Gauss–Seidel methods converge or diverge simultaneously, and Gauss–Seidel is faster in the former case. Hint: Prove that

$$(\rho(G) \geq 1) \Longrightarrow (\rho(J) \geq 1) \Longrightarrow (\rho(G) \geq \rho(J))$$

and

$$(\rho(G) \leq 1) \Longrightarrow (\rho(J) \geq \rho(G)).$$

4. Let $n \geq 2$ and $A \in \mathbf{HPD}_n$ be given. Assume that A is tridiagonal.

   a. Verify that the spectrum of J is real and even.
   b. Show that the eigenvalues of J satisfy $\lambda < 1$.
   c. Deduce that the Jacobi method is convergent.

5. Let $A \in \mathbf{HPD}_n$, $A = D - E - E^*$. Use the Hermitian norm $\|\cdot\|_2$.

   a. Show that $|((E + E^*)v, v)| \leq \rho(J)\|D^{1/2}v\|^2$ for every $v \in \mathbb{C}^n$. Deduce that

   $$\mathrm{cond}(A) \leq \frac{1 + \rho(J)}{1 - \rho(J)}\mathrm{cond}(D).$$

   b. Let us define a function by

   $$g(x) := \frac{\sqrt{x} - 1}{\sqrt{x} + 1}.$$

   Verify that

   $$g\left(\frac{1 + \rho(J)}{1 - \rho(J)}\right) = \frac{1 - \sqrt{1 - \rho(J)^2}}{\rho(J)}.$$

   c. Deduce that if A is tridiagonal and if $D = dI_n$, then the convergence rate θ of the conjugate gradient is at least the half of that of SOR with optimal parameter.

6. Here is another proof of Theorem 12.1, when ω is real. Let $A \in \mathbf{HPD}_n$.

   a. Suppose we are given $\omega \in (0, 2)$.

i. Assume that $\lambda = e^{2i\theta}$ (θ real) is an eigenvalue of $\mathscr{L}_\omega$. Show that $(1 - \omega - \lambda)e^{-i\theta} \in \mathbb{R}$.

ii. Deduce that $\lambda = 1$; then show that this case is impossible too.

iii. Let $m(\omega)$ be the number of eigenvalues of $\mathscr{L}_\omega$ of modulus less than or equal to one (counted with multiplicities). Show that m is constant on $(0,2)$.

   b.   i. Compute

$$\lim_{\omega \to 0} \frac{1}{\omega}(\mathscr{L}_\omega - I_n).$$

ii. Deduce that $m = n$, hence that the SOR converges for every $\omega \in (0,2)$.

7. (Extension of Theorem 12.2 to complex values of ω). We still assume that A is tridiagonal, that the Jacobi method converges, and that the spectrum of J is real. We retain the notation of Section 12.4.

   a. Given an index a such that $\lambda_a > 0$, verify that $\Delta(\lambda_a)$ vanishes for two real values of ω, of which only one, denoted by ω_a, belongs to the open disk $D = D(1;1)$. Show that $1 < \omega_a < 2$.

   b. Show that if $\omega \in D \setminus [\omega_a, 2)$, then the roots of $X^2 + \omega - 1 - \omega\lambda_a X$ have distinct moduli, with one and only one of them, denoted by $\mu_a(\omega)$, of modulus larger than $|\omega - 1|^{1/2}$.

   c. Show that $\omega \mapsto \mu_a$ is holomorphic on its domain, and that

$$\lim_{|\omega-1| \to 1} |\mu_a(\omega)|^2 = 1,$$

$$\lim_{\omega \to \gamma} |\mu_a(\omega)|^2 = \gamma - 1 \quad \text{if } \gamma \in [\omega_a, 2).$$

   d. Deduce that $|\mu_a(\omega)| < 1$ (use the maximum principle), then that the relaxation method converges for every $\omega \in D$.

   e. Show, finally, that the spectral radius of $\mathscr{L}_\omega$ is minimal for $\omega = \omega_r$, which was previously denoted by ω_J.

8. Let B be a cyclic matrix of order three. With square diagonal blocks, it reads blockwise as

$$B = \begin{pmatrix} 0 & 0 & M_1 \\ M_2 & 0 & 0 \\ 0 & M_3 & 0 \end{pmatrix}.$$

We wish to compare the Jacobi and Gauss–Seidel methods for the matrix $A := I - B$. Compute the matrix G. Show that $\rho(G) = \rho(J)^3$. Deduce that both methods converge or diverge simultaneously and that, in the case of convergence, Gauss–Seidel is three times faster than Jacobi. Show that for A^T, the convergence or the divergence still holds simultaneously, but that Gauss–Seidel is only one and a half times faster. Generalize to cyclic matrices of any order p.

9. (Preconditionned conjugate gradient method.)

Let $Ax = b$ be a linear system whose matrix A is symmetric positive-definite (all entries are real.) Recall that the convergence rate of the conjugate gradient method is the number

$$\tau_{GC} = -\log \frac{\sqrt{K(A)} - 1}{\sqrt{K(A)} + 1},$$

that behaves like $2/\sqrt{K(A)}$ when $K(A)$ is large, as it used to be in real life. The number $K(A) := \lambda_{\max}(A)/\lambda_{\min}(A)$ is the condition number of A.

Preconditioning is a technique that reduces the condition number, hence increases the convergence rate, through a change of variables. Say that a new unknown is $y := B^T x$, so that the system is equivalent to $\tilde{A}y = \tilde{b}$, where

$$\tilde{A} := B^{-1}AB^{-T}, \quad b = B\tilde{b}.$$

With a given preconditioning, we associate the matrices $C := BB^T$ and $T := I_n - C^{-1}A$. Notice that preconditioning with C or $\alpha^{-1}C$ is essentially the same trick if $\alpha > 0$, although $T = T(\alpha)$ differs significantly. Thus we merely associate with C the whole family

$$\{T(\alpha) = I_n - \alpha C^{-1}A \mid \alpha > 0\}.$$

a. Show that $\tilde{A}$ is similar to $C^{-1}A$.
b. Consider the decomposition $A = M - N$ with $M = \alpha^{-1}C$ and $N = \alpha^{-1}C - A$. This yields an iterative method

$$C(x^{k+1} - x^k) = b - \alpha Ax^k,$$

whose iteration matrix is $T(\alpha)$. Show that there exist values of α for which the method is convergent. Show that the optimal parameter (the one that maximizes the convergence rate) is

$$\alpha_{\text{opt}} = \frac{2}{\lambda_{\min}(\tilde{A}) + \lambda_{\max}(\tilde{A})},$$

with the convergence rate

$$\tau_{\text{opt}} = -\log \frac{K(\tilde{A}) - 1}{K(\tilde{A}) + 1}.$$

c. When $K(\tilde{A})$ is large, show that

$$\frac{\tau_{GCP}}{\tau_{\text{opt}}} \approx \sqrt{K(\tilde{A})},$$

where τ_{GCP} stands for the preconditioned conjugate gradient, that is, the conjugate gradient applied to $\tilde{A}$.

Conclusion?

10. (Continuation.) We now start from a decomposition $A = M - N$ and wish to construct a preconditioning.

    Assume that $M + N^T$, obviously a symmetric matrix, is positive-definite. We already know that $\|M^{-1}N\|_A < 1$, where $\|\cdot\|_A$ is the Euclidean norm associated with A (Lemma 9.3.1.)

    a. Define $T := (I_n - M^{-T}A)(I_n - M^{-1}A)$. Prove that $\|T\|_A < 1$. Deduce that the "symmetric" method

$$Mx^{k+1/2} = Nx^k + b, \quad M^T x^{k+1} = N^T x^{k+1/2} + b$$

    is convergent (remark that $A = M^T - N^T$.)

    This method is called *symmetric S.O.R.*, or *S.S.O.R.* when M is as in the relaxation method.

    b. From the identity $T = I_n - M^{-T}(M + N^T)M^{-1}A$, we define $C = M(M + N^T)^{-1}M^T$. Express the corresponding preconditioning $C(\omega)$ when M and N come from the S.O.R. method:

$$M = \frac{1}{\omega}D - E, \qquad \omega \in (0, 2).$$

    This is the *S.S.O.R.* preconditioning.

    c. Show that $\lambda_{\max}(C(\omega)^{-1}A) \leq 1$, with equality when $\omega = 1$.

    d. Compute $\rho(T)$ and $K(\tilde{A})$ when A is tridiagonal with $a_{ii} = 2$, $a_{i,i\pm 1} = -1$ and $a_{ij} = 0$ otherwise. Compare the S.S.O.R. method and the S.S.O.R. preconditioned conjugate gradient method.

Chapter 13
Approximation of Eigenvalues

13.1 General Considerations

The computation of the eigenvalues of a square matrix is a problem of considerable difficulty. The naive idea, according to which it is enough to compute the characteristic polynomial and then find its roots, turns out to be hopeless because of Abel's theorem, which states that the general equation $P(x) = 0$, where P is a polynomial of degree $d \geq 5$, is not solvable using algebraic operations and roots of any order. For this reason, there exists no direct method, even an expensive one, for the computation of $\mathrm{Sp}(M)$.

Dropping half of that program, one could compute the characteristic polynomial exactly, then compute an approximation of its roots. But the cost and the instability of the computation are prohibitive. Amazingly, the opposite strategy is often used: a standard algorithm for computing the roots of a polynomial $P \in \mathbb{C}[X]$ of high degree consists in forming its companion matrix[1] B_p and then applying to this matrix the QR algorithm to compute its eigenvalues with good accuracy.

Hence, all the methods are iterative and use the matrices directly. We need a notion of convergence, thus we limit ourselves to the cases $K = \mathbb{R}$ or $\mathbb{C}$. The general strategy consists in constructing a sequence of matrices

$$ M^{(0)}, M^{(1)}, \ldots, M^{(m)}, \ldots, $$

pairwise similar. Each method is conceived in such a way that the sequence converges to a simple form, triangular or diagonal, because then the eigenvalues can be read on the diagonal. Such a convergence is not always possible. For example, an algorithm in $\mathbf{M}_n(\mathbb{R})$ cannot converge to a triangular form when the matrix under consideration possesses a pair of nonreal eigenvalues.

---

[1] Fortunately, the companion matrix is a Hessenberg matrix; see below for this notion and its practical aspects.

13.1.1 Stability

In the course of the calculations, it is fundamental that the sequence

$$M^{(0)}, M^{(1)}, \ldots, M^{(m)}, \ldots$$

remain bounded, in order to keep away from overflow, as well as to be allowed to apply Theorem 5.2. This is not guaranteed a priori, because the set of matrices similar to M is unbounded in general. For instance, the following matrices are pairwise similar for all values of $a \in K^*$:

$$\begin{pmatrix} 0 & a \\ a^{-1} & 0 \end{pmatrix}.$$

This boundedness is one important issue among others. When passing from $M^{(k)}$ to $M^{(k+1)}$, the conjugation by a matrix Q yields to an amplification of the round-off errors by a factor that can be estimated as the *condition number* of Q, namely $\kappa(Q) := \|Q\|_2 \|Q^{-1}\|_2$. We recall that $\kappa(Q) \geq 1$, with equality if and only if Q is the matrix of a similitude. In order to keep control of the roundoff error, it thus seems necessary that the *product* of the numbers $\kappa(Q^{(k)})$ remain bounded. Because this is an infinite product, we need that $\kappa(Q^{(k)}) \to 1$ as $k \to +\infty$. In other words, the distance from $Q^{(k)}$ to $\mathbb{C} \cdot \mathbf{U}_n$ must tend to zero. Notice that a scalar factor in Q is harmless inasmuch as it cancels with the inverse factor in Q^{-1}. For the sake of simplicity, we thus ask that each iteration be a unitary conjugation: each $M^{(k)}$ is unitary similar to M, thus remains bounded because $\mathbf{U}_n$ is compact. When dealing with matrices in $\mathbf{M}_n(\mathbb{R})$, we employ orthogonal conjugation instead.

13.1.2 Expected Convergence

We thus assume that $M^{(k+1)} = Q_k^{-1} M^{(k)} Q_k$ for a unitary Q_k. Set $P_j := Q_0 \cdots Q_{j-1}$, which is unitary too. We have $M^{(j)} = P_j^{-1} M^{(0)} P_j$. Because $\mathbf{U}_n$ is compact, the sequence $(P_j)_{j \in \mathbb{N}}$ possesses cluster values. Let P be one of them. Then $M' := P^{-1} M^{(0)} P = P^* M^{(0)} P$ is a cluster point of $(M^{(j)})_{j \in \mathbb{N}}$ and is conjugated to M. If the sequence $(M^{(j)})_j$ converges, its limit is therefore (unitarily) similar to M, and hence has the same spectrum.

This argument shows that in general, the sequence $(M^{(j)})_j$ does not converge to a diagonal matrix, because then the eigenvectors of M would be the columns of P. In other words, M would have an orthonormal eigenbasis: M would be normal. Except in this special case, one expects merely that the sequence $(M^{(j)})_j$ converges to a triangular matrix, an expectation that is compatible with Theorem 5.1. But even this hope is too optimistic in general.

13.1.3 Initialization

Given $M \in \mathbf{M}_n(\mathbb{C})$, there are two strategies for the choice of $M^{(0)}$. One can naively take $M^{(0)} = M$. But because an iteration on a generic matrix is rather costly, one often uses a preliminary reduction to a simple form (e.g., the Hessenberg form, in the QR algorithm), which is preserved throughout the iterations. With a few such tricks, certain methods can be astonishingly efficient.

13.2 Hessenberg Matrices

We recall the notion of *Hessenberg* matrices.

Definition 13.1 *A square matrix $M \in \mathbf{M}_n(K)$ is called* upper Hessenberg *(one speaks simply of a Hessenberg matrix) if $m_{jk} = 0$ for every pair (j,k) such that $j - k \geq 2$.*

A Hessenberg matrix thus has the form

$$\begin{pmatrix} x & \cdots & \cdots & & \\ y & \ddots & & & \\ 0 & \ddots & \ddots & & \vdots \\ \vdots & \ddots & \ddots & \ddots & \vdots \\ 0 & \cdots & 0 & z & t \end{pmatrix}.$$

In particular, an upper-triangular matrix is a Hessenberg matrix.

From the point of view of matrix reduction by conjugation, one can attribute two advantages to the Hessenberg class, compared with the class of triangular matrices. First of all, if $K = \mathbb{R}$, many matrices are not trigonalizable in $\mathbb{R}$, although all are trigonalizable in $\mathbb{C}$. Even within complex numbers, the trigonalization cannot be done in practice, because it would require the computation of the eigenvalues. On the contrary, we show that every square matrix with real or complex entries is similar to a Hessenberg matrix over the real or complex numbers, respectively. This is obtained after a finite number of operations.

13.2.1 Stability of the Hessenberg Form

If M is Hessenberg and T upper-triangular, the products TM and MT are still Hessenberg.[2] For example, if M admits an LU factorization, then L is Hessenberg, and

---

[2] But the product of two Hessenberg matrices is not Hessenberg in general.

thus has only two nonzero diagonals, because $L = MU^{-1}$. Likewise, if $M \in \mathbf{GL}_n(\mathbb{C})$ is Hessenberg, then the factor Q in the factorization $M = QR$ is again Hessenberg, because $Q = MR^{-1}$. An elementary compactness and continuity argument shows that the same fact holds true for every $M \in \mathbf{M}_n(\mathbb{C})$.

13.2.2 Hessenberg Form versus Irreducibility

We have seen in Proposition 3.26 that a Hessenberg matrix such that the $m_{j+1,j}$s are nonzero has geometrically simple eigenvalues. The algebraic multiplicity can, however, be arbitrary, as shown in the following example

$$ M = \begin{pmatrix} 1 & 1 \\ -1 & -1 \end{pmatrix}. $$

13.2.3 Transforming a Matrix into a Hessenberg One

Theorem 13.1 *For every matrix $M \in \mathbf{M}_n(\mathbb{C})$ there exists a unitary transformation U such that $U^{-1}MU$ is a Hessenberg matrix. If $M \in \mathbf{M}_n(\mathbb{R})$, one may take $U \in \mathbf{O}_n$.*

Moreover, the matrix U is computable in $4n^3/3 + O(n^2)$ multiplications and $4n^3/3 + O(n^2)$ additions.

Proof. Let $X \in \mathbb{C}^m$ be a unit vector: $X^*X = 1$. The matrix of the unitary (orthogonal) symmetry with respect to the hyperplane $X^\perp$ is $S = I_m - 2XX^*$. In fact, $SX = X - 2X = -X$, and $Y \in X^\perp$ (i.e., $X^*Y = 0$) implies $SY = Y$.

We construct a sequence $M_1 = M, \ldots, M_{n-1}$ of unitarily similar matrices. The matrix M_{n-r} is of the form

$$ \begin{pmatrix} H & B \\ 0_{r,n-r-1} & Z \; N \end{pmatrix}, $$

where $H \in \mathbf{M}_{n-r}(\mathbb{C})$ is Hessenberg and Z is a vector in $\mathbb{C}^r$. Hence, M_{n-1} is Hessenberg.

One passes from M_{n-r} to M_{n-r+1}, that is, from r to $r-1$ as follows. Let $\mathbf{e}^1$ be the first vector of the canonical basis of $\mathbb{C}^r$. If Z is already colinear to $\mathbf{e}^1$, one does nothing besides defining $M_{n-r+1} = M_{n-r}$. Otherwise, one chooses $X \in \mathbb{C}^r$ so that SZ is parallel to $\mathbf{e}^1$ (we discuss below the possible choices for X). Then one sets

$$ V = \begin{pmatrix} I_{n-r} & 0_{n-r,r} \\ 0_{r,n-r} & S \end{pmatrix}, $$

which is a unitary matrix, with $V^* = V^{-1} = V$ (such a matrix is called a *Householder matrix*). We then have

$$V^{-1}M_{n-r}V = \begin{pmatrix} H & BS \\ 0_{n,n-r-1} & SZ \; SNS \end{pmatrix}.$$

We thus define $M_{n-r+1} = V^{-1}M_{n-r}V$.

There are two possible choices for S, given by

$$X_{\pm} := \frac{1}{\|Z \pm \|Z\|_2 \mathbf{q}\|_2} (Z \pm \|Z\|_2 \mathbf{q}), \quad \mathbf{q} = \frac{z_1}{|z_1|} \mathbf{e}^1.$$

It is always advantageous to choose the sign that gives the largest denominator, namely the positive sign. One thus optimizes the roundoff errors in the case where Z is almost aligned with $\mathbf{e}^1$.

13.2.4 Complexity

Let us consider now the complexity of the $(n - r)$th step. Only the terms of order r^2 and $r(n - r)$ are meaningful. The computation of X, in $O(r)$ operations, is thus negligible, like that of X^* and of $2X$. The computation of $BS = B - (BX)(2X^*)$ needs about $4r(n - r)$ operations. Then $2NX$ needs $2r^2$ operations, as does $2X^*N$. We next compute $4X^*NX$, and then form the vector $T := 4(X^*NX)X - 2NX$ at the cost $O(r)$. The product TX^* takes r^2 operations, as $2X(X^*N)$. Then $N + TX^* - X(2X^*N)$ needs $2r^2$ additions. The complete step is thus accomplished in $2r^2 + 4rn + O(n)$ operations. A sum from $r = 1$ to $n - 2$ yields a complexity of $\frac{8}{3}n^3 + O(n^2)$, in which one recognizes $\frac{4}{3}n^3 + O(n^2)$ multiplications, $\frac{4}{3}n^3 + O(n^2)$ additions, and $O(n)$ square roots. $\quad\square$

13.2.5 The Hermitian Case

When M is Hermitian, the matrix $U^{-1}MU$ is still Hermitian. Because it is Hessenberg, it is tridiagonal, with $a_{j,j+1} = \bar{a}_{j+1,j}$ and $a_{jj} \in \mathbb{R}$. The symmetry reduces the complexity to $2n^3/3 + O(n^2)$ multiplications. One can then use the Hessenberg form of M in order to localize its eigenvalues.

Proposition 13.1 *If M is tridiagonal Hermitian and if the entries $m_{j+1,j}$ are nonzero (i.e., if M is irreducible), then the eigenvalues of M are real and simple. Furthermore, if M_j is the (Hermitian, tridiagonal, irreducible) matrix obtained by keeping only the j last rows and columns of M, the eigenvalues of M_j strictly separate those of M_{j+1}.*

The separation, not necessarily strict, of the eigenvalues of M_{j+1} by those of M_j has already been proved, in a more general framework, in Theorem 6.5.

Proof. The geometric simplicity of the eigenvalues has been stated in Proposition 3.26. Because M is Hermitian, it is diagonalizable: geometric multiplicity equals the algebraic one. Thus the eigenvalues are simple. In addition, an Hermitian matrix has a real spectrum.

We proceed by induction on j. If $j \geq 1$, we decompose the matrix M_{j+1} block-wise:

$$\begin{pmatrix} m & \bar{a} & 0 & \cdots & 0 \\ a & & & & \\ 0 & & M_j & & \\ \vdots & & & & \\ 0 & & & & \end{pmatrix},$$

where $a \neq 0$ and $m \in \mathbb{R}$. Let P_ℓ be the characteristic polynomial of M_ℓ. We compute that of M_{j+1} by expanding the determinant with respect to the first column:

$$P_{j+1}(X) = (X - m)P_j(X) - |a|^2 P_{j-1}(X), \tag{13.1}$$

where $P_0 \equiv 1$ by convention.

The induction hypothesis is as follows. The polynomials P_j and P_{j-1} have real coefficients and have, respectively, j and $j-1$ real roots $\mu_1, \ldots, \mu_j$ and $\sigma_1, \ldots, \sigma_{j-1}$, with

$$\mu_1 < \sigma_1 < \mu_2 < \cdots < \sigma_{j-1} < \mu_j.$$

In particular, they have no other roots, and their roots are simple. The signs of the values of P_{j-1} at points μ_j thus alternate. Because P_{j-1} is positive over $(\sigma_{j-1}, +\infty)$, we have $(-1)^{j-k}P_{j-1}(\mu_k) > 0$.

This hypothesis clearly holds at step $j = 1$. If $j \geq 2$ and if it holds at step j, then (13.1) shows that $P_{j+1} \in \mathbb{R}[X]$. Furthermore,

$$(-1)^{j-k}P_{j+1}(\mu_k) = -|a|^2(-1)^{j-k}P_{j-1}(\mu_k) < 0.$$

From the intermediate value theorem, P_{j+1} possesses a root λ_k in (μ_{k-1}, μ_k). Furthermore, $P_{j+1}(\mu_j) < 0$, and $P_{j+1}(x)$ is positive for $x \gg 1$; hence there is another root in $(\mu_j, +\infty)$. Likewise, P_{j+1} has a root in $(-\infty, \mu_1)$. Hence, P_{j+1} possesses $j+1$ distinct real roots λ_k, with

$$\lambda_1 < \mu_1 < \lambda_2 < \cdots < \mu_j < \lambda_{j+1}.$$

Because P_{j+1} has degree $j+1$, it has no root other than the λ_ks, and these are simple.
$\square$

The sequence of polynomials P_j is a *Sturm sequence*, which allows us to compute the number of roots of P_n in a given interval (a, b). A Sturm sequence is a finite sequence of real polynomials $Q_0, \ldots, Q_n$, with Q_0 a nonzero constant such that

- If $Q_j(x) = 0$ and $0 < j < n$, then $Q_{j+1}(x)Q_{j-1}(x) < 0$. In particular, Q_j and Q_{j+1} do not share a common root.

- Likewise, if $Q_0(c) = 0$ for some $c \in (a,b)$, then

$$\frac{Q_0(x)Q_1(x)}{x - c} < 0, \qquad \forall x \in (c - \varepsilon, c + \varepsilon)$$

  for some $\varepsilon > 0$.

If $a \in \mathbb{R}$ is not a root of Q_n, we denote by $V(a)$ the number of sign changes in the sequence $(Q_0(a), \ldots, Q_n(a))$, in which the zeroes play no role and can be ignored.

Proposition 13.2 *If $Q_n(a) \neq 0$ and $Q_n(b) \neq 0$, and if $a < b$, then the number of roots of Q_n in (a,b) is equal to $V(a) - V(b)$.*

Let us remark that it is not necessary to compute the polynomials P_j in order to apply them to this proposition. Given $a \in \mathbb{R}$, it is enough to compute the sequence of values $P_j(a)$.

Once an interval (a,b) is known to contain an eigenvalue λ and only that one (by means of Proposition 13.2 or Theorem 5.7), one can compute an approximate value of λ, either by dichotomy, or by computing the numbers $V((a+b)/2), \ldots$, or by the secant or Newton method. In the latter case, one must compute P_n itself. The last two methods are convergent, provided that we have a good initial approximation at our disposal, because $P_n'(\lambda) \neq 0$.

13.3 The *QR* Method

The *QR* method is considered the most efficient one for the approximate computation of the whole spectrum of a general square matrix $M \in \mathbf{M}_n(\mathbb{C})$. One employs it only after having reduced M to Hessenberg form, because this form is preserved throughout the algorithm, whereas each iteration is much cheaper than it would be for an arbitrary matrix.

13.3.1 Description of the **QR** Method

Let $A \in \mathbf{M}_n(K)$ be given, with $K = \mathbb{R}$ or $\mathbb{C}$. We construct a sequence of matrices $(A_j)_{j \in \mathbb{N}}$, with $A_0 = A$. The induction $A_j \mapsto A_{j+1}$ consists in performing the *QR* factorization of A_j, $A_j = Q_j R_j$, and then defining $A_{j+1} := R_j Q_j$. We have

$$A_{j+1} = Q_j^{-1} A_j Q_j,$$

which shows that A_{j+1} is unitarily similar to A_j. Hence,

$$A_j = (Q_0 \cdots Q_{j-1})^{-1} A (Q_0 \cdots Q_{j-1}) \tag{13.2}$$

is conjugate to A by a unitary transformation.

13.3.1.1 Obstructions

- If A is unitary, then $A_j = A$ for every j, with $Q_j = A$ and $R_j = I_n$. The convergence occurs but is useless, because the limit A is not simpler than the data. We show later on that the reason for this bad behavior is that the eigenvalues of a unitary matrix have the same modulus. The QR method does not do a good job of separating the eigenvalues of close modulus.
- Another bad situation is when our matrix has at least two eigenvalues of the same modulus. This happens in particular if A has real entries. In the latter case, then each Q_j is real orthogonal, R_j is real, and A_j is real. This is seen by induction on j. A limit A' is not triangular if some eigenvalues of A are nonreal, namely if A possesses a pair of complex conjugate eigenvalues.

Let us sum up what can be expected in a brave new world. If all the eigenvalues of $A \in \mathbf{M}_n(\mathbb{C})$ have distinct moduli, the sequence $(A_j)_j$ might converge to a triangular matrix, or at least its lower-triangular part might converge to

$$\begin{pmatrix} \lambda_1 & & & \\ 0 & \lambda_2 & & \\ \vdots & \ddots & \ddots & \\ 0 & \cdots & 0 & \lambda_n \end{pmatrix}.$$

When $A \in \mathbf{M}_n(\mathbb{R})$, let us make the following assumption. Let p be the number of real eigenvalues and $2q$ that of nonreal eigenvalues ; then there are $p + q$ distinct eigenvalue moduli. In that case, $(A_j)_j$ might converge to a block-triangular form, the diagonal blocks being 2×2 or 1×1. The limits of the diagonal blocks trivially provide the eigenvalues of A.

Herebelow, we treat the complex case with eigenvalues of pairwise distinct moduli. The case with real entries and pairs of complex conjugate eigenvalues has been treated in [23].

13.3.2 The Case of a Singular Matrix

When A is not invertible, the QR factorization is not unique, raising a difficulty in the definition of the algorithm. The computation of the determinant would immediately detect the case of noninvertibility, but would not provide any cure. However, if the matrix has been first reduced to the Hessenberg form, then a single QR iteration makes a diagnosis and does provide a cure. Indeed, if A is Hessenberg and singular, then in $A = QR$, Q is Hessenberg and R is singular. If $a_{21} = 0$, the matrix A is block-triangular and we may reduce our calculations to the case of a matrix of size $(n-1) \times (n-1)$ by deleting the first row and the first column. Otherwise, there exists $j \geq 2$ such that $r_{jj} = 0$. The matrix $A_1 = RQ$ is then block-triangular, because it is

Hessenberg and $(A_1)_{j,j-1} = r_{jj}q_{j,j-1} = 0$. Again, we may reduce our calculations to that of the spectra of two matrices of sizes $j \times j$ and $(n-j) \times (n-j)$, the diagonal blocks of A_1. After finitely many such steps (not larger than the multiplicity of the null eigenvalue), there remain only Hessenberg invertible matrices to deal with. We assume therefore from now on that $A \in \mathbf{GL}_n(K)$.

13.3.3 Complexity of an Iteration

An iteration of the *QR* method requires the factorization $A_j = Q_j R_j$ and the computation of $A_{j+1} = R_j Q_j$. Each part costs $O(n^3)$ operations if it is done on a generic matrix (using the naive way of multiplying matrices). The reduction to the Hessenberg form has a comparable cost, therefore we loose nothing by reducing A to this form. Actually, we make considerable gains in two aspects. First of all, the cost of each *QR* iteration is reduced to $O(n^2)$. Secondly, the cluster values of the sequence $(A_j)_j$ must have the Hessenberg form too.

Let us first examine the Householder method of *QR* factorization for a generic matrix A. In practice, one computes only the factor R and matrices of unitary symmetries whose product is Q. One then multiplies these unitary matrices by R on the left to obtain $A' = RQ$.

Let $\mathbf{a}_1 \in \mathbb{C}^n$ be the first column vector of A. We begin by determining a unit vector $v_1 \in \mathbb{C}^n$ such that the hyperplane symmetry $H_1 := I_n - 2v_1 v_1^*$ sends $\mathbf{a}_1$ to $\|\mathbf{a}_1\|_2 e^1$. The matrix $H_1 A$ has the form

$$
\tilde{A} = \begin{pmatrix} \|\mathbf{a}_1\|_2 & x & \cdots \\ 0 & \vdots & \\ \vdots & \vdots & \\ 0 & y & \cdots \end{pmatrix}.
$$

We then perform these operations again on the matrix extracted from $\tilde{A}$ by deleting the first rows and columns, and so on. At the kth step, H_k is a matrix of the form

$$
\begin{pmatrix} I_k & 0 \\ 0 & I_{n-k} - 2v_k v_k^* \end{pmatrix},
$$

where $v_k \in \mathbb{C}^{n-k}$ is a unit vector. The computation of v_k requires $O(n-k)$ operations. The product $H_k A^{(k)}$, where $A^{(k)}$ is block-triangular, amounts to that of two square matrices of size $n - k$, one of them $I_{n-k} - 2v_k v_k^*$. We thus compute a matrix $N - 2vv^*N$ from v and N, which costs about $4(n-k)^2$ operations. Summing from $k = 1$ to $k = n - 1$, we find that the complexity of the computation of R alone is $4n^3/3 + O(n^2)$. As indicated above, we do not compute the factor Q, but compute all the matrices $RH_{n-1} \cdots H_k$. That necessitates $2n^3 + O(n)$ operations. The complexity of one step of the *QR* method on a generic matrix is thus $10n^3/3 + O(n^2)$.

Let us now analyze the situation when A is a Hessenberg matrix. By induction on k, we see that v_k belongs to the plane spanned by e^k and e^{k+1}. Its computation needs $O(1)$ operations. Then the product of H_k and $A^{(k)}$ can be obtained by simply recomputing the rows of indices k and $k+1$, about $6(n-k)$ operations. Summing from $k=1$ to $n-1$, we find that the complexity of the computation of R alone is $3n^2 + O(n)$. The computation of the product $(RH_{n-1}\cdots H_{k+1})H_k$ needs about $6k$ operations. Finally, the complexity of the QR iteration on a Hessenberg matrix is $6n^2 + O(n)$, in which there are $4n^2 + O(n)$ multiplications.

To sum up, the cost of the preliminary reduction of a matrix to Hessenberg form is less than or equal to what is saved during the first iteration of the QR method.

13.3.4 Convergence of the QR Method

As explained above, the best convergence statement assumes that the eigenvalues have distinct moduli.

Let us recall that the sequence A_k is not always convergent. For example, if A is already triangular, its QR factorization is $Q = D$, $R = D^{-1}A$, with $d_j = a_{jj}/|a_{jj}|$. Hence, $A_1 = D^{-1}AD$ is triangular, with the same diagonal as that of A. By induction, A_k is triangular, with the same diagonal as that of A. We have thus $Q_k = D$ for every k, so that $A_k = D^{-k}AD^k$. The entry of index (ℓ, m) is thus multiplied at each step by a unit number $z_{\ell m}$, which is not necessarily equal to one if $\ell < m$. Hence, the part above the diagonal of A_k may not converge.

Summing up, a convergence theorem may concern only the diagonal of A_k and what lies below it.

Lemma 24. *Let $A \in \mathbf{GL}_n(K)$ be given, with $K = \mathbb{R}$ or $\mathbb{C}$. Let $A_k = Q_k R_k$ be the sequence of matrices given by the QR algorithm. Let us define $P_k = Q_0 \cdots Q_{k-1}$ and $U_k = R_{k-1} \cdots R_0$. Then $P_k U_k$ is the QR factorization of the kth power of A:*

$$A^k = P_k U_k.$$

Proof. From (13.2), we have $A_k = P_k^{-1} A P_k$; that is, $P_k A_k = A P_k$. Then

$$P_{k+1} U_{k+1} = P_k Q_k R_k U_k = P_k A_k U_k = A P_k U_k.$$

By induction, $P_k U_k = A^k$. However, $P_k \in \mathbf{U}_n$ and U_k is triangular, with a positive real diagonal, as a product of such matrices. $\square$

Theorem 13.2 *Let $A \in \mathbf{GL}_n(\mathbb{C})$ be given. Assume that the moduli of the eigenvalues of A are distinct:*

$$|\lambda_1| > |\lambda_2| > \cdots > |\lambda_n| \qquad (>0).$$

In particular, the eigenvalues are simple, and thus A is diagonalizable:

$$A = Y^{-1}\mathrm{diag}(\lambda_1, \ldots, \lambda_n)Y.$$

Assume also that Y admits an LU factorization. Then the strictly lower-triangular part of A_k converges to zero, and the diagonal of A_k converges to

$$D := \text{diag}(\lambda_1, \ldots, \lambda_n).$$

Proof. Let $Y = LU$ be the factorization of Y. We also make use of the QR factorization of Y^{-1} : $Y^{-1} = QR$. Because $A^k = Y^{-1}D^kY$, we have $P_kU_k = Y^{-1}D^kY = QRD^kLU$.

The matrix D^kLD^{-k} is lower-triangular with unit numbers on its diagonal. Each term is multiplied by $(\lambda_i/\lambda_j)^k$, therefore its strictly lower-triangular part tends to zero, because $|\lambda_i/\lambda_j| < 1$ for $i > j$. Therefore, $D^kLD^{-k} = I_n + E_k$ with $E_k \to 0_n$ as $k \to +\infty$. Hence, $P_kU_k = QR(I_n + E_k)D^kU = Q(I_n + RE_kR^{-1})RD^kU = Q(I_n + F_k)RD^kU$, where $F_k \to 0_n$. Let $O_kT_k = I_n + F_k$ be the QR factorization of $I_n + F_k$. By continuity, O_k and T_k both tend to I_n. Then

$$P_kU_k = (QO_k)(T_kRD^kU).$$

The first product is a unitary matrix, whereas the second is a triangular one. Let $|D|$ be the "modulus" matrix of D (whose entries are the moduli of those of D), and let D_1 be $|D|^{-1}D$, which is unitary. We also define $D_2 = \text{diag}(u_{jj}/|u_{jj}|)$ and $U' = D_2^{-1}U$. Then D_2 is unitary and the diagonal of U' is positive real. From the uniqueness of the QR factorization of an invertible matrix we obtain

$$P_k = QO_kD_1^kD_2, \quad U_k = (D_1^kD_2)^{-1}T_kRD_1^kD_2|D|^kU',$$

which yields

$$Q_k = P_k^{-1}P_{k+1} = D_2^{-1}D_1^{-k}O_k^{-1}O_{k+1}D_1^{k+1}D_2,$$
$$R_k = U_{k+1}U_k^{-1} = D_2^{-1}D_1^{-k-1}T_{k+1}RDR^{-1}T_k^{-1}D_1^kD_2.$$

Because D_1^{-k} and D_1^{k+1} are bounded, we deduce that Q_k converges, to D_1. Likewise, $R_k - R_k' \to 0_n$, where

$$R_k' = D_2^{-1}D_1^{-k}RDR^{-1}D_1^{k-1}D_2. \tag{13.3}$$

The fact that the matrix R_k' is upper-triangular shows that the strictly lower-triangular part of $A_k = Q_kR_k$ tends to zero (observe that the sequence $(R_k)_{k\in\mathbb{N}}$ is bounded, because the set of matrices unitarily conjugate to A is bounded). Likewise, the diagonal of R_k' is $|D|$, which shows that the diagonal of A_k converges to $D_1|D| = D$. $\quad\square$

Remark

Formula (13.3) shows that the sequence A_k does not converge, at least when the eigenvalues have distinct complex arguments. However, if the eigenvalues have equal complex arguments, for example, if they are real and positive, then $D_1 = \alpha I_n$

and $R_k \to T := D_2^{-1} R |D| R^{-1} D_2$; hence A_k converges. Note that the limit αT is not diagonal in this case.

The odd assumption about Y (LU factorization) in Theorem 13.2 is fulfilled in most practical situations:

Theorem 13.3 *Let $A \in \mathbf{GL}_n(\mathbb{C})$ be an irreducible Hessenberg matrix whose eigenvalues are of distinct moduli:*

$$|\lambda_1| > \cdots > |\lambda_n| \quad (> 0).$$

Then the QR *method converges; that is, the lower-triangular part of A_k converges to*

$$\begin{pmatrix} \lambda_1 & & & \\ 0 & \lambda_2 & & \\ \vdots & \ddots & \ddots & \\ 0 & \cdots & 0 & \lambda_n \end{pmatrix}.$$

Proof. In the light of Theorem 13.2, it is enough to show that the matrix Y in the previous proof admits an *LU* factorization. We have $YA = \text{diag}(\lambda_1, \ldots, \lambda_n)Y$. The rows of Y are thus the left eigenvectors: $\ell_j A = \lambda_j \ell_j$.

If $x \in \mathbb{C}^n$ is nonzero, there exists a unique index r such that $x_r \neq 0$, and $j > r$ implies $x_j = 0$. By induction, quoting the Hessenberg form and the irreducibility of A, we obtain $(A^m x)_{r+m} \neq 0$, while $j > r + m$ implies $(A^m x)_j = 0$. Hence, the vectors $x, Ax, \ldots, A^{n-r}x$ are linearly independent. A linear subspace, invariant for A and containing x, is thus of dimension greater than or equal to $n - r + 1$.

Let F be a linear subspace, invariant for A, of dimension $p \geq 1$. Let r be the smallest integer such that F contains a nonzero vector x with $x_{r+1} = \cdots = x_n = 0$. The minimality of r implies that $x_r \neq 0$. Hence, we have $p \geq n - r + 1$. By construction, the intersection of F and of linear subspace $[\mathbf{e}^1, \ldots, \mathbf{e}^{r-1}]$ spanned by $\mathbf{e}^1, \ldots, \mathbf{e}^{r-1}$ reduces to $\{0\}$. Thus we also have $p + (r - 1) \leq n$. Finally, $r = n - p + 1$, and we see that

$$F \oplus [\mathbf{e}^1, \ldots, \mathbf{e}^{n-p}] = \mathbb{C}^n.$$

Let us choose $F = [\ell_1, \ldots, \ell_q]^\perp$, which is invariant for A. Then $p = n - q$, and we have

$$[\ell_1, \ldots, \ell_q]^\perp \oplus [\mathbf{e}^1, \ldots, \mathbf{e}^q] = \mathbb{C}^n.$$

This amounts to saying that $\det(\ell_j \mathbf{e}^k)_{1 \leq j,k \leq q} \neq 0$. In other words, the leading principal minor of order q of Y is nonzero. By Theorem 11.1, Y admits an *LU* factorization. $\square$

13.3.5 The Case of Hermitian Matrices

The situation is especially favorable for tridiagonal Hermitian matrices. To begin with, we may assume that A is positive-definite, up to the change of A into $A + \mu I_n$

with $\mu > -\rho(A)$. Next, we can write A in block-diagonal form, where the diagonal blocks are tridiagonal irreducible Hermitian matrices. The QR method then treats each block separately. We are thus reduced to the case of an Hermitian positive-definite, tridiagonal, and irreducible matrix. Its eigenvalues are real, strictly positive, and simple, from Proposition 13.1: we have $\lambda_1 > \cdots > \lambda_n > 0$. Theorems 13.2 and 13.3 can then be applied.

Corollary 13.1 *If $A \in \mathbf{HPD}_n$ and if A_0 is a Hessenberg matrix, unitarily similar to A (e.g., a matrix obtained by Householder's method), then the sequence A_k defined by the* QR *method converges to a diagonal matrix whose diagonal entries are the eigenvalues of A.*

Indeed, the lower-triangular part converges, hence the whole matrix, because it is Hermitian.

13.3.6 Implementing the QR Method

The QR method converges faster as λ_n, or merely λ_n/λ_{n-1}, becomes smaller. We can obtain this situation by translating $A_k \mapsto A_k - \alpha_k I_n$. The strategies for the choice of α_k are described in [27]. This procedure is called *Rayleigh translation*. It yields a significant improvement of the convergence of the QR method. If the eigenvalues of A are simple, a suitable translation places us into the case of eigenvalues of distinct moduli. This trick has a nonnegligible cost if A is a real matrix with a pair of complex conjugate eigenvalues, inasmuch as it requires a translation by a nonreal number α. As mentioned above, the computations become much more costly in $\mathbb{C}$ than they are in $\mathbb{R}$.

As k increases, the triangular form of A_k shows up first at the last row. As a by-product, the sequence $(A_k)_{nn}$ converges more rapidly than other sequences $(A_k)_{jj}$. When the last row is sufficiently close to $(0, \dots, 0, \lambda_n)$, the Rayleigh translation must be selected in such a way as to bring λ_{n-1}, instead of λ_n, to the origin; and so on.

With a clever choice of Rayleigh translations, the QR method, when it converges, is of order two for a generic matrix, and is of order three for an Hermitian matrix.

13.4 The Jacobi Method

The Jacobi method gives an approximate value of the whole spectrum of a real symmetric matrix $A \in \mathbf{Sym}_n$. As in the QR method, one constructs a sequence of matrices, unitarily similar to A. In particular, the roundoff errors are not amplified. Each iteration is cheap ($O(n)$ operations), and the convergence may be quadratic or even faster when the eigenvalues are distinct. It is thus a rather efficient method.

13.4.1 Conjugating by a Rotation Matrix

Let $1 \leq p,q \leq n$ be two distinct indices and $\theta \in [-\pi, \pi)$ an angle. We denote by $R_{p,q}(\theta)$ the matrix of rotation of angle θ in the plane spanned by e^p and e^q. For example, if $p < q$, then

$$
R = R_{p,q}(\theta) := \begin{pmatrix} I_{p-1} & \vdots & 0 & \vdots & 0 \\ \cdots & \cos\theta & \cdots & \sin\theta & \cdots \\ 0 & \vdots & I_{q-p-1} & \vdots & 0 \\ \cdots & -\sin\theta & \cdots & \cos\theta & \cdots \\ 0 & \vdots & 0 & \vdots & I_{n-q} \end{pmatrix}.
$$

If H is a symmetric matrix, we compute $K := R^{-1}HR = R^T HR$, which is also symmetric, with the same spectrum. Setting $c = \cos\theta$, $s = \sin\theta$ the following formulæ hold.

$$
\begin{aligned}
k_{ij} &= h_{ij} && \text{if } i,j \neq p,q, \\
k_{ip} &= ch_{ip} - sh_{iq} && \text{if } i \neq p,q, \\
k_{iq} &= ch_{iq} + sh_{ip} && \text{if } i \neq p,q, \\
k_{pp} &= c^2 h_{pp} + s^2 h_{qq} - 2csh_{pq}, \\
k_{qq} &= c^2 h_{qq} + s^2 h_{pp} + 2csh_{pq}, \\
k_{pq} &= cs(h_{pp} - h_{qq}) + (c^2 - s^2)h_{pq}.
\end{aligned}
$$

The cost of the computation of entries k_{ij} for $i,j \neq p,q$ is zero; that of k_{pp}, k_{qq}, and k_{pq} is $O(1)$. The cost of this conjugation is thus $6n + O(1)$ operations, keeping in mind the symmetry $K^T = K$.

Let us remark that the conjugation by the rotation of angle $\theta \pm \pi$ yields the same matrix K, up to signs. For this reason, we limit ourselves to angles $\theta \in [-\pi/2, \pi/2)$.

13.4.2 Description of the Method

One constructs a sequence $A^{(0)} = A, A^{(1)}, \ldots$ of symmetric matrices, each one conjugate to the previous one by a rotation as above: $A^{(k+1)} = (R^{(k)})^T A^{(k)} R^{(k)}$. At step k, we choose two distinct indices p and q (in fact, p_k, q_k) in such a way that $a_{pq}^{(k)} \neq 0$ (if it is not possible, $A^{(k)}$ is already a diagonal matrix similar to A). We then choose θ (in fact θ_k) in such a way that $a_{pq}^{(k+1)} = 0$. From the formulæ above, this is equivalent to

$$
cs(a_{pp}^{(k)} - a_{qq}^{(k)}) + (c^2 - s^2)a_{pq}^{(k)} = 0.
$$

This amounts to solving the equation

$$
\cot 2\theta = \frac{a_{qq}^{(k)} - a_{pp}^{(k)}}{2a_{pq}^{(k)}} =: \sigma_k. \tag{13.4}
$$

This equation possesses two solutions in $[-\pi/2, \pi/2)$, namely $\theta_k \in [-\pi/4, \pi/4)$ and $\theta_k \pm \pi/2$. There are thus two possible rotation matrices, which yield to two distinct results. Once the angle has been selected, its computation is useless (it would actually be rather expensive). In fact, $t := \tan \theta_k$ solves

$$\frac{2t}{1-t^2} = \tan 2\theta;$$

that is,

$$t^2 + 2t\sigma_k - 1 = 0.$$

The two angles correspond to the two possible roots of this quadratic equation. We then obtain

$$c = \frac{1}{\sqrt{1+t^2}}, \quad s = tc.$$

We show below that the stablest choice is the angle $\theta_k \in [-\pi/4, \pi/4)$, which corresponds to the unique root t in $[-1, 1)$.

The computation of c, s needs only $O(1)$ operations, so that the cost of an iteration of the Jacobi method is still $6n + O(1)$. Observe that an entry that has vanished at a previous iteration becomes in general nonzero after a few more iterations.

13.4.3 The Choice of the Pair $(\mathbf{p}, \mathbf{q})$

We use here the Schur norm $\|M\| = (\operatorname{Tr} M^T M)^{1/2}$, also called the Frobenius norm, denoted elsewhere by $\|M\|_F$. We wish to show that $A^{(k)}$ converges to a diagonal matrix, therefore we decompose $A^{(k)} = D_k + E_k$, where $D_k = \operatorname{diag}(a_{11}^{(k)}, \ldots, a_{nn}^{(k)})$. To begin with, because the sequence is formed of unitarily similar matrices, we have $\|A^{(k)}\| = \|A\|$.

Lemma 25. *We have*

$$\|E_{k+1}\|^2 = \|E_k\|^2 - 2\left(a_{pq}^{(k)}\right)^2.$$

Proof. It suffices to redo the calculations of Section 13.4.1, noting that

$$k_{ip}^2 + k_{iq}^2 = h_{ip}^2 + h_{iq}^2$$

whenever $i \neq p, q$, whereas $k_{pq}^2 = 0$. $\square$

We deduce from the lemma that $\|D_{k+1}\|^2 = \|D_k\|^2 + 2\left(a_{pq}^{(k)}\right)^2$. The convergence of the Jacobi method then depends on the choice of the pair (p, q) at each step. Notice that the choice of the same pair at two consecutive iterations is inadvisable, inasmuch as it yields $A^{(k+1)} = A^{(k)}$.

There are essentially three strategies for chosing the pair (p, q) at a given step.

Optimal choice.   One chooses a pair (p,q) for which the modulus of a_{pq} is maximal among off-diagonal entries of $A^{(k)}$. At first glance, this looks to be the most
efficient choice, but needs a comparison procedure whose cost is about $n^2 \log n$.
If a careful storage of the order of moduli at previous steps is made, the comparison reduces to about n^2 operations, still costly enough, compared to the $6n$
operations needed in the conjugation.

Sequential choice.   Here the pair is a periodic function of k. Typically, one chooses
first $(1,2)$ then $(2,3), \ldots, (n-1,n), (1,3), (2,4), \ldots, (1,n)$. Variant: because the
position $(2,3)$ was affected by the operations made around $(1,2)$, it might be better to find an order beginning with $(1,2), (3,4), \ldots,$ in such a way that an index
p is not present in two consecutive pairs, in order to treat all the entries as fast as
possible.

Random choice.   The set of pairs (p,q) with $1 \leq p < q \leq n$ is equipped with
the uniform probability. The pair (p,q) is taken at random at step k, and independently of the previous choices. Some variants of the random choice can be
elaborated.

13.4.4 Convergence with the Optimal Choice

Theorem 13.4 *With the "optimal choice" of (p_k, q_k) and with the choice $\theta_k \in$
$[-\pi/4, \pi/4)$, the Jacobi method converges in the following sense. There exists a
diagonal matrix D such that*

$$\|A^{(k)} - D\| \leq \frac{\sqrt{2}\|E_0\|}{1-\rho} \rho^k, \qquad \rho := \sqrt{1 - \frac{2}{n^2 - n}}.$$

*In particular, the spectrum of A consists of the diagonal terms of D and the limit of
D_k; the Jacobi method is of order one at least.*

This kind of convergence is called *linear*, because it is typical of methods in
which the error obeys a linear inequality $\varepsilon_{k+1} \leq \rho \varepsilon_k$, with $\rho < 1$. We also say that
the convergence is of order one at least. This is a rather slow convergence that we
already encountered in iterative methods for linear systems (Chapter 12).

Proof. With the optimal choice of (p,q), we have

$$(n^2 - n)\left(a_{pq}^{(k)}\right)^2 \geq \|E_k\|^2.$$

Hence,

$$\|E_{k+1}\|^2 \leq \left(1 - \frac{2}{n^2 - n}\right)\|E_k\|^2 = \rho^2\|E_k\|^2.$$

It follows that $\|E_k\| \leq \rho^k\|E_0\|$. In particular, E_k tends to zero as $k \to +\infty$.

It remains to show that D_k converges too. A calculation using the notation of Section 13.4.1 and the fact that $k_{pq} = 0$ yields

$$k_{pp} - h_{pp} = -th_{pq}.$$

Because $|\theta_k| \leq \pi/4$, we have $|t| \leq 1$, so that $|a_{pp}^{(k+1)} - a_{pp}^{(k)}| \leq |a_{pq}^{(k)}|$. Likewise, $|a_{qq}^{(k+1)} - a_{qq}^{(k)}| \leq |a_{pq}^{(k)}|$. The other diagonal entries are unchanged, thus we have $\|D_{k+1} - D_k\| \leq \|E_k\|$.

We therefore have

$$\|D_\ell - D_k\| \leq \|E_0\|(\rho^{\ell-1} + \cdots + \rho^k) \leq \|E_0\| \frac{\rho^k}{1-\rho}, \quad \ell > k.$$

The sequence $(D_k)_{k\in\mathbb{N}}$ is thus Cauchy, hence convergent. Because E_k tends to zero, $A^{(k)}$ converges to the same limit D. This matrix is diagonal, with the same spectrum as A, because this is true for each $A^{(k)}$. Finally, we obtain

$$\|A^{(k)} - D\|^2 = \|D_k - D\|^2 + \|E_k\|^2 \leq \frac{2}{(1-\rho)^2} \|E_k\|^2.$$

$\square$

We analyze, in Exercise 10, the (bad) behavior of D_k when we make the opposite choice $\pi/4 \leq |\theta_k| \leq \pi/2$.

13.4.5 Optimal Choice: Super-Linear Convergence

The following statement shows that the Jacobi method compares rather well with other methods.

Theorem 13.5 *The Jacobi method with optimal choice of (p,q) converges super-linearly when the eigenvalues of A are simple, in the following sense. Let $N = n(n-1)/2$ be the number of elements under the diagonal. Then there exists a number $c > 0$ such that*

$$\|E_{k+N}\| \leq c\|E_k\|^2,$$

for every $k \in \mathbb{N}$.

In the present setting, the order of the Jacobi method can be estimated at least as $\nu := 2^{1/N}$, which is slightly larger than 1. This is typical of a method where the error obeys an inequality of the form $\varepsilon_{k+1} < \text{cst} \cdot (\varepsilon_k)^\nu$ (mind, however, that the inequality given in the theorem is not exactly of this form). The convergence is much faster than a linear one. We expect in practice that the order be even larger than ν. For instance, Exercise 15 gives the order $(1 + \sqrt{5})/2$ when $n = 3$. The exact order for a general n is still unknown.

Proof. We first remark that if $i \neq j$ with $\{i, j\} \neq \{p_\ell, q_\ell\}$, then

$$|a_{ij}^{(\ell+1)} - a_{ij}^{(\ell)}| \leq |t_\ell| \sqrt{2} \|E_\ell\|, \tag{13.5}$$

where $t_\ell = \tan \theta_\ell$. To see this, observe that $1 - c \leq t$ and $|s| \leq t$ whenever $|t| \leq 1$. However, Theorem 13.4 ensures that D_k converges to $\mathrm{diag}(\lambda_1, \ldots, \lambda_n)$, where the λ_js are the eigenvalues of A. Because these are distinct, there exist $K \in \mathbb{N}$ and $\delta > 0$ such that, if $k \geq K$, then

$$\min_{i \neq j} |a_{ii}^{(k)} - a_{jj}^{(k)}| \geq \delta$$

for $k \geq K$. We have therefore

$$|\sigma_k| \geq \frac{\delta}{\sqrt{2}\|E_k\|} \xrightarrow{k \to +\infty} +\infty.$$

It follows that t_k tends to zero and, more precisely, that

$$t_k \approx -\frac{1}{2\sigma_k}.$$

Finally, there exists a constant c_1 such that

$$|t_k| \leq c_1 \|E_k\|.$$

Let us then fix k larger than K, and let us denote by J the set of pairs (p_ℓ, q_ℓ) when $k \leq \ell \leq k + N - 1$. For such an index, we have $\|E_\ell\| \leq \rho^{\ell-k}\|E_k\| \leq \|E_k\|$. In particular, $|t_\ell| \leq c_1 \|E_k\|$.

If $(p, q) \in J$ and if $\ell < k + N$ is the largest index such that $(p, q) = (p_\ell, q_\ell)$, a repeated application of (13.5) shows that

$$|a_{pq}^{(k+N)}| \leq c_1 N \sqrt{2} \|E_k\|^2.$$

If J is equal to the whole set of pairs (i, j) such that $i < j$, these inequalities ensure that $\|E_{k+N}\| \leq c_2 \|E_k\|^2$. Otherwise, there exists a pair (p, q) that one sets to zero twice: $(p, q) = (p_\ell, q_\ell) = (p_m, q_m)$ with $k \leq \ell < m < k + N$. In that case, the same argument as above shows that

$$\|E_{k+N}\| \leq \|E_m\| \leq \sqrt{2N}|a_{pq}^{(m)}| \leq 2\sqrt{N}c_1(m - \ell)\|E_k\|^2.$$

$\square$

Remarks

We show in Exercise 13 that when the eigenvalues of A are simple, the distance between the diagonal and the spectrum of A is $O(\|E_k\|^2)$, and not $O(\|E_k\|)$ as expected from Theorem 5.7.

13.4.6 Convergence with the Random Choice

Recall that we choose the pair (p, q) independently of those chosen at previous steps, according to the uniform distribution. The matrix $A^{(k)}$ is therefore a function of $A^{(k)}$ and of the random variable (p_k, q_k); as such, it is a random variable.

We are interested in the *expectation* of the norm of the error $\|E_{k+1}\|^2$. To begin with, we consider the *conditional* expectation, knowing $\|E_k\|^2$. We have

$$\mathrm{e}\left[\|E_{k+1}\|^2 \mid \|E_k\|^2\right] = \frac{2}{n^2 - n} \sum_{1 \le p < q \le n} \|E_{k+1}(p, q)\|^2.$$

Because of Lemma 25, we obtain

$$\mathrm{e}\left[\|E_{k+1}\|^2 \mid \|E_k\|^2\right] = \frac{2}{n^2 - n} \sum_{1 \le p < q \le n} \left(\|E_k\|^2 - |a_{pq}^{(k)}|^2\right)$$

$$= \|E_k\|^2 - \frac{2}{n^2 - n} \sum_{1 \le p < q \le n} |a_{pq}^{(k)}|^2$$

$$= \left(1 - \frac{2}{n^2 - n}\right) \|E_k\|^2 = \rho^2 \|E_k\|^2.$$

Taking now the expectation with respect to the previous choices, we obtain

$$\mathrm{e}\left[\|E_{k+1}\|^2\right] = \rho^2 \mathrm{e}\left[\|E_k\|^2\right].$$

By induction, this yields

$$\mathrm{e}\left[\|E_k\|^2\right] = \rho^{2k} \mathrm{e}\left[\|E_0\|^2\right]. \tag{13.6}$$

Let β be a number given in the interval $(\rho, 1)$. Let us denote $c_0 := \mathrm{e}\left[\|E_0\|^2\right]$. Then the probability that $\|E_k\|$ is larger than β^k is less than $c_0(\rho/\beta)^{2k}$, according to (13.6). We therefore have

$$\sum_{k=0}^{\infty} \mathbb{P}\left(\|E_k\| > \beta^k\right) < \infty. \tag{13.7}$$

Thanks to the theorem of Borel–Cantelli, this implies that for almost every choice of the sequence $(p_k, q_k)_{k \in \mathbb{N}}$, the inequality $\|E_k\| \le \beta^k$ is true for all but finitely many indices k ; in other words, $\|E_k\| \le \beta^k$ is true for large enough k. When this happens, we may apply the same analysis of the diagonal part D_k as that made in Section 13.4.4. Finally, we have the following theorem.

Theorem 13.6 *Consider the Jacobi method with random choice, the pairs (p_k, q_k) being independent and chosen according to the uniform distribution.*

For every $\varepsilon > 0$, the error $\|E_k\|$ decays almost surely as an $O\left((\rho + \varepsilon)^k\right)$, with

$$\rho := \sqrt{1 - \frac{2}{n^2 - n}}.$$

Provided the angle θ_k is chosen in the interval $(-\pi/4, \pi/4]$, the diagonal converges as soon as the error tends to zero, and the diagonal entries of its limit are the eigenvalues of A.

13.5 The Power Methods

The power methods are designed for the approximation of a single eigenvalue. Consequently, their cost is significantly lower than that of the QR or the Jacobi methods. The standard power method is used in particular when searching for the optimal parameter in the SOR method for a tridiagonal matrix, where we have to compute the spectral radius of the Jacobi iteration matrix (Theorem 12.2).

13.5.1 The Standard Method

Let $M \in \mathbf{M}_n(\mathbb{C})$ be a matrix. We search for an approximation of its eigenvalue of maximum modulus, whenever only one such exists. The standard method consists in choosing a norm on $\mathbb{C}^n$, a unit vector $x^0 \in \mathbb{C}^n$, and then successively computing the vectors x^k by the formula

$$x^{k+1} := \frac{1}{\|Mx^k\|} Mx^k.$$

The justification of this method is given in the following theorem.

Theorem 13.7 *One assumes that* $\mathrm{Sp}\, M$ *contains only one element* λ *of maximal modulus (that modulus is thus equal to* $\rho(M)$*).*

If $\rho(M) = 0$*, the method stops because* $Mx^k = 0$ *for some* $k < n$.

Otherwise, let $\mathbb{C}^n = E \oplus F$ *be the decomposition of* $\mathbb{C}^n$*, where* E, F *are invariant linear subspaces under M, with* $\mathrm{Sp}(M|_E) = \{\lambda\}$ *and* $\lambda \notin \mathrm{Sp}(M|_F)$*. Assume that* $x^0 \notin F$*. Then* $Mx^k \neq 0$ *for every* $k \in \mathbb{N}$ *and*

$$\lim_{k \to +\infty} \|Mx^k\| = \rho(M). \tag{13.8}$$

In addition,

$$V := \lim_{k \to +\infty} \left(\frac{\bar{\lambda}}{\rho(M)} \right)^k x^k$$

is a unit eigenvector of M, associated with the eigenvalue λ*. If* $V_j \neq 0$*, then*

$$\lim_{k \to +\infty} \frac{(Mx^k)_j}{x_j^k} = \lambda.$$

Proof. The case $\rho(M) = 0$ is obvious because M is then nilpotent.

Assume otherwise that $\rho(M) > 0$. Let $x^0 = y^0 + z^0$ be the decomposition of x^0 with $y^0 \in E$ and $z^0 \in F$. By assumption, $y^0 \neq 0$. Because $M|_E$ is invertible, $M^k y^0 \neq 0$. Because $M^k x^0 = M^k y^0 + M^k z^0$, $M^k y^0 \in E$, and $M^k z^0 \in F$, we have $M^k x^0 \neq 0$. The algorithm may be rewritten as[3]

$$x^k = \frac{1}{\|M^k x^0\|} M^k x^0.$$

We therefore have $x^k \neq 0$.

If $F \neq \{0\}$, then $\rho(M|_F) < \rho(M)$ by construction. Hence there exist (from Theorem 7.1) $\eta < \rho(M)$ and $C > 0$ such that $\|(M|_F)^k\| \leq C\eta^k$ for every k. Then $\|(M|_F)^k z^0\| \leq C_1 \eta^k$. On the other hand, $\rho((M|_E)^{-1}) = 1/\rho(M)$, and the same argument as above ensures that $\|(M|_E)^{-k}\| \leq 1/C_2 \mu^k$, for some $\mu \in (\eta, \rho(M))$, so that $\|M^k y^0\| \geq C_3 \mu^k$. Hence,

$$\|M^k z^0\| \ll \|M^k y^0\|,$$

so that

$$x^k \approx \frac{1}{\|M^k y^0\|} M^k y^0.$$

We are thus led to the analysis of the case where $z^0 = 0$, namely when M has no eigenvalue but λ. That is assumed from now on.

Let r be the degree of the minimal polynomial of M. The vector space spanned by the vectors $x^0, Mx^0, \ldots, M^{r-1}x^0$ contains all the x^ks. Up to the replacement of $\mathbb{C}^n$ by this linear subspace, we may assume that it equals $\mathbb{C}^n$. Then we have $r = n$. Furthermore, because $\ker(M - \lambda)^{n-1}$, a nontrivial linear subspace, is invariant under M, we see that $x^0 \notin \ker(M - \lambda)^{n-1}$.

The vector space $\mathbb{C}^n$ then admits the basis

$$\{v^1 = x^0, v^2 = (M - \lambda)x^0, \ldots, v^n = (M - \lambda)^{n-1}x^0\}.$$

With respect to this basis, M becomes the Jordan matrix

$$\tilde{M} = \begin{pmatrix} \lambda & 0 & \cdots & \cdots & \\ 1 & \ddots & \ddots & & \vdots \\ 0 & \ddots & \ddots & \ddots & \vdots \\ \vdots & \ddots & \ddots & \ddots & 0 \\ & \cdots & 0 & 1 & \lambda \end{pmatrix}.$$

---

[3] One could normalize x^k at the end of the computation, but we prefer doing it at each step in order to avoid overflows, and also to ensure (13.8).

The matrix $\lambda^{-k}\tilde{M}^k$ depends polynomially on k. The coefficient of highest degree, as $k \to +\infty$, is at the intersection of the first column and the last row. It equals

$$\binom{k}{n-1}\lambda^{1-n},$$

which is equivalent to $\dfrac{(k/\lambda)^{n-1}}{(n-1)!}$. We deduce that

$$M^k x^0 \sim \frac{k^{n-1}\lambda^{k-n+1}}{(n-1)!}v^n.$$

Hence,

$$x^k \sim \left(\frac{\lambda}{|\lambda|}\right)^{k-n+1}\frac{v^n}{\|v^n\|}.$$

Because v^n is an eigenvector of M, the claims of the theorem have been proved. $\quad\square$

The case where the algebraic and geometric multiplicities of λ are equal (i.e., $M|_E = \lambda I_E$), for example, if λ is a simple eigenvalue, is especially favorable. Indeed, $M^k y^0 = \lambda^k y^0$, and therefore

$$x^k = \left(\frac{\lambda}{|\lambda|}\right)^k\frac{1}{\|y^0\|}y^0 + O\left(\frac{\|M^k z^0\|}{|\lambda|^k}\right).$$

Theorem 7.1 thus shows that the error

$$x^k - \left(\frac{\lambda}{|\lambda|}\right)^k\frac{1}{\|y^0\|}y^0$$

tends to zero faster than

$$\left(\frac{\rho(M|_F)+\varepsilon}{\rho(M)}\right)^k,$$

for every $\varepsilon > 0$. The convergence is thus of order one, and becomes faster as the ratio $|\lambda_2|/|\lambda_1|$ becomes smaller (arranging the eigenvalues by nonincreasing moduli). However, the convergence is much slower when the Jordan blocks of M relative to λ are nontrivial. The error then behaves like $1/k$ in general.

The situation is more delicate when $\rho(M)$ is the modulus of several distinct eigenvalues. The vector x^k, suitably normalized, does not converge in general but "spins" closer and closer to the sum of the corresponding eigenspaces. The observation of the asymptotic behavior of x^k allows us to identify the eigendirections associated with the eigenvalues of maximal modulus. The sequence $\|Mx^k\|$ does not converge and depends strongly on the choice of the norm. However, $\log\|Mx^k\|$ converges in the Cesaro sense, that is, in the mean, to $\log\rho(M)$ (Exercise 12).

Remark

The hypothesis on x_0 is generic, in the sense that it is satisfied for every choice of x_0 in an open dense subset of $\mathbb{C}^n$. If by chance x^0 belongs to F, the power method theoretically furnishes another eigenvalue, of smaller modulus. In practice, a large enough number of iterations always leads to the convergence to λ. In fact, the number λ is rarely exactly representable in a computer. When it is not, the linear subspace F does not contain any nonzero representable vector. Thus the vector x^0, or its computer representation, does not belong to F, and Theorem 13.7 applies.

13.5.2 The Inverse Power Method

Let us assume that M is invertible. The standard power method, applied to M^{-1}, furnishes the eigenvalue of least modulus, whenever it is unique, or at least produces its modulus in the general case. The inversion of a matrix is a costly operation, therefore we involve ourselves with that idea only if M has already been inverted, for example if we had previously had to make an LU or a QR factorization. That is typically the situation when one begins to implement the QR algorithm for M. It might look strange to involve a method giving only one eigenvalue in the course of a method that is expected to compute the whole spectrum.

The inverse power method is thus subtle. Here is how it works. One begins by implementing the QR method until one gets coarse approximations $\mu_1, \ldots, \mu_n$ of the eigenvalues $\lambda_1, \ldots, \lambda_n$. If one persists in the QR method, the proof of Theorem 13.2 shows that the error is at best of order σ^k with $\sigma = \max_j |\lambda_{j+1}/\lambda_j|$. When n is large, σ is in general close to 1 and this convergence is rather slow. Likewise, the method with Rayleigh translations, for which σ is replaced by $\sigma(\eta) := \max_j |(\lambda_{j+1} - \eta)/(\lambda_j - \eta)|$, is not satisfactory. However, if one wishes to compute a *single* eigenvalue, say λ_p, with full accuracy, the power method, applied to $M - \mu_p I_n$, produces an error on the order of θ^k, where $\theta := |\lambda_p - \mu_p| / \min_{j \neq p} |\lambda_j - \mu_p|$ is a small number, since $\lambda_p - \mu_p$ is small.

In practice, the inverse power method is used mainly to compute an approximate eigenvector, associated with an eigenvalue for which one already had a good approximate value.

Exercises

1. Given a polynomial $P \in \mathbb{R}[X]$, use Euclidean division in order to define a sequence of nonzero polynomials P_j in the following way. Set $P_0 = P$, $P_1 = P'$. If P_j is not constant, $-P_{j+1}$ is the remainder of the division of P_{j-1} by P_j: $P_{j-1} = Q_j P_j - P_{j+1}$, $\deg P_{j+1} < \deg P_j$.

a. Assume that P has only simple roots. Show that the sequence $(P_j)_j$ is well defined, that it has only finitely many terms, and that it is a Sturm sequence.

b. Use Proposition 13.2 to compute the number of real roots of the real polynomials $X^2 + aX + b$ or $X^3 + pX + q$ in terms of their discriminants.

2. (Wilkinson [40], Section 5.45.) Let $n = 2p - 1$ be an odd number and $W_n \in \mathbf{M}_n(\mathbb{R})$ be the symmetric tridiagonal matrix

$$
\begin{pmatrix}
p & 1 & & & \\
1 & \ddots & \ddots & & \\
& \ddots & 1 & \ddots & \\
& & \ddots & \ddots & 1 \\
& & & 1 & p
\end{pmatrix}.
$$

The diagonal entries are thus $p, p - 1, \ldots, 2, 1, 2, \ldots, p - 1, p$, and the subdiagonal entries are equal to 1.

a. Show that the linear subspace

$$
E' = \{X \in \mathbb{R}^n \mid x_{p+j} = x_{p-j}, 1 \le j < p\}
$$

is invariant under W_n. Likewise, show that the linear subspace

$$
E'' = \{X \in \mathbb{R}^n \mid x_{p+j} = -x_{p-j}, 0 \le j < p\}
$$

is stable under W_n.

b. Deduce that the spectrum of W_n is the union of the spectra of the matrices

$$
W_n' =
\begin{pmatrix}
p & 1 & & & \\
1 & \ddots & \ddots & & \\
& \ddots & \ddots & \ddots & \\
& & & 1 & 2 & 1 \\
& & & & 2 & 1
\end{pmatrix}, \qquad
W_n'' =
\begin{pmatrix}
p & 1 & & & \\
1 & \ddots & \ddots & & \\
& \ddots & \ddots & \ddots & \\
& & & 1 & 3 & 1 \\
& & & & 1 & 2
\end{pmatrix}
$$

(we have $W_n' \in \mathbf{M}_p(\mathbb{R})$ and $W_n'' \in \mathbf{M}_{p-1}(\mathbb{R})$).

c. Show that the eigenvalues of W_n'' strictly separate those of W_n'.

3. For $a_1, \ldots, a_n \in \mathbb{R}$, with $\sum_j a_j = 1$, form the matrix

$$
M(a) := \begin{pmatrix}
a_1 & a_2 & a_3 & a_4 & & a_n \\
a_2 & b_2 & a_3 & \vdots & \vdots & \vdots \\
a_3 & a_3 & b_3 & & \vdots & \vdots \\
a_4 & \cdots & & & & \vdots \\
& \cdots & \cdots & & & a_n \\
a_n & \cdots & \cdots & \cdots & a_n & b_n
\end{pmatrix},
$$

where $b_j := a_1 + \cdots + a_{j-1} - (j-2)a_j$.

    a. Compute the eigenvalues and the eigenvectors of $M(a)$.

    b. We limit ourselves to n-uplets a that belong to the simplex S defined by $0 \le a_n \le \cdots \le a_1$ and $\sum_j a_j = 1$. Show that for $a \in S$, $M(a)$ is bistochastic and $b_2 - a_2 \le \cdots \le b_n - a_n \le 1$.

    c. Let $\mu_1, \ldots, \mu_n$ be an n-uplet of elements in $[0, 1]$ with $\mu_n = 1$. Show that there exists a unique a in S such that $\{\mu_1, \ldots, \mu_n\}$ is equal to the spectrum of $M(a)$ (counting with multiplicity).

4. Show that the cost of an iteration of the QR method for an Hermitian tridiagonal matrix is $20n + O(1)$.

5. Show that the reduction to the Hessenberg form (in this case, tridiagonal form) of an Hermitian matrix costs $7n^3/6 + O(n^2)$ operations.

6. (Invariants of the algorithm QR.) For $M \in \mathbf{M}_n(\mathbb{R})$ and $1 \le k \le n-1$, let us denote by $(M)_k$ the matrix of size $(n-k) \times (n-k)$ obtained by deleting the first k rows and the last k columns. For example, $(I)_1$ is the Jordan matrix $J(0; n-1)$. We also denote by $K \in \mathbf{M}_n(\mathbb{R})$ the matrix defined by $k_{1n} = 1$ and $k_{ij} = 0$ otherwise.

    a. For an upper-triangular matrix T, explicitly compute KT and TK.

    b. Let $M \in \mathbf{M}_n(\mathbb{R})$. Prove the equality

$$
\det(M - \lambda I - \mu K) = (-1)^n \mu \det(M - \lambda I)_1 + \det(M - \lambda I).
$$

    c. Let $A \in \mathbf{GL}_n(\mathbb{R})$ be given, with factorization $A = QR$. Prove that

$$
\det(A - \lambda I)_1 = \frac{\det R}{r_{nn}} \det(Q - \lambda R^{-1})_1.
$$

    d. Let $A' := RQ$. Show that

$$
r_{nn} \det(A' - \lambda I)_1 = r_{11} \det(A - \lambda I)_1.
$$

    e. Generalize the previous calculation by replacing the index 1 by k. Deduce that the roots of the polynomial $\det(A - \lambda I)_k$ are conserved throughout the QR algorithm. How many such roots do we have for a general matrix? How many for a Hessenberg matrix?

7. (Invariants; continuing.) For $M \in \mathbf{M}_n(\mathbb{R})$, let us define $P_M(h;z) := \det((1 - h)M + hM^T - zI_n)$.

   a. Show that $P_M(h;z) = P_M(1 - h;z)$. Deduce that there exists a polynomial Q_M such that $P_M(h;z) = Q_M(h(1 - h);z)$.

   b. Show that Q_M remains constant throughout the QR algorithm: if $Q \in \mathbf{O}_n(\mathbb{R})$, R is upper-triangular, and $M = QR$, $N = RQ$, then $Q_M = Q_N$.

   c. Deduce that there exist polynomial functions J_{rk} on $\mathbf{M}_n(\mathbb{R})$, defined by

$$P_M(h;z) = \sum_{r=0}^{n} \sum_{k=0}^{[r/2]} (h(1 - h))^k z^{n-r} J_{rk}(M),$$

that are invariant throughout the QR algorithm. Verify that the J_{r0}s can be expressed in terms of invariants that we already know.

   d. Compute explicitly J_{21} when $n = 2$. Deduce that in the case where Theorem 13.2 applies and $\det A > 0$, the matrix A_k converges.

   e. Show that for $n \geq 2$,

$$J_{21}(M) = -\frac{1}{2} \mathrm{Tr}\left((M - M^T)^2\right).$$

Deduce that if A_k converges to a diagonal matrix, then A is symmetric.

8. In the Jacobi method, show that if the eigenvalues are simple, then the product $R^1 \cdots R^m$ converges to an orthogonal matrix R such that R^*AR is diagonal.

9. Extend the Jacobi method to Hermitian matrices. **Hint:** Replace the rotation matrices

$$\begin{pmatrix} \cos\theta & \sin\theta \\ -\sin\theta & \cos\theta \end{pmatrix}$$

by unitary matrices

$$\begin{pmatrix} z_1 & z_2 \\ z_3 & z_4 \end{pmatrix}.$$

10. Let $A \in \mathbf{Sym}_n(\mathbb{R})$ be a matrix whose eigenvalues, of course real, are simple. Apply the Jacobi method, but selecting the angle θ_k so that $\pi/4 \leq |\theta_k| \leq \pi/2$.

   a. Show that E_k tends to zero, that the sequence D_k is relatively compact, and that its cluster values are diagonal matrices whose diagonal terms are the eigenvalues of A.

   b. Show that an iteration has the effect of permuting, asymptotically, $a_{pp}^{(k)}$ and $a_{qq}^{(k)}$, where $(p,q) = (p_k,q_k)$. In other words

$$\lim_{k\to+\infty} |a_{pp}^{(k+1)} - a_{qq}^{(k)}| = 0,$$

and vice versa, permuting p and q.

11. The Bernoulli method computes an approximation of the root of largest modulus for a polynomial $a_0 X^n + \cdots + a_n$, when that root is unique. To do so, one defines a sequence by a linear induction of order n:

$$z_k = -\frac{1}{a_0}(a_1 z_{k-1} + \cdots + a_n z_{k-n}).$$

Compare this method with the power method for a suitable matrix.

12. Consider the power method for a matrix $M \in \mathbf{M}_n(\mathbb{C})$ of which several eigenvalues are of modulus $\rho(M) \neq 0$. Again, $\mathbb{C}^n = E \oplus F$ is the decomposition of $\mathbb{C}^n$ into linear subspaces stable under M, such that $\rho(M|_F) < \rho(M)$ and the eigenvalues of $M|_E$ are of modulus $\rho(M)$. Finally, $x^0 = y^0 + z^0$ with $y^0 \in E$, $z^0 \in F$, and $y^0 \neq 0$.

  a. Express

  $$\frac{1}{m}\sum_{k=0}^{m-1} \log \|Mx^k\|$$

  in terms of $\|M^m x^0\|$.

  b. Show that if $0 < \mu < \rho(M) < \eta$, then there exist constants C, C' such that

  $$C\mu^k \leq \|M^k x^0\| \leq C' \eta^k, \quad \forall k \in \mathbb{N}.$$

  c. Deduce that $\log \|Mx^k\|$ converges in the mean to $\log \rho(M)$.

13. Let $M \in \mathbf{M}_n(\mathbb{C})$ be given. Assume that the Gershgorin disk D_ℓ is disjoint from the other disks D_m, $m \neq \ell$. Show that the inverse power method, applied to $M - m_{\ell\ell}I_n$, provides an approximate computation of the unique eigenvalue of M that belongs to D_ℓ.

14. The ground field is $\mathbb{R}$.

  a. Let P and Q be two monic polynomials of respective degrees n and $n-1$ ($n \geq 2$). We assume that P has n real and distinct roots, strictly separated by the $n-1$ real and distinct roots of Q. Show that there exist two real numbers d and c, and a monic polynomial R of degree $n-2$, such that

  $$P(X) = (X - d)Q(X) - c^2 R(X).$$

  b. Let P be a monic polynomial of degree n ($n \geq 2$). We assume that P has n real and distinct roots. Build sequences $(d_j, P_j)_{1 \leq j \leq n}$ and $(c_j)_{1 \leq j \leq n-1}$, where d_j, c_j are real numbers and P_j is a monic polynomial of degree j, with

  $$P_n = P, \quad P_j(X) = (X - d_j)P_{j-1}(X) - c_{j-1}^2 P_{j-2}(X), \quad (2 \leq j \leq n).$$

  Deduce that there exists a tridiagonal matrix A, which we can obtain by algebraic calculations (involving square roots), whose characteristic polynomial is P.

c. Let P be a monic polynomial. We assume that P has n real roots. Prove that one can factorize $P = Q_1 \cdots Q_r$, where each Q_j has simple roots, and the factorization requires only finitely many operations. Deduce that there is a finite algorithm, involving no more than square roots calculations, which provides a tridiagonal symmetric matrix A, whose characteristic polynomial is P (a *tridiagonal symmetric companion matrix*).

15. We apply the Jacobi method to a real 3×3 matrix A. Our strategy is one that we have called "optimal choice".

a. Let (p_1, q_1), (p_2, q_2), $\ldots$, (p_k, q_k), $\ldots$ be the sequence of index pairs that are chosen at consecutive steps (recall that one vanishes the off-diagonal entry of largest modulus). Prove that this sequence is cyclic of order three: it is either the sequence

$$\ldots, (1,2), (2,3), (3,1), (1,2), \ldots,$$

or

$$\ldots, (1,3), (3,2), (2,1), (1,3), \ldots.$$

b. Assume now that A has simple eigenvalues. At each step, one of the three off-diagonal entries is null, and the two other ones are small, because the method converges. Say that they are $0, x_k, y_k$ with $0 < |x_k| \le |y_k|$ (if x_k vanishes then one diagonal entry is an eigenvalue and the method ends one step further). Show that $y_{k+1} \sim x_k$ and $x_{k+1} \sim 2x_k y_k / \delta$, where δ is a gap between two eigenvalues. Deduce that the method is of order $\omega = (1 + \sqrt{5})/2$, the golden ratio, meaning that the error ε_k at step k satisfies

$$\varepsilon_{k+1} = O(\varepsilon_k \varepsilon_{k-1}).$$

c. Among the class of Hessenberg matrices, we distinguish the *unit* ones, which have 1s below the diagonal:

$$M = \begin{pmatrix} * & \cdots & & \cdots & * \\ 1 & \ddots & & & \vdots \\ 0 & \ddots & & & \\ \vdots & \ddots & \ddots & \ddots & \vdots \\ 0 & \cdots & 0 & 1 & * \end{pmatrix}.$$

i. Let $M \in \mathbf{M}_n(k)$ be a unit Hessenberg matrix. We denote by M_k the submatrix obtained by retaining the first k rows and columns. For instance, $M_n = M$ and $M_1 = (m_{11})$. We set P_k the characteristic polynomial of M_k. Show that

$$P_n(X) = (X - m_{nn}) P_{n-1}(X) - m_{n-1,n} P_{n-2}(X) - \cdots - m_{2n} P_1(X) - m_{1n}.$$

ii. Let $Q_1, \ldots, Q_n \in k[X]$ be monic polynomials, with $\deg Q_k = k$. Show that there exists one and only one unit Hessenberg matrix M such that, for every $k = 1, \ldots, n$, the characteristic polynomial of M_k equals Q_k.
Hint: Argue by induction over n.

Note: The roots of the polynomials $P_1, \ldots, P_n$ are called the *Ritz values* of M.

References

1. Jacques Baranger. *Analyse Numérique*. Hermann, Paris, 1991.
2. Genrich R. Belitskii and Yurii. I. Lyubich. *Matrix Norms and Their Applications*, volume 36 of *Operator theory: advances and applications*. Birkhäuser, Basel, 1988.
3. Marcel Berger and Bernard Gostiaux. *Differential Geometry: Manifold, Curves and Surfaces*, volume 115 of *Graduate Texts in Mathematics*. Springer-Verlag, New York, 1988.
4. Rajendra Bhatia. *Matrix Analysis*, volume 169 of *Graduate Texts in Mathematics*. Springer-Verlag, Heidelberg, 1996.
5. Rajendra Bhatia. Pinching, trimming, truncating, and averaging of matrices. *Amer. Math. Monthly*, 107(7):602–608, 2000.
6. Rajendra Bhatia. Linear algebra to quantum cohomology: The story of Alfred Horn's inequalities. *Amer. Math. Monthly*, 108(4):289–318, 2001.
7. Peter Bürgisser, Michael Clausen, and M. Amin Shokrollahi. *Algebraic Complexity Theory*. Springer-Verlag, Berlin, 1997. With the collaboration of Thomas Lickteig.
8. Philippe Ciarlet. *Introduction to Numerical Linear Algebra and Optimisation*. Cambridge Texts in Applied Mathematics. Cambridge University Press, Cambridge, England, 1989.
9. Philippe Ciarlet and Jean-Marie Thomas. *Exercices d'analyse numérique matricielle et d'optimisation*. Mathématiques appliquées pour la maîtrise. Masson, Paris, 1982.
10. Harvey Cohn. *Advanced Number Theory*. Dover, New York, 1980. Reprint of *A Second Course in Number Theory*, 1962, Dover Books on Advanced Mathematics.
11. Don Coppersmith and Shmuel Winograd. Matrix multiplication via arithmetic progressions. *J. Symbolic Comput.*, 9(3):251–280, 1990.
12. Philip J. Davis. *Circulant Matrices*. Chelsea, New York, 1979.
13. Miroslav Fiedler and Vlastimil Pták. On matrices with non-positive off-diagonal elements and positive principal minors. *Czech. Math. Journal*, 12:382–400, 1962.
14. Edward Formanek. Polynomial identities and the Cayley–Hamilton theorem. *Math. Intelligencer*, 11(1):37–39, 1989.
15. Edward Formanek. *The polynomial identities and invariants of $n \times n$ matrices*. Number 78 in CBMS Regional Conf. Ser. Math. Amer. Math. Soc., Providence, RI, 1991.
16. William Fulton. Eigenvalues, invariant factors, highest weights, and Schubert calculus. *Bull. Amer. Math. Soc. (N.S.)*, 37(3):209–249 (electronic), 2000.
17. Felix R. Gantmacher. *The Theory of Matrices. Vol. 1*. Chelsea, New York, 1959.
18. Felix R. Gantmacher. *The Theory of Matrices. Vol. 2*. Chelsea, New York, 1959.
19. Gene H. Golub and Charles F. Van Loan. *Matrix Computations*, volume 3 of *Series in the Mathematical Sciences*. Johns Hopkins University Press, Baltimore, MD, 1983.
20. Nicholas Higham. *Accuracy and Stability of Numerical Algorithms*. SIAM, Philadelphia, 1996.
21. Roger A. Horn and Charles R. Johnson. *Matrix Analysis*. Cambridge University Press, Cambridge, England, 1985.

22. Alston S. Householder. *The Theory of Matrices in Numerical Analysis*. Dover, New York, 1975.
23. Huajun Huang and Tin-Yau Tam. On the QR iterations of real matrices. *Linear Algebra Appl.*, 408:161–176, 2005.
24. Tosio Kato. *Perturbation Theory for Linear Operators*. Springer-Verlag, Berlin, 1995. Reprint of the 1980 edition.
25. Nicholas M. Katz and Peter Sarnak. *Random Matrices, Frobenius Eigenvalues and Monodromy*. Number 45 in Colloquium Publ. Amer. Math. Soc., Providence, RI, 1999.
26. Anthony W. Knapp. *Representation of Semisimple Groups. An Overview Based on Examples*. Princeton Mathematical Series. Princeton University Press, Princeton, NJ, 1986.
27. Patrick Lascaux and R. Théodor. *Analyse Numérique Matricielle Appliquée à l'Art de l'Ingénieur*. Masson, Paris, 1987.
28. Chi-Wang Li and Roy Mathias. Extremal characterization of the Schur complement and resulting inequalities. *SIAM Review*, 42:233–246, 2000.
29. Helmut Lütkepohl. *Handbook of Matrices*. J. Wiley & Sons, New York, 1996.
30. Rached Mneimné and Frédéric Testard. *Introduction à la Théorie des Groupes de Lie Classiques*. Hermann, Paris, 1986.
31. F. P. Preparata and D. V. Sarwate. An improved parallel processor bound in fast matrix inversion. *Inf. Process. Lett.*, 7(3):148–150, 1978.
32. Shmuel Rosset. A new proof of the Amitsur-Levitski identity. *Israel J. Math.*, 23(2):187–188, 1976.
33. Walter Rudin. *Real and Complex Analysis*. McGraw-Hill, New York, third edition, 1987.
34. Walter Rudin. *Functional Analysis*. McGraw–Hill, New York, second edition, 1991.
35. Eugene Seneta. *Non-Negative Matrices and Markov Chains*. Springer Series in Statistics. Springer-Verlag, New York–Berlin, 1981.
36. Joseph Stoer and Christoph Witzgall. Transformations by diagonal matrices in a normed space. *Numer. Math.*, 4:158–171, 1962.
37. Volker Strassen. Gaussian elimination is not optimal. *Numer. Math.*, 13:354–356, 1969.
38. Lloyd N. Trefethen and Mark Embree. *Spectra and Pseudospectra*. Princeton University Press, Princeton, NJ, 2005. The behavior of nonnormal matrices and operators.
39. Joseph H. M. Wedderburn. *Lectures on Matrices*, volume XVII of *Colloquium*. Amer. Math. Soc., New York, 1934.
40. James H. Wilkinson. *The Algebraic Eigenvalue Problem*. Oxford Science, Oxford, UK, 1965.
41. V. A. Yakubovich and V. M. Starzhinskii. *Linear Differential Equations with Periodic Coefficients*. Wiley & Sons, New York, 1975.

Index of Notations

$|A|$, 149
$a|b$, 163
$A \circ B$, 122
$A^\dagger$, 218
$A \geq 0$, 149
$a \prec b$, 117
$a \sim b$, 163
A^*, 31, 163

$B \otimes C$, 29
$(b_1, \ldots, b_r)$, 163
B_p, 172

$\mathrm{charc}(K)$, 1
$\mathrm{circ}(a_0, \ldots, a_{n-1})$, 61
$\mathscr{C}_n$, 61
$\mathrm{cond}(A)$, 239
C_r, 152

δ_i^j, 16
$\det M$, 31
$D_i(A)$, 102
$\mathrm{diag}(d_1, \ldots, d_n)$, 17
$\dim E$, 3
$\dim_K F$, 4
$D_k(N)$, 167
$\mathbf{DS}_n$, 156

$\mathbf{e}$, 156
$\mathbf{e}^i$, 4
$E_K(\lambda)$, 48

E_λ, 51
$\mathrm{End}(E)$, 6
$\varepsilon(\sigma)$, 31
$\exp A$, 185

$F + G$, 2
$F \otimes G$, 69
$F \oplus G$, 4
$\mathbb{F}_p$, 1

G, 227
$\mathscr{G}$, 190
$\mathscr{G}(A)$, 102
G_α, 193
$G \otimes_{\mathbb{R}} \mathbb{C}$, 5
$\gcd$, 164
$\mathbf{GL}_n(A)$, 39
$\mathbf{GL}_n(K)$, 22
G_0, 194

$\mathscr{H}(A)$, 98
$\mathbf{H}_n$, 84
$\mathbf{H}_n^+$, 85
$\mathbf{HPD}_n$, 85

$\Im$, imaginary part, 104
$\Im M$, 84
I_n, 16

J, 227
$J(a; r)$, 176

J_{ik}, 166
J_2, 198
J_3, 198
J_4, 198

$\overline{K}$, 2
$\ker u$, 5
K^I, 2
$K[M]$, 22
K_n, 122
$K[X]$, 31
$k[X,Y]$, 165

$\Lambda(E)$, 76
$\Lambda^2(E)$, 72
$\Lambda^k(E)$, 73
$\mathscr{L}(E;F)$, 5
$\mathscr{L}_\omega$, 228
$L^p(\Omega)$, 135

$\bar{M}$, 83
$M\begin{pmatrix} i_1 & i_2 & \cdots & i_p \\ j_1 & j_2 & \cdots & j_p \end{pmatrix}$, 34
M^k, 22
M^{-1}, 39
M^{-k}, 39
M^{-T}, 39
M^{-1}, 22
$\mathbf{M}_n(K)$, 16
$\mathbf{M}_{n \times m}(A)$, 16
M^*, 83
M^{-*}, 83

$\| A \|$, 130
$\| A \|_p$, 131
$\| x \|_2$, 13, 98, 109
$\| x \|_p$, 127
$\| x \|_A$, 230
$\| x \|_\infty$, 127
$\| \cdot \|'$, 130
$\||| \cdot \|||$, 131

ω_J, 234
0_n, 16
$\mathbf{O}_n(K)$, 23, 39
0_{nm}, 16

$\mathbf{O}_n^-$, 192
$\mathbf{O}^+(1,3)$, 195
$\mathbf{O}(p,q)$, 188
$x \perp y$, 24
$\perp_A$, 236

Pf, 55
P_G, 232
π_0, 194
P_J, 232
P_M, 43
P_ω, 232
p', 128
$\mathbf{PSL}_2(\mathbb{R})$, 104

rk M, 17
$r_A(x)$, 98
$\mathfrak{R}$, real part, 129
$\mathfrak{R}M$, 84
$R(h;A)$, 205
$\rho(A)$, 127
rk M, 23
$R(u)$, 5
$r(x)$, 236
$R(z;A)$, 96

σ_r, 56
$s_j(A)$, 140
$s_k(a)$, 117
$\mathbf{SL}_n(A)$, 39
$\mathscr{S}_\ell(X_1,\ldots,X_\ell)$, 78
s_m, 57
S_n, 31
$\mathbf{SO}_n(K)$, 39
S^1, 155
$\mathbf{SO}^+(1,3)$, 195
Sp(M), 43
Sp(u), 6
Sp$_K(M)$, 43
$\mathbf{SPD}_n$, 85
$\mathbf{Sp}_m$, 188
S^2, 104, 195
$\mathbf{SU}_n$, 84
$\mathbf{Sym}_n(K)$, 18
$\mathbf{Sym}^2(E)$, 72
$\mathbf{Sym}_n^+$, 85

τ, 227
τ_{CG}, 240
T_k, 238
$T(E)$, 75
$T^k(E)$, 73
$\mathrm{Tr}\,M$, 43

$\mathbf{U}_n$, 84
$\mathscr{U}_p$, 154
$\mathbf{U}(p,q)$, 188

$V(a)$, 253

$|x|$, 149
$x \le y$, 149
$x > 0$, 149
$x \ge 0$, 149

$\mathbb{Z}[i]$, 1

General Index

algebra
    Lie, 200

Abel
    theorem, 176, 247
adjugate, 35
algebra
    exterior, 76
    tensor, 75
algebraically closed field, 2
alternate
    matrix, 28
alternating
    form, 10
antilinear, 13

Bézout identity, 164
basis, 3
    change of, 26
    dual, 6
bidual, 7
bilinear space, 11
blockwise
    diagonal, 19
    triangular, 19

Campbell–Hausdorff formula, 200
canonical form, 173, 176
Cauchy–Binet formula, 37
Cauchy–Schwarz inequality, 12, 13
Cayley–Hamilton theorem, 44

central subspace, 97
change of basis, 26
characteristic
    of a field, 1
    polynomial, 43
Choleski factorization, 213
cofactor, 35
commutator, 22
complexity, 209
compression, 115
condition number, 239, 248
congruent matrices, 29
conjugate
    exponents, 128
    gradient, 235
contraction, 137
contractive subspace, 98
convergence
    linear, 262
    rate, 227, 240
    super-linear, 263
convergent
    method, 226
Cotlar lemma, 142
Cramer's formulae, 42

degenerate
    bilinear form, 10
    Hermitian matrix, 85
determinant, 31

of n vectors, 33
diagonal
    blockwise, 19
    matrix, 17
    quasi-, 167
diagonalizable, 49
    orthogonally, 93
diagonally dominant, 103
    strictly, 104
    strongly, 103
dimension, 3
direct solving method, 207
domain
    Euclidean, 165
    principal ideal, 163
dual
    basis, 6
    space, 6
Dunford decomposition, 52
Dunford–Taylor integral, 95

eigenbasis, 49
eigenprojector, 97
eigenspace, 48
    generalized, 51
eigenvalue, 6, 42
    multiplicity
        algebraic, 44
        geometric, 44
    semisimple, 46
    simple, 44
eigenvector, 6, 42
elementary divisor, 175
endomorphism, 6
equivalent
    matrices, 26
    norms, 129
Euclidean
    division, 165
    domain, 165
    space, 12
expansive subspace, 98
exponential, 185
exterior
    algebra, 76

    product, 74, 76
extremal point, 144, 156

Fredholm principle, 25
Frobenius
    canonical form, 173
    norm, 261

Gauss
    integers, 1
    method, 225
Gauss–Seidel
    method, 227
gcd, 164
Gershgorin
    disk, 102, 107
    domain, 102
Greville's algorithm, 221
group
    linear, 39
    modular, 104
    orthochronous Lorentz, 195
    orthogonal, 39, 188
        special, 39
    special linear, 39
    special orthogonal, 192
    symmetric, 31
    symplectic, 188
    topological, 201
    unitary, 188

Hadamard product, 122
Hermitian
    form, 13
    space, 14
Hessenberg matrix, 60, 249
Householder
    matrix, 250
    method, 255
    theorem, 133

ideal, 163
    principal, 163
idempotent matrix, 23
inequality
    Cauchy–Schwarz, 129

Hölder, 127
Minkowski, 127
Weyl, 111
integral domain, 31
invariant
  subspace, 96
invariant factor, 167
inverse
  generalized, 218
  left, 218
  right, 218
inverse matrix, 22
irreducibility, 59
isometry, 11
  q-, 12
isotropic
  cone, 12
  vector, 24

Jacobi
  identity, 200
  method, 227, 260
Jordan
  block, 176
  decomposition, 177

kernel, 5
  bilinear form, 10
  quadratic form, 11

matrices
  commuting, 22
  product of, 20
matrix, 15
  adjugate, 35
  band, 220
  companion, 172
  cyclic, 154
  diagonal, 17
  diagonalizable, 49
  elementary, 166
  Hermitian, 84
    positive-definite, 85
  Hermitian adjoint, 83
  Hessenberg, 60, 249
  Householder, 250

idempotent, 23
identity, 16
imaginary part, 84
inverse, 22, 39
invertible, 22, 39
Jordan, 176
nilpotent, 23
nonnegative, 149
nonsingular, 22, 39
normal, 84
of a bilinear form, 28
of a linear map, 19
orthogonal, 23, 84
orthostochastic, 158
Pascal's, 197
permutation, 17, 156
projection, 61
quasi-diagonal, 167
rank-one, 25
real part, 84
singular, 39
skew-Hermitian, 84
skew-symmetric, 18
square, 16
stochastic, 156
  bi-, 156
symmetric, 18
totally positive, 222
triangular, 17
  strictly, 17
tridiagonal, 231
unitary, 84
Vandermonde, 63
maximal compact subgroup, 190
method
  conjugate gradient, 235
  Gauss–Seidel, 227
  Jacobi, 227, 260
  power, 266
    inverse, 269
  QR, 253
  relaxation, 228
minimal polynomial, 46
minor, 34
  leading principal, 35

principal, 35, 208
Moore–Penrose inverse, 218

neutral subspace, 98
nilpotent matrix, 23
nondegenerate
    bilinear form, 10
    Hermitian matrix, 85
nonsingular, 22
norm
    l^p, 127
    algebra, 131
    Frobenius, 261
    induced, 131
    matrix, 131
    Schur, 105, 197, 261
    subordinated, 131
    superstable, 101
norms
    equivalent, 129
numerical
    radius, 101
    range, 98

operator monotone, 114
orthogonal
    group, 23, 39, 188
    matrix, 23
    of a set, 24
    sets, 24
    vectors, 24
orthogonally
    diagonalizable, 93

Perron–Frobenius theorem, 150, 151
Pfaffian, 55
polar
    decomposition, 184
    form, 11
polynomial
    invariant, 170
positive definite
    Hermitian form, 13
    quadratic form, 12
positive-semidefinite, 85
preconditioning, 241

product
    Hadamard, 122
    of matrices, 20
    scalar, 12
projector, 61, 96

QR
    method, 253
QR
    factorization, 215
quadratic form, 11

range, 5
rank, 17
    decomposition, 170
    of a matrix, 23
    of a tensor, 70
rank-one
    affine, 53
    matrix, 25
Rayleigh
    ratio, 110
    translation, 259
reducibility, 59
relaxation
    method, 228
    over-, 228
residue, 236
resolvant
    of a matrix, 198
resolvent
    of a matrix, 96
    set, 96
resolvent set, 147
Riesz–Thorin theorem, 135
ring
    factorial, 165
    Noetherian, 164
    principal ideal domain, 163

scalar product, 12, 14, 24
Schur
    complement, 41, 115
    lemma, 62
    norm, 261
    theorem, 86

semisimple eigenvalue, 47
sesquilinear form, 13
signature, 93
similar
      matrices, 27
      unitarily, 85
similarity invariant, 44, 170
singular value, 140, 216
solving method
      direct, 207
      iterative, 225
spectral radius, 113, 127
spectrum, 43
square root
      Hermitian, 110
stable subspace, 97
standard polynomial, 78
Strassen algorithm, 213
Sturm sequence, 252
subspace, 2
      central, 97
      contractive, 98
      expansive, 98
      neutral, 98
      stable, 97
      unstable, 97
Sylvester equation, 65
Sylvester index, 93
symmetric
      bilinear form, 9
      group, 31
      matrix, 18
symplectic
      group, 188

tensor
      algebra, 75
      product, 69
tensor product
      of matrices, 77
trace, 43
triangular
      blockwise, 19
trigonalizable matrix, 49
trigonalization

unitary, 86

unitarily similar, 85
unitary
      diagonalization, 91
      group, 188
      trigonalization, 86
unstable subspace, 97

Vandermonde matrix, 63
vector
      column, 17
      nonnegative, 149
      positive, 149
      row, 17
vector space, 2

Weyl inequalities, 111

Cited Names

Aluthge, 205
Ando, 205

Banach, 145
Boyd, 161
Brouwer, 203

Cauchy, 67, 199

Diaconis, 161
Duncan, 199

Euler, 180

Flanders, 180
Frobenius, 107, 162

Hadamard, 123
Hessenberg, 68, 274
Horn
    Roger, 161
Householder, 147
Hua, Loo-Keng, 67

Jacobi, 274
John, 145
Jordan (Camille), 203
Jung, 205

Ko, Eungil, 205

Loewner, 114

Lyapunov, 202

Mazur, 145
Moore, 222

Oppenheim (Sir A.), 123

Parrott, 146
Pearcy, 205
Penrose, 222
Perron, 162
Pusz, 123

Riccati, 125
Ritz, 275
Rota, 147

Schur, 67, 106, 107, 123, 161
Schwarz, 199
Smith, 181
Starzhinskii, 202
Strang, 147
Sun, Jun, 161

Wimmer, 180
Woronowicz, 123

Xiao, Lin, 161

Yakubovich, 202